Springer
Proceedings in Physics 55

Springer Proceedings in Physics

Managing Editor: H. K. V. Lotsch

44 *Optical Fiber Sensors*
Editors: H. J. Arditty, J. P. Dakin, and R. Th. Kersten

45 *Computer Simulation Studies in Condensed Matter Physics II: New Directions*
Editors: D. P. Landau, K. K. Mon, and H.-B. Schüttler

46 *Cellular Automata and Modeling of Complex Physical Systems*
Editors: P. Manneville, N. Boccara, G. Y. Vichniac, and R. Bidaux

47 *Number Theory and Physics*
Editors: J.-M. Luck, P. Moussa, and M. Waldschmidt

48 *Many-Atom Interactions in Solids*
Editors: R .M. Nieminen, M. J. Puska, and M. J. Manninen

49 *Ultrafast Phenomena in Spectroscopy*
Editors: E. Klose and B. Wilhelmi

50 *Magnetic Properties of Low-Dimensional Systems II: New Developments*
Editors: L. M. Falicov, F. Mejía-Lira, and J. L. Morán-López

51 *The Physics and Chemistry of Organic Superconductors*
Editors: G. Saito and S. Kagoshima

52 *Dynamics and Patterns in Complex Fluids: New Aspects of the Physics-Chemistry Interface*
Editors: A. Onuki and K. Kawasaki

53 *Computer Simulation Studies in Condensed Matter Physics III*
Editors: D. P. Landau, K. K. Mon, and H.-B. Schüttler

54 *Polycrystalline Semiconductors II*
Editors: J. H. Werner and H. P. Strunk

55 *Nonlinear Dynamics and Quantum Phenomena in Optical Systems*
Editors: R. Vilaseca and R. Corbalán

56 *Amorphous and Crystalline Silicon Carbide III*
Editors: G. L. Harris, M. G. Spencer, and C. Y.-W. Yang

57 *Evolutionary Trends in the Physical Sciences*
Editors: M. Suzuki and R. Kubo

58 *New Trends in Nuclear Collective Dynamics*
Editors: Y. Abe, H. Horiuchi, and K. Matsuyanagi

59 *Exotic Atoms in Condensed Matter*
Editors: G. Benedek and H. Schneuwly

60 *The Physics and Chemistry of Oxide Superconductors*
Editors: Y. Iye and H. Yasuoka

61 *Surface X-Ray and Neutron Scattering*
Editors: H. Zabel and I. K. Robinson

62 *Surface Science: Lectures on Basic Concepts and Applications*
Editors: F. A. Ponce and M. Cardona

63 *Coherent Raman Spectroscopy: Recent Advances*
G. Marowsky and V. V. Smirnov

Volumes 1–43 are listed on the back inside cover

R. Vilaseca R. Corbalán (Eds.)

Nonlinear Dynamics and Quantum Phenomena in Optical Systems

Proceedings of the Third International Workshop
Blanes (Girona, Spain), October 1–3, 1990

With 150 Figures

Springer-Verlag
Berlin Heidelberg New York
London Paris Tokyo
Hong Kong Barcelona
Budapest

Professor Dr. Ramon Vilaseca

Universitat de Valencia, Departament Interuniversitari D'Optica
46100 Burjassot (Valencia), Spain

Professor Dr. Ramon Corbalán

Universitat Autònoma de Barcelona, Departament de Física
08193 Bellaterra (Barcelona), Spain

ISBN-13:978-3-642-76375-5 e-ISBN-13:978-3-642-76373-1
DOI: 10.1007/978-3-642-76373-1

Softcover reprint of the hardcover 1st edition 1991

Typesetting: Camera ready by authors

54/3140-5 4 3 2 1 0 – Printed on acid-free paper

Preface

This volume contains contributions based on the lectures delivered at the Third International Workshop on Nonlinear Dynamics and Quantum Phenomena in Optical Systems, which was held in Blanes, Girona, Spain, 1–3 October 1990. Blanes is a charming small town located on the well-known Costa Brava. With the convenient facilities of the Centre for Advanced Studies (CEAB), Blanes followed the tradition of the previous meetings of this series, which were held at Palma de Mallorca and Santander.

We aimed to provide an opportunity for scientists active in the broad field of quantum optics to meet in an informal atmosphere, thus promoting the exchange of ideas and allowing a search for interconnections between seemingly unrelated topics and a cross fertilization of the different subfields of quantum optics. We encouraged contributions dealing with the newest and most important developments in quantum optics. The main topics included instabilities, chaos, spatiotemporal dynamics, and pattern formation in lasers and nonlinear optical devices; phase dynamics; generation and detection of squeezed and other nonclassical states of light; coherent interaction of light with atomic systems; and multiphoton processes and above-threshold ionization.

The meeting brought together a group of 72 optical scientists from Belgium, France, Germany, Israel, Italy, Spain, the United Kingdom, USA, and USSR. In the technical program, nine invited papers were included, presented by eight distinguished specialists.

The workshop was very successful due to the high scientific level of the contributions and the active participation of the audience. We wish to express our gratitude to all of the participants for their collaboration and interest and to the contributors for the timely preparation of their outstanding manuscripts, which are now available to a wider audience. We would also like to thank our colleagues F.J. Bermejo, L. Pesquera, and M. San Miguel for help in the organization of the meeting, and the various bodies that provided financial support. These include: Ministerio de Educación y Ciencia (DGICYT); Generalitat de Catalunya (Department d'Ensenyament, CIRIT); Consejo Superior de Investigaciones Científicas (CSIC); CEE (Directorate-General for Science, Research and Development); British Council; Universitat de València; Universitat Autònoma de Barcelona, and Universidad de Cantabria.

We look forward to another productive meeting in the near future.

Barcelona, June 1991

R. Vilaseca
R. Corbalán

Contents

Part I

Squeezed and Nonclassical Light

Discrete-Mode Theory of Quantum-Optical Processing

R. Loudon

Physics Department, Essex University, Colchester CO4 3SQ, UK

Abstract. The main features of discrete cavity-mode theory are reviewed for a system in which balanced-homodyne detection is used to measure the phase-dependent electric field of an optical signal. The effects of attenuation and amplification are modelled by suitable optical beam splitters, and the amounts of added noise are determined. The advantages and disadvantages of discrete-mode theory are summarized. It is concluded that most practical optical systems are better described by a continuous-mode theory based on free space quantization.

1. Introduction

The increasing availability of beams of nonclassical light [1-3] necessitates treatments of the main optical processing elements within the framework of quantum optics. The main properties of the discrete-mode quantization of the electromagnetic field are summarized in Sect. 2, and the theory is applied to detection, attenuation and amplification respectively in Sects. 3, 4 and 5. Sect. 6 reviews the advantages and disadvantages of discrete-mode theory.

2. Discrete-Mode Quantized Field

A one-dimensional cavity of length L has the travelling-wave mode spacing

$$\Delta k = 2\pi/L \quad \text{or} \qquad \Delta\omega = 2\pi c/L \tag{1}$$

where k is the wavevector and ω is the angular frequency. The harmonic oscillator operators associated with modes i and j have the usual commutator

$$\left[\hat{a}_i, \hat{a}_j^{\dagger}\right] = \delta_{ij} \,. \tag{2}$$

The electric field operator for waves of given polarization propagating in the positive z-direction parallel to the cavity axis is (see for example [4])

Springer Proceedings in Physics, Vol. 55
Nonlinear Dynamics and Quantum Phenomena in Optical Systems
Editors: R. Vilaseca, R. Corbalán

$$\hat{E}(z,t) = i \sum_i (\hbar\omega_i/2\varepsilon_0 AL)^{1/2} \left\{\hat{a}_i \exp[-i\omega_i(t - z/c)] + \text{h.c.}\right\}, \tag{3}$$

where A is the cavity cross-sectional area.

It is sufficient for many calculations to consider only a single mode of the quantized field, when the mode label i can be dropped and the field operator is conveniently expressed in units of

$$2(\hbar\omega/2\varepsilon_0 AL)^{1/2} \tag{4}$$

in the form

$$\hat{E}(\chi) = \frac{1}{2}\hat{a} \exp(-i\chi) + \frac{1}{2}\hat{a}^{\dagger} \exp(i\chi) . \tag{5}$$

Here χ represents the phase of the field, as given by the usual travelling-wave exponent in eqn (3). The field operator can alternatively be expressed in terms of the quadrature operators $\hat{X}$ and $\hat{Y}$ defined by

$$\hat{X} = \frac{1}{2} (\hat{a}^{\dagger} + \hat{a}) \qquad \text{and} \qquad \hat{Y} = \frac{1}{2} i(\hat{a}^{\dagger} - \hat{a}) . \tag{6}$$

These are Hermitian operators with commutator

$$\left[\hat{X},\hat{Y}\right] = i/2 \tag{7}$$

and uncertainty relation

$$\Delta X \Delta Y \geq \frac{1}{4}. \tag{8}$$

The field operator from eqn (5) can be written

$$\hat{E}(\chi) = \hat{X} \cos\chi + \hat{Y} \sin\chi . \tag{9}$$

Measurements of the field at different phase angles χ_1 and χ_2 are not in general compatible since it follows from eqns (5) or (9) with use of the appropriate commutators that

$$\left[\hat{E}(\chi_1),\hat{E}(\chi_2)\right] = -\frac{1}{2} i \sin(\chi_1 - \chi_2) . \tag{10}$$

The field measurements are therefore subject to an uncertainty relation, or minimum noise limitation,

$$\Delta E(\chi_1)\Delta E(\chi_2) \geq \frac{1}{4} \left| \sin(\chi_1 - \chi_2) \right| . \tag{11}$$

Some important excitations have a field variance that is independent of the phase, when the minimum noise is given by

$$\Delta E(\chi) \geq \frac{1}{2} . \tag{12}$$

On the other hand, excitations with phase-dependent noise may have

$$\Delta E(\chi) < \frac{1}{2} \tag{13}$$

for some range of phase angles χ, in which cases the field is said to be squeezed.

The coherent state defined by

$$|\alpha\rangle = \exp(\alpha\hat{a}^{\dagger} - \alpha^{*}\hat{a})\,|\,0\,\rangle , \tag{14}$$

where

$$\alpha = |\alpha| \exp(i\phi) \tag{15}$$

is any complex number and $|0\rangle$ is the vacuum state, is a particularly important single-mode excitation. It has the usual properties

$$\hat{a}\,|\,\alpha\,\rangle = \alpha\,|\,\alpha\,\rangle \qquad \text{and} \qquad \langle\,\alpha\,|\,\hat{a}^{\dagger} = \langle\,\alpha\,|\,\alpha^{*} , \tag{16}$$

leading to the quadrature operator means

$$\langle\, X \,\rangle = |\alpha| \cos\phi \qquad \text{and} \qquad \langle\, Y \,\rangle = |\alpha| \sin\phi , \tag{17}$$

and uncertainties

$$\Delta X = \Delta Y = \frac{1}{2} . \tag{18}$$

The field operator from eqn (9) has the corresponding properties

$$\langle\, E(\chi) \,\rangle = |\alpha| \cos(\chi - \phi) \tag{19}$$

and

$$\Delta E(\chi) = \frac{1}{2} . \tag{20}$$

The coherent state has phase-independent noise and it has the minimum uncertainties allowed by the relations (8) and (12).

3. Balanced Homodyne Detection

Fig. 1 represents the beam splitter that superposes the signal beam (mode operator $\hat{a}$) with the local oscillator (mode operator $\hat{a}_L$) in a balanced homodyne detector [5]. The output mode operators are given by

$$\hat{d}_1 = r\hat{a} + t\hat{a}_L \qquad \text{and} \qquad \hat{d}_2 = t\hat{a} + r\hat{a}_L, \tag{21}$$

where r and t are the complex reflection and transmission coefficients of the beam splitter. The transformation from input to output operators is unitary if

$$|r|^2 + |t|^2 = 1 \qquad \text{and} \qquad r^*t + rt^* = 0, \tag{22}$$

and independent boson input operators then produce independent boson output operators. The output photon number operators are

$$\hat{d}_1^\dagger\hat{d}_1 = |r|^2\hat{a}^\dagger\hat{a} + r^*t\hat{a}^\dagger\hat{a}_L + rt^*\hat{a}_L^\dagger\hat{a} + |t|^2\hat{a}_L^\dagger\hat{a}_L$$

$$\hat{d}_2^\dagger\hat{d}_2 = |t|^2\hat{a}^\dagger\hat{a} + t^*r\hat{a}^\dagger\hat{a}_L + tr^*\hat{a}_L^\dagger\hat{a} + |r|^2\hat{a}_L^\dagger\hat{a}_L, \tag{23}$$

and conservation of photon number is ensured by eqn (22).

A 50/50 beam splitter is used in balanced homodyne detection, with the amplitudes and phases of r and t given by

$$|r| = |t| = 2^{-1/2} \qquad \text{and} \qquad \phi_r - \phi_t = \pi/2. \tag{24}$$

The measured quantity is the difference between the output photon numbers, represented by

$$\hat{n}_{21} = \hat{d}_2^\dagger\hat{d}_2 - \hat{d}_1^\dagger\hat{d}_1 = i\left(\hat{a}^\dagger\hat{a}_L - \hat{a}_L^\dagger\hat{a}\right), \tag{25}$$

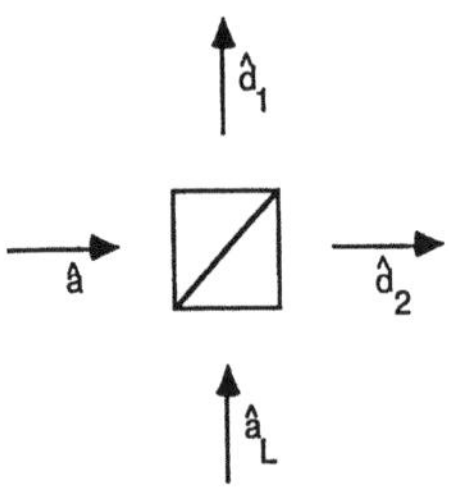

Fig. 1 Beam splitter component in a balanced homodyne detector

where eqns (23) have been used. The local oscillator light is assumed to be in a coherent state $|\alpha_L\rangle$ with properties analogous to those in eqns (16). Then if the local oscillator is much more intense than the signal, it is a good approximation to replace the local oscillator operators in eqn (25) by the equivalent c-numbers, and the measurement operator can be written

$$\hat{n}_{21} = 2|\alpha_L|\hat{E}(\chi), \tag{26}$$

where the field operator is given by eqns (5) or (9) with phase angle

$$\chi = \phi_L + \frac{\pi}{2}. \tag{27}$$

Balanced homodyne detection therefore measures the signal-mode electric-field operator at a phase angle determined by the local oscillator.

4. Linear Attenuation

The effects of attenuation on the quantum properties of light can be determined by considering the passage of light through a beam splitter. Fig. 2 shows a beam splitter with input light (mode operator $\hat{a}_{in}$) whose energy is partly scattered (mode operator $\hat{a}_S$) and partly transmitted to form the output beam (mode operator $\hat{a}_{out}$). The unitary properties of the beam splitter require the inclusion of the second input mode with operator $\hat{b}$, and the input-output relations are

$$\hat{a}_S = r\hat{a}_{in} + t\hat{b} \qquad \text{and} \qquad \hat{a}_{out} = t\hat{a}_{in} + r\hat{b}, \tag{28}$$

similar to eqns (21). The conditions (22) on the reflection and transmission coefficients ensure conservation of photon number, with

$$\hat{a}^\dagger_{out}\hat{a}_{out} + \hat{a}^\dagger_S\hat{a}_S = \hat{a}^\dagger_{in}\hat{a}_{in} + \hat{b}^\dagger\hat{b}. \tag{29}$$

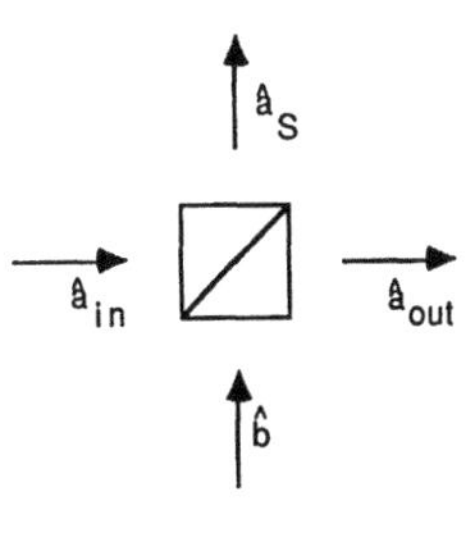

Fig. 2 Attenuating beam splitter

For an attenuating environment in thermal equilibrium at temperature T, the input mode operator $\hat{b}$ is assumed to have the expectation values

$$\langle \hat{b} \rangle = \langle \hat{b}^{\dagger} \rangle = 0 \tag{30}$$

and

$$\langle \hat{b}^{\dagger} \hat{b} \rangle = n(T) = 1/\{\exp(\hbar\omega/k_B T) - 1\} . \tag{31}$$

The relations (28) can be used to determine the effect of the beam splitter on the electric field operator of eqn (5) that controls the response in balanced homodyne detection. Thus

$$\hat{E}_{out}(\chi) = |t| \, \hat{E}_{in}(\chi - \phi_t) + |r| \, \hat{E}_T(\chi - \phi_r), \tag{32}$$

where $\hat{E}_T$ is the field of the $\hat{b}$ mode with the thermal properties given by eqns (30) and (31). The mean and the variance of the output field are therefore

$$\langle E_{out}(\chi) \rangle = |t| \langle E_{in}(\chi - \phi_t) \rangle \tag{33}$$

and

$$\left(\Delta E_{out}(\chi)\right)^2 = |t|^2 \left(\Delta E_{in}(\chi - \phi_t)\right)^2 + \frac{1}{2}(1 - |t|^2)\left[n(T) + \frac{1}{2}\right]. \tag{34}$$

These results can be written in more compact forms by ignoring the phase angles and defining the attenuation coefficient K to be

$$K = |t|^2 \leq 1. \tag{35}$$

Then

$$\langle E_{out} \rangle = K^{1/2} \langle E_{in} \rangle \tag{36}$$

and

$$(\Delta E_{out})^2 = K(\Delta E_{in})^2 + \frac{1}{2}(1 - K)\left[n(T) + \frac{1}{2}\right]. \tag{37}$$

This last result for the effect of attenuation on the noise conforms to the standard fluctuation - dissipation theorem [6].

The added noise represented by the second term on the right-hand side of eqn (37) takes its minimum value for a zero-temperature attenuator, when the result can be written

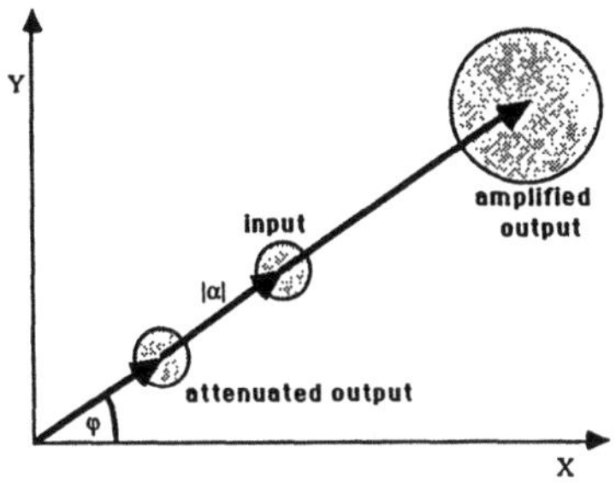

Fig. 3 Representation of the effects of fourfold attenuation and amplification on coherent light.

$$(\Delta E_{out})^2 - \frac{1}{4} = K\left\{(\Delta E_{in})^2 - \frac{1}{4}\right\}. \tag{38}$$

It follows that the noise tends to the coherent-state value in eqn (20) under severe attenuation, and that input coherent light has its noise unchanged. The phase-independent noise of coherent light is represented by the circular disc labelled 'input' in Fig. 3, and the attenuation with $K = 1/4$ maintains the input noise but reduces the amplitude of the state by a factor of 2. The properties of the electric field measured in balanced homodyne detection are obtained by projection of the state representation in Fig. 3 on to an axis inclined at angle χ to the horizontal. The signal-to-noise ratio is reduced by attenuation according to

$$\langle E_{out} \rangle / \Delta E_{out} = K^{1/2} \langle E_{in} \rangle / \Delta E_{in}. \tag{39}$$

5. Linear Amplification

The attenuator theory of the previous section can be formally converted to an amplifier theory by changing the loss to a gain,

$$K \rightarrow G \geq 1, \tag{40}$$

and changing the temperature of the input mode $\hat{b}$ from positive to negative,

$$n(T) \rightarrow n(-|T|) = -[n(|T|) + 1], \tag{41}$$

where the form of the thermal distribution of eqn (31) has been used. The amplifier input-output relations obtained by conversion of eqns (36) and (37) are

$$\langle E_{out} \rangle = G^{1/2} \langle E_{in} \rangle \tag{42}$$

and

$$(\Delta E_{out})^2 = G(\Delta E_{in})^2 + \frac{1}{2}(G-1)\left[n(|T|) + \frac{1}{2}\right]. \tag{43}$$

The added noise represented by the second term on the right-hand side of (43) is again minimized for zero temperature, but the result now takes the form

$$(\Delta E_{out})^2 + \frac{1}{4} = G\left\{(\Delta E_{in})^2 + \frac{1}{4}\right\}. \tag{44}$$

Thus, unlike attenuation, the noise is always increased by amplification, and its effect on coherent light for a gain of G = 4 is represented in Fig. 3. The signal-to-noise ratio is reduced by amplification according to

$$\langle E_{out} \rangle / \Delta E_{out} = \{G/(2G-1)\}^{1/2} \langle E_{in} \rangle / \Delta E_{in} \tag{45}$$

for a coherent input. Eqn (44) also shows that a highly-squeezed input with essentially zero noise loses its squeezed character, defined by eqn (13), for gains of G = 2 or larger.

The conversions (40) and (41) from attenuator to amplifier imply a change in the character of the beam splitter used to model the former. The negative temperature is synonymous with inverted harmonic oscillators in the upwards travelling modes in Fig. 2, and the mode operator relations (28) are converted to [7]

$$\hat{a}_S = r\,\hat{a}_{in}^{\dagger} + t\,\hat{b} \qquad \text{and} \qquad \hat{a}_{out} = t\,\hat{a}_{in} + r\,\hat{b}^{\dagger}. \tag{46}$$

The coupling coefficients must now satisfy

$$|t|^2 - |r|^2 = 1, \tag{47}$$

in contrast to the conventional beam-splitter properties in eqn (22). The resulting photon-number conservation relation

$$\hat{a}_{out}^{\dagger}\,\hat{a}_{out} - \hat{a}_S^{\dagger}\,\hat{a}_S = \hat{a}_{in}^{\dagger}\,\hat{a}_{in} - \hat{b}^{\dagger}\hat{b} \tag{48}$$

has negative contributions in accordance with the oscillator inversions. With the identifications

$$|t|^2 = G \quad \text{and} \quad |r|^2 = G - 1, \tag{49}$$

the electric-field input-output relations obtained by application of eqns (46) are the same as found in eqns (42) and (43).

The theory outlined above applies to the main varieties of phase-insensitive amplifier [8]. Thus the atomic population factor P that occurs in theories of inverted population amplifiers or attenuators [9] is equivalent to -n(T) in the current theory. Again, the standard operator relation in the theories of Raman or non-degenerate parametric amplifiers,

$$\hat{a}_{out} = \hat{a}_{in} \cosh s - \hat{a}_i^{\dagger} e^{i\theta} \sinh s, \tag{50}$$

has a form identical to eqn (46), with the identifications

$$t = \cosh s = G^{1/2} \quad \text{and} \quad r = -e^{i\theta} \sinh s. \tag{51}$$

Here $\hat{a}_i$ describes the thermal input in the form of material excitation or idler field mode.

The input modes labelled by $\hat{a}_{in}$ and $\hat{b}$ are considered to be independent in the above analysis. Quite different amplifier behaviour is obtained when the two input modes become identical, and the output operator is given by

$$\hat{a}_{out} = \hat{a}_{in} \cosh s - \hat{a}_{in}^{\dagger} e^{i\theta} \sinh s \tag{52}$$

in the same notation as eqn (50). This is the standard form for a degenerate parametric amplifier, and in conjunction with the definition of the field operator in eqn (5) it leads to

$$\hat{E}_{out}(\chi) = G(\chi)^{1/2} \hat{E}_{in}(\chi_o), \tag{53}$$

where

$$G(\chi) = e^{2s} \sin^2\left(\chi - \frac{1}{2}\theta\right) + e^{-2s} \cos^2\left(\chi - \frac{1}{2}\theta\right) \tag{54}$$

and the phase angles are related by

$$\tan\left(\chi_o - \frac{1}{2}\theta\right) = e^{2s} \tan\left(\chi - \frac{1}{2}\theta\right). \tag{55}$$

The input-output relations obtained from eqn (53) are

$$\langle E_{out} \rangle = G(\chi)^{1/2} \langle E_{in} \rangle \tag{56}$$

and

$$(\Delta E_{out})^2 = G(\chi)(\Delta E_{in})^2. \tag{57}$$

The additive noise terms found in eqns (37) and (43) are now absent, and the signal-to-noise ratio is unchanged by degenerate parametric amplification. However, this improvement is achieved at the expense of a gain that varies with the phase angle χ, and indeed turns into a loss over part of the range, with

$$G(\chi) < 1 \qquad \text{for} \qquad |\chi - \frac{1}{2}\theta| < \tan^{-1}(e^{-s}). \tag{58}$$

6. Conclusions

The discrete-mode theory outlined above has the advantage of extreme simplicity in systems that involve a small number of modes. It provides models for the main processes of attenuation, amplification, and detection. The results derived above refer to the means and variances of field operators expressed in the quantization-cavity-dependent units of eqn (4), but the cavity length L often disappears in the final theoretical expressions used for comparison with experiment. Of course some experiments are in practice performed in optical cavities, when the discrete-mode theory is entirely appropriate.

There is however no real cavity in most optical experiments, where the light flows one way from sources to detectors with no significant reflection or recycling. The detected photons are steadily replenished by the light sources without the associated reduction in optical energy that typically occurs in closed-cavity theories [10,11]. The discrete-mode formalism is awkward to apply to optical pulses and to systems that produce spatial variations in beam intensity. In addition the cavity length L sometimes survives in the final results of calculations for optical systems without even a notional cavity to provide an identifiable length.

It is therefore often preferable to work with a theory that is quantized in free space, with a set of travelling-wave modes of the electromagnetic field characterized by a continuous wavevector [12]. Such a theory overcomes the problems mentioned in the previous paragraph, while confirming some of the simple results of discrete-mode theory in appropriate limits.

References

1. R. Loudon and P.L. Knight, J. Mod. Opt. 34, 709, (1987).
2. M.C. Teich and B.E.A. Saleh, in Progress in Optics, edited by E. Wolf (North-Holland, Amsterdam, 1988) vol. 26, p. 1.
3. E. Giacobino, this volume.
4. R. Loudon, The Quantum Theory of Light, 2nd ed. (Clarendon Press, Oxford, 1983).
5. M.J. Collett, R. Loudon and C.W. Gardiner, J. Mod. Opt. 34, 881 (1987).
6. L.D. Landau and E.M. Lifshitz, Statistical Physics (Pergamon, Oxford, 1980), vol. 1.
7. R.J. Glauber, in Frontiers in Quantum Optics, edited by E.R. Pike and S. Sarkar (Adam Hilger, Bristol, 1986) p. 534.
8. C.M. Caves, Phys. Rev. D 26, 1817 (1982).
9. R. Loudon and T.J. Shepherd, Optica Acta 31, 1243 (1984).
10. T.J. Shepherd, Optica Acta 28, 567 (1981).
11. M.D. Srinivas and E.B. Davies, Optica Acta 28, 981 (1981).
12. J.R. Jeffers, R. Loudon, K.J. Blow and S.J.D. Phoenix, this volume, following paper.

Continuous-Mode Theory of Quantum-Optical Processing

J.R. Jeffers[1], *R. Loudon*[1,2], *K.J. Blow*[2], *and S.J.D. Phoenix*[2]

[1]Physics Department, Essex University, Colchester CO4 3SQ, UK
[2]British Telecom Research Laboratories,
Martlesham Heath, Ipswich IP5 7RE, UK

Abstract. We formulate the quantum theory of the electromagnetic field in a dispersive dielectric optical fibre of finite cross-section but infinite length without recourse to cavity quantization. Operator representations are derived for the optical fields and flux in the fibre. The forms of field excitation that describe coherent light and thermal noise are identified. The effects of propagation down attenuating and amplifying fibres on the quadrature field means and variances detected by balanced homodyne detection are calculated. Results for the specific case of exact compensation of fibre loss by amplification are presented. The way in which the continuous-mode quantum theory accommodates pulse spreading effects in propagation through group-velocity dispersive fibres is demonstrated.

1. Introduction

Many optical experiments take place in an open system geometry with no easily identifiable cavity in which the electromagnetic field should be quantized. This is the case for example in pulse propagation along optical fibres, where the cross-section is well defined but the total length plays little or no role in governing the natures of the excitations and their propagation characteristics. Such systems are conveniently treated by a one-dimensional continuous-mode field that is quantized in a space of finite cross section but infinite length. Sect. 2 summarizes the main features of the quantized field [1], while Sects. 3, 4 and 5 respectively cover the theories of balanced homodyne detection, travelling-wave attenuation, and amplification. The effects of dielectric dispersion on quantum-optical pulse propagation are treated in Sect. 6. Much reference is made to the corresponding results that are found in the cavity-quantized discrete-mode theory of reference [2], and equations from this paper are identified by the prefix I.

2. Continuous-Mode Quantized Field

The discrete-mode quantized electric-field operator of eqn I(3) is readily converted to continuous-mode form by allowing the cavity length L to tend

Springer Proceedings in Physics, Vol. 55
Nonlinear Dynamics and Quantum Phenomena in Optical Systems
Editors: R. Vilaseca, R. Corbalán

to infinity. The mode spacing $\Delta\omega$ of I(1) then tends to zero, the discrete operators convert into continuum operators and the Kronecker delta into a Dirac delta function according to

$$\hat{a}_i \rightarrow (\Delta\omega)^{1/2}\,\hat{a}(\omega) \quad \text{and} \qquad \delta_{ij} \rightarrow \Delta\omega\,\delta(\omega-\omega'), \tag{1}$$

and the commutation relation I(2) therefore becomes

$$\left[\hat{a}(\omega), \hat{a}^{+}(\omega')\right] = \delta(\omega-\omega')\,. \tag{2}$$

The summation in I(3) is converted to an integration in the usual way, and the electric-field operator can be written

$$\hat{E}(z,t) = \hat{E}^{+}(z,t) + \hat{E}^{-}(z,t) \tag{3}$$

where

$$\hat{E}^{+}(z,t) = i\int_{0}^{\infty} d\omega(\hbar\omega/4\pi\varepsilon_0 cA)^{1/2}\,\hat{a}(\omega)\exp[-i\omega(t-z/c)]$$

$$= c\,\hat{B}^{+}(z,t), \tag{4}$$

$$\hat{E}^{-}(z,t) = \left[\hat{E}^{+}(z,t)\right]^{\dagger} \tag{5}$$

and the magnetic field operator $\hat{B}$ is defined in a similar fashion.

The integration in eqn (4) strictly covers only positive frequencies, but the mode excitation in most cases of practical interest is restricted to a narrow bandwidth $\mathcal{B}$ centred on a frequency ω_0 such that

$$\mathcal{B} << \omega_0\,. \tag{6}$$

The integration in these cases can be extended to cover the complete range from $-\infty$ to ∞ to a very good approximation. It is then convenient to define Fourier transformed operators

$$\hat{a}(t) = (2\pi)^{-1/2}\int d\omega\,\hat{a}(\omega)\exp(-i\omega t) \tag{7}$$

with a commutator

$$\left[\hat{a}(t), \hat{a}^{+}(t')\right] = \delta(t-t') \tag{8}$$

that is easily proved with the use of eqn (2).

The most convenient representation of the optical field energy in the continuous-mode description is provided by the normal-order Poynting vector

$$\hat{S}(z,t) = \frac{1}{\mu_o}\left\{\hat{E}^-(z,t)\hat{B}^+(z,t) + \hat{B}^-(z,t)\hat{E}^+(z,t)\right\}. \tag{9}$$

This expression takes a somewhat complicated form when the field-operator expansions are substituted from eqn (4), but the result simplifies to

$$\hat{S}(z,t) = (\hbar\omega_o/A)\hat{a}^\dagger(t-z/c)\hat{a}(t-z/c) \tag{10}$$

in the narrow-band limit of eqn (6). It is clear from this expression that the flux of a light beam, in units of photons per unit time, is represented by

$$\hat{f}(t) = \hat{a}^\dagger(t)\hat{a}(t) = (1/2\pi)\int d\omega \int d\omega'\, \hat{a}^\dagger(\omega)\hat{a}(\omega')\exp[i(\omega-\omega')t]\ . \tag{11}$$

The continuous-mode formalism can handle light beams whose fluxes have arbitrary time dependences, including pulse trains and beams of constant mean intensity. A beam of constant mean flux F has a frequency-dependent correlation function of the form

$$\langle\hat{a}^\dagger(\omega)\hat{a}(\omega')\rangle = 2\pi\, f(\omega)\, \delta(\omega-\omega'), \tag{12}$$

where $f(\omega)$ is the mean flux per unit ω bandwidth. The total mean flux from eqn (11) is then

$$f(t) = \langle\hat{f}(t)\rangle = \int d\omega\, f(\omega) \equiv F. \tag{13}$$

An example of a light beam of constant mean flux is provided by thermally-excited light with a bandwidth that is narrow in accordance with eqn (6), where the total mean flux is

$$F = \mathcal{B}\, f(\omega_o), \tag{14}$$

with

$$2\pi f(\omega_o) = n(T) = 1/\{\exp(\hbar\omega_o/k_BT) - 1\}. \tag{15}$$

The total mean photon number in the field, given by

$$N = \int dt\, f(t), \tag{16}$$

is clearly infinite for a beam of constant flux.

A continuous-mode coherent state is defined by

$$|\{\alpha\}\rangle = \exp\left\{\int d\omega\left[\alpha(\omega)\hat{a}^{\dagger}(\omega) - \alpha^{*}(\omega)\hat{a}(\omega)\right]\right\}|0\rangle$$

$$= \exp\left\{\int dt\left[\alpha(t)\hat{a}^{\dagger}(t) - \alpha^{*}(t)\hat{a}(t)\right]\right\}|0\rangle \tag{17}$$

analogous to the discrete-mode definition in eqn I(14). The definitions in terms of frequency or time integrals are equivalent, where

$$\alpha(t) = (2\pi)^{-1/2}\int d\omega\, \alpha(\omega)\exp(-i\omega t) \tag{18}$$

is any complex function and $|0\rangle$ is the continuous-mode vacuum state. The coherent states defined in this way have the properties

$$\hat{a}(\omega)\,|\{\alpha\}\rangle = \alpha(\omega)\,|\{\alpha\}\rangle \tag{19}$$

and

$$\hat{a}(t)\,|\{\alpha\}\rangle = \alpha(t)\,|\{\alpha\}\rangle \tag{20}$$

together with the conjugate relations, similar to eqn I(16). It follows that the mean flux of the coherent state is

$$f(t) = |\alpha(t)|^2 \tag{21}$$

and this function can have any desired time dependence. 'Single-mode' coherent light of frequency ω_o, constant mean flux F and phase θ is represented by

$$\alpha(\omega) = (2\pi F)^{1/2}\exp(i\theta)\delta(\omega-\omega_o) \tag{22}$$

and

$$\alpha(t) = F^{1/2}\exp(i\theta - i\omega_o t). \tag{23}$$

3. Balanced Homodyne Detection

The general scheme of a balanced homodyne detector is the same as that outlined in Sect. 3 of reference [2] (see [3] for a more complete description). Thus the measured difference between the photon numbers in the two output arms of the beam splitter is still represented by an operator similar to that of eqn I(25). However, for a continuous-mode light beam whose flux may vary with the time, it is important to take account of the detection period. If the detection commences at time τ and has an integration time $\mathcal{T}$, the appropriate generalization of I(25) is the dimensionless photon-number difference operator

$$\hat{n}_{21} = i \int_{\tau}^{\tau+\mathcal{T}} dt \left\{ \hat{a}^{\dagger}(t)\hat{a}_L(t) - \hat{a}_L^{\dagger}(t)\hat{a}(t) \right\}. \tag{24}$$

The local oscillator is assumed to be in a coherent state $|\{\alpha_L\}\rangle$ with a very narrow spread of frequencies so that a 'single-mode' form analogous to eqn (23) can be used, with

$$\alpha_L(t) = F_L^{1/2} \exp(i\phi_L - i\omega_o t). \tag{25}$$

Then if the local oscillator is much more intense than the signal, it is a good approximation to replace the local oscillator operators in eqn (24) by the corresponding c-numbers, and the measurement operator can be written

$$\hat{n}_{21} = 2(F_L\mathcal{T})^{1/2}\, \hat{E}(\chi,\tau), \tag{26}$$

where

$$\hat{E}(\chi,\tau) = \mathcal{T}^{-1/2} \int_{\tau}^{\tau+\mathcal{T}} dt \left\{ \frac{1}{2}\hat{a}(t)\exp(i\omega_o t - i\chi) + \frac{1}{2}\hat{a}^{\dagger}(t)\exp(-i\omega_o t + i\chi) \right\} \tag{27}$$

and

$$\chi = \phi_L + \frac{\pi}{2}. \tag{28}$$

It is seen that eqn (26) is very similar to the discrete-mode measurement operator in eqn I(26). The electric field operator defined in eqn (27) represents the property of the signal field that is measured in balanced homodyne detection with a narrow-band local oscillator of constant mean flux, and it provides the continuous-mode generalization of the single-mode field operator defined in eqn I(5).

Measurements of the signal field with different local-oscillator phase angles are not in general compatible since it follows from eqns (8) and (27) that

$$\left[\hat{E}(\chi_1,\tau),\hat{E}(\chi_2,\tau)\right] = -\frac{1}{2}\,i\,\sin(\chi_1-\chi_2), \tag{29}$$

with the resulting uncertainty relation

$$\Delta E(\chi_1,\tau)\Delta E(\chi_2,\tau) \geq \frac{1}{4}\,\left|\sin(\chi_1-\chi_2)\right|. \tag{30}$$

More generally, the field commutator is easily calculated for homodyne measurements with different start times τ_1 and τ_2, and its value falls off linearly with decreasing overlap of the detection periods, to become zero for nonoverlapping periods. Analogous to eqn I(13), the signal is said to be squeezed when

$$\Delta E(\chi,\tau) < \frac{1}{2} \tag{31}$$

for some range of local oscillator phase angles.

4. Travelling-Wave Attenuation

Fig. 1 shows a model that can be used to determine the effects of attenuation on the quantum properties of a light beam that is propagated along an optical fibre over a distance z. The attenuation is caused by scattering centres that can be modelled by beam splitters. These are taken to be discrete components in the initial stages of the calculation, but their number

$$\mathcal{N} = z/\Delta z \tag{32}$$

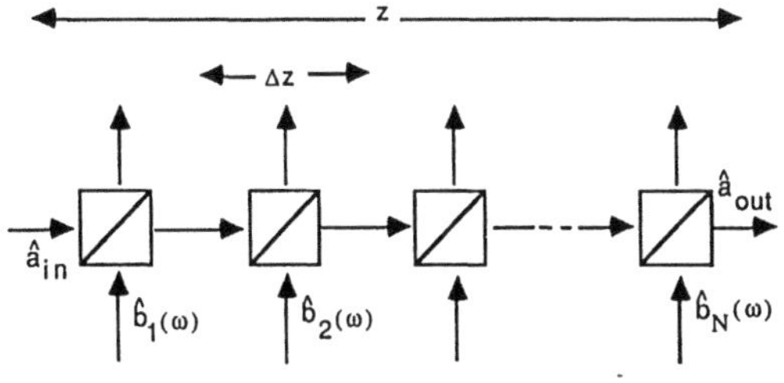

Fig. 1 Beam-splitter representation of scattering in an optical fibre

is later taken to infinity in order to model a continuous attenuating medium. Each beam splitter has the unitary properties described in reference [2], with reflection and transmission coefficients that satisfy

$$|r|^2 + |t|^2 = 1 \qquad \text{and} \qquad r^*t + rt^* = 0. \tag{33}$$

The phases of the coefficients are here chosen such that

$$r = i\,|r| \text{ and } t = |t|. \tag{34}$$

The input operators $\hat{b}_n(\omega)$ refer to independent thermally-excited modes, and their commutator is

$$\left[\hat{b}_m(\omega), \hat{b}_n^{\dagger}(\omega')\right] = \delta_{mn}\,\delta(\omega - \omega'). \tag{35}$$

The input signal-mode operator $\hat{a}_{in}(\omega)$ has the commutator

$$\left[\hat{a}_{in}(\omega), \hat{a}_{in}^{\dagger}(\omega')\right] = \delta(\omega - \omega') \tag{36}$$

similar to eqn (2). The calculation of the effects of attenuation requires an expression for the output signal operator $\hat{a}_{out}(\omega)$ in terms of the input signal operator. This is readily obtained by iteration of the result found in Sect. 4 of reference [2] for the scattering by a single beam splitter. Thus, taking account of the beam propagation phase with wavevector k, the required expression is

$$\hat{a}_{out}(\omega) = \left(|t|\,e^{ik\Delta z}\right)^{\mathcal{N}} \hat{a}_{in}(\omega) + i\,|r| \sum_{m=1}^{\mathcal{N}} \left(|t|\,e^{ik\Delta z}\right)^{\mathcal{N}-m} \hat{b}_m(\omega). \tag{37}$$

The beam splitters are now converted to a continuous array by taking the limits

$$\mathcal{N} \to \infty, \qquad \Delta z \to 0 \text{ and } \qquad |r|^2 \to 0 \tag{38}$$

such that the attenuation index

$$\kappa = |r|^2/\Delta z \tag{39}$$

remains finite. The usual exponential limit then provides the result

$$|t|^{2\mathcal{N}} = (1 - |r|^2)^{\mathcal{N}} = (1 - \kappa z/\mathcal{N})^{\mathcal{N}} \to e^{-\kappa z}. \tag{40}$$

The input operators and their commutator are converted to continuous spatial dependences by the replacements

$$\hat{b}_m(\omega) \rightarrow (\Delta z)^{1/2}\, \hat{b}(\omega,z') \quad \text{and} \quad \delta_{mn} \rightarrow \Delta z\, \delta(z'-z''), \tag{41}$$

with the result

$$\left[\hat{b}(\omega,z'),\hat{b}^{\dagger}(\omega',z'')\right] = \delta(\omega-\omega')\delta(z'-z''). \tag{42}$$

The summation in eqn (37) is converted to an integration in the usual way, and with the above conversions the output operator takes the form

$$\begin{aligned} \hat{a}_{out}(\omega) &= \exp\left(ikz - \tfrac{1}{2}\kappa z\right)\hat{a}_{in}(\omega) \\ &+ i\kappa^{1/2}\int_0^z dz' \exp\left\{(ik-\tfrac{1}{2}\kappa)(z-z')\right\}\hat{b}(\omega,z'). \end{aligned} \tag{43}$$

It is not difficult to verify with the use of eqns (36) and (42) that the output operators have the proper boson commutator

$$\left[\hat{a}_{out}(\omega),\hat{a}^{\dagger}_{out}(\omega')\right] = \delta(\omega-\omega'). \tag{44}$$

It is seen from eqn (43) that the effect of the attenuation is to reduce the input component of the output operator and to introduce contributions from the noise operators.

The theory of balanced homodyne detection of an attenuated signal requires the evaluation of moments of the field operator defined in eqn (27). In the simple case where the wavevector has the free-space value

$$k = \omega/c, \tag{45}$$

the Fourier transform of eqn (43) is

$$\hat{a}_{out}(t) = e^{-\kappa z/2}\hat{a}_{in}\left(t - \frac{z}{c}\right) + i\kappa^{1/2}\int_0^z dz'\, e^{-\kappa(z-z')/2}\, \hat{b}\left(t - \frac{z-z'}{c}, z'\right). \tag{46}$$

Insertion of this relation and its Hermitian conjugate into eqn (27) provides an expression for the output field in terms of the input field and the noise operators, which is somewhat complicated but leads to simple results for the mean and the variance. Thus if the attenuation coefficient is defined to be

$$K = e^{-\kappa z}, \tag{47}$$

the mean signal field measured in balanced homodyne detection is

$$\langle E_{out}(\chi,\tau)\rangle = K^{1/2}\left\langle E_{in}\left(\chi - \frac{\omega_0 z}{c}, \tau - \frac{z}{c}\right)\right\rangle \tag{48}$$

and its variance is

$$\left(\Delta E_{out}\right)^2 = K(\Delta E_{in})^2 + \frac{1}{2}(1-K)\left[n(T) + \frac{1}{2}\right], \tag{49}$$

where phase angles identical to those in eqn (48) have been suppressed and the thermal input modes have been assumed to have the time-dependent correlation function

$$\langle \hat{b}^{\dagger}(t,z')\hat{b}(t',z'')\rangle = n(T)\delta(t-t')\delta(z'-z'') . \tag{50}$$

The results (48) and (49) are identical in form to those derived by discrete-mode theory in Sect. 4 of reference [2], and the discussion that follows I(36) and I(37) applies equally in the continuous-mode theory. However, despite these similarities, the results derived above represent a substantial generalization since they allow for the effects of propagation and detector integration time that are absent from the discrete-mode theory.

5. Travelling-Wave Amplification

Optical signals can be continuously amplified as they propagate along a fibre, for example by stimulated Raman scattering or by interaction with an inverted atomic population. The process can be modelled by a line of beam splitters similar to that shown in Fig.1 but where each beam splitter now has the amplifying properties described in Sect.5 of reference [2]. The calculation is very similar to that for the travelling-wave attenuator outlined in the previous section. Thus a gain index g is defined by

$$g = |r|^2/\Delta z, \tag{51}$$

analogous to eqn (39), while the continuum limit analogous to that defined in eqns (38),(40) and (41) produces the output operator relation

$$\hat{a}_{out}(t) = e^{gz/2}\,\hat{a}_{in}\left(t - \frac{z}{c}\right) + i g^{1/2} \int_{0}^{z} dz'\, e^{g(z-z')/2}\, \hat{b}^{\dagger}\left(t - \frac{z-z'}{c}, z'\right). \tag{52}$$

The field operator at the output from the amplifier is obtained by substitution of eqn (52) into eqn (27). The thermal input modes are assumed to have a correlation function similar to eqn (50) but with the mean photon number of eqn (15) evaluated at $|T|$, since the temperature T itself is now negative. Then with the gain coefficient defined by

$$G = e^{gz}, \tag{53}$$

the mean signal field measured in balanced homodyne detection is

$$\langle E_{out}(\chi,\tau)\rangle = G^{1/2}\left\langle E_{in}\left(\chi - \frac{\omega_0 z}{c}, \tau - \frac{z}{c}\right)\right\rangle \tag{54}$$

and the variance is

$$(\Delta E_{out})^2 = G(\Delta E_{in})^2 + \frac{1}{2}(G-1)\left[n(|T|) + \frac{1}{2}\right], \tag{55}$$

where the suppressed χ and τ variables are identical to those in eqn (54). These results have the same forms as those derived by discrete-mode theory, and the discussion that follows eqns I(42) and I(43) also applies to the continuous-mode theory.

The generalization of eqns (46) and (52) for an optical fibre that simultaneously has an attenuation index κ and a gain index g is

$$\hat{a}_{out}(t) = e^{(g-\kappa)z/2}\,\hat{a}_{in}\left(t - \frac{z}{c}\right)$$

$$+ i\int_{0}^{z} dz'\, e^{(g-\kappa)(z-z')/2}\left\{g^{1/2}\,\hat{b}_{+}^{\dagger}\left(t - \frac{z-z'}{c}, z'\right) + \kappa^{1/2}\,\hat{b}_{-}\left(t - \frac{z-z'}{c}, z'\right)\right\}, \tag{56}$$

where the +(-) subscripts denote the independent thermal input modes associated with the amplification (attenuation). The relation between the input and output field operators is readily calculated as before. There is particular practical interest in conditions where the fibre loss is exactly compensated by the gain, so that $g = \kappa$, and the input and output signal-field means and variances are related by

$$\langle E_{out} \rangle = \langle E_{in} \rangle \tag{57}$$

and

$$(\Delta E_{out})^2 = (\Delta E_{in})^2 + \frac{1}{2}\kappa z \qquad (g = \kappa). \tag{58}$$

The thermal-mode temperatures have here been set equal to zero, and the χ and τ variables in eqns (57) and (58) are the same as before. Thus for gain-compensated low-loss optical fibre with

$$\kappa = 0.046\ \mathrm{km}^{-1} \quad (0.2\ \mathrm{dB\ km}^{-1}) \tag{59}$$

and for a highly-squeezed input with essentially zero noise, the squeezing defined by the inequality (31) is lost after a propagation distance of about 11 km.

6. Field Quantization in a Dielectric

The quantized field expressions of eqn (4) are generalized in a dielectric material to [1]

$$\hat{E}^+(z,t) = i \int_0^\infty d\omega \left(\frac{\hbar\omega}{4\pi\varepsilon_0 c A n(\omega)} \right)^{1/2} \hat{a}(\omega) \exp\left[-i\omega\left(t - \frac{n(\omega) z}{c} \right) \right] \tag{60}$$

and a similar expression for $c\hat{B}^+(z,t)$ except that $n(\omega)$ is moved to the numerator of the square-root factor. Here $n(\omega)$ is the frequency-dependent refractive index, related to the dielectric function $\varepsilon(\omega)$ by

$$n(\omega)^2 = \varepsilon(\omega) \tag{61}$$

and the optical wavevector is now

$$k = \omega\, n(\omega)/c. \tag{62}$$

The definition (9) of the Poynting vector remains valid with these modifications, and the beam photon flux operator is given by eqn (11).

However the optical dispersion in a fibre causes progressive distortions of propagating optical pulses. Consider for example an input Gaussian pulse represented by the coherent-state excitation of eqn (17) with

$$\alpha_{in}(\omega) = (2\pi\delta^2)^{-1/4} \exp\left\{i(\omega-\omega_o)t_o - (\omega-\omega_o)^2/4\delta^2\right\} \tag{63}$$

and

$$\alpha_{in}(t) = (2\delta^2/\pi)^{1/4} \exp\left\{-i\omega_o t_o + \delta^2(t-t_o)^2\right\}. \tag{64}$$

The pulse properties at the output from the fibre are obtained with the use of eqn (43), but the free-space wavevector of eqn (45) must now be replaced by the dispersive wavevector of eqn (62) in determining the time dependence of the output field. A Taylor expansion around the central frequency of the pulse, given by

$$k = k_o + (\omega-\omega_o)k_o' + \frac{1}{2}(\omega-\omega_o)^2 k_o'' + \ldots, \tag{65}$$

leads to the output mean value

$$\langle a_{out}(t)\rangle = (2\pi)^{-1/4}(2\Delta^2/\delta)^{1/2}\exp\left\{-i\omega_o t + ik_o z - \frac{1}{2}\kappa z - \Delta^2(t-t_o-k_o' z)^2\right\} \tag{66}$$

where

$$\Delta^2 = \delta^2/(1 - 2i\delta^2 k_o'' z) \tag{67}$$

and the attenuation index has been assumed to be independent of the frequency. It is seen that the pulse travels with the group velocity $1/k_o'$, and its shape remains Gaussian, but with a duration increased by a factor

$$\left(1 + 4\delta^4 k_o''^2 z^2\right)^{1/2} \tag{68}$$

resulting from the effects of the group velocity dispersion represented by the parameter k_o''. Eqn (68) reproduces a well-known result (see for example [4]).

7. Conclusions

The electromagnetic field is straightforwardly quantized in a one-dimensional medium of finite cross-sectional area but infinite length. The resulting theory is appropriate for calculations of quantum-optical effects associated with propagation in long optical fibres. The concepts of discrete-mode theory carry over to the continuous-mode formalism, but the Poynting vector replaces the total energy as the most appropriate measure of field excitation. The formal results of calculations of balanced homodyne detection, attenuation, and amplification resemble those of discrete-mode theory, but the continuous-mode formalism embraces optical pulse propagation, including the effects of dispersive media. The theory can also be applied to determine the effects of nonlinear processes such as self-phase modulation on quantum optical propagation in fibres [5].

References

1. K.J. Blow, R. Loudon, S.J.D. Phoenix and T.J. Shepherd, Phys. Rev. A 42, 4102 (1990).
2. R. Loudon, this volume, preceding paper.
3. M.J. Collett, R. Loudon and C.W. Gardiner, J. Mod. Opt. 34, 881 (1987).
4. G.P. Agrawal, Nonlinear Fiber Optics (Academic Press, San Diego, 1989).
5. K.J. Blow, R. Loudon and S.J.D. Phoenix, to be published.

Generation of Squeezed Light

E. Giacobino

Laboratoire de Spectroscopie Hertzienne de L'ENS,
Université Pierre et Marie Curie, F-75252 Paris Cedex 05, France

After a brief presentation of squeezed states, the various techniques used for generation of squeezed light are reviewed with an emphasis on one of the most efficient methods to date, parametric interaction, either in the degenerate or in the non-degenerate operation.

1. Introduction

Random fluctuations of various origins limit the accuracy of any measurement. In optics, several kinds of techniques have been successfully implemented to reduce the "technical noise", or "instrumental noise" and the suppression of these causes of noise can be so efficient that the basic limit due to the quantum nature of light is actually encountered in many situations. In the last decade, theoretical studies have shown that this limit could be circumvented, and numerous experimental demonstrations of quantum noise reduction have been done. The amount of noise reduction in several experiments now ranges between 70% and 80%.

2. The quantum fluctuations

While a full description of the quantum of the electromagnetic field and its fluctuations requires the formalism of quantum optics, many properties of quantum fields can be understood simply in terms of mean square fluctuations of their components. A particular mode of the electromagnetic field, corresponding to a plane wave of given frequency, direction and polarization, is described by the electric field operator:

$$E = E_1 \cos\omega t + E_2 \sin\omega t \qquad (1)$$

where E_1 et E_2 are the two quadrature components (defined with respect to a given phase reference). In quantum optics, the two quadrature components are conjugate variables, as the position q and the momentum p of an oscillator. Their dispersions in any quantum state obey a Heisenberg inequality:

$$\Delta E_1 \, \Delta E_2 \geq \varepsilon^2 \qquad (2)$$

Springer Proceedings in Physics, Vol. 55
Nonlinear Dynamics and Quantum Phenomena in Optical Systems
Editors: R. Vilaseca, R. Corbalán

where ε is a constant corresponding to the electric field of a single photon. In the case of a particle, the Heisenberg inequality concerns two variables q and p which are of different physical nature. In contrast, the two conjugate variables of the electromagnetic field, E_1 and E_2, are of the same nature, and most of the usual light sources have no preferred quadrature component. In other words, the fluctuations of the electromagnetic field usually have no natural phase reference. As a result, the dispersions ΔE_1 and ΔE_2 are equal and independently fulfill an inequality:

$$\Delta E_1 = \Delta E_2 \geq \varepsilon . \tag{3}$$

The factor ε gives a standard scale for the fluctuations of the electromagnetic field and for the quantum noise. In particular, it gives the scale of the *vacuum fluctuations*, which correspond to the state with the lowest energy. They are equivalent to the zero point fluctuations for a matter oscillator. Since there is no phase reference in the vacuum, the corresponding field is a minimal state fulfilling $\Delta E_1 = \Delta E_2 = \varepsilon$.

It can often be considered that the vacuum fluctuations are the source of quantum noise in optical measurements. For example, the *coherent* state is a superposition of a classical field and of the vacuum fluctuations.

To surpass the limit set by eq. (3), it is necessary to break the symmetry between the two quadratures. The Heisenberg inequality does not forbid one to decrease ΔE_1 below ε, provided that ΔE_2 is increased. Such states are called *squeezed states*. For example

$$\Delta E_1 < \varepsilon \qquad \Delta E_2 \geq \varepsilon^2 / \Delta E_1 . \tag{4}$$

The photon noise is associated with the quantum fluctuations in the field intensity. If we suppose that the mean field is aligned along the E_1 axis, the intensity fluctuations can be written as

$$\Delta I^2 = 4 \langle E_1 \rangle^2 \, \Delta E_1{}^2 . \tag{5}$$

The intensity fluctuations are proportional to the fluctuations of the field amplitude. The standard photon noise corresponds to the case in which $\Delta E_1{}^2 = \varepsilon^2$. It is characterized by a variance proportional to the mean intensity.

This noise is reduced by squeezing the fluctuations of the E_1 component of the field. More generally, in every optical measurement, it can be shown that the fluctuations correspond to some component of the field, which can be a quadrature component of one field or a superposition of several fields. As an example, let us consider the case of two-field squeezing. We assume that the two fields have equal non zero mean values aligned along the E_1 axis. We will study the quantity : $\mathcal{E} = (E^1{}_1 - E^2{}_1)/\sqrt{2}$. For uncorrelated coherent fields, the mean square fluctuations in $\mathcal{E}$ are

$$(\Delta \mathcal{E})^2 = ((\Delta E^1{}_1)^2 + (\Delta E^2{}_1)^2) / 2 = \varepsilon^2 . \tag{6}$$

If the fluctuations of the field amplitudes are correlated, $(\Delta\mathcal{E})^2$ is smaller than ε^2. $\mathcal{E}$ is a squeezed quantity. In that case the fluctuations in the intensity difference $\mathcal{J} = |E^1|^2 - |E^2|^2$ are

$$(\Delta\mathcal{J})^2 = 4\,(2\bar{I})\,(\Delta\mathcal{E})^2 \quad , \tag{7}$$

where $\bar{I}$ is the common mean intensity of the two fields. If the fields are uncorrelated, $(\Delta\mathcal{E})^2 = \varepsilon^2$, and comparison with eq. (5) shows that the fluctuations in $\mathcal{J}$ are just equal to the standard photon noise of a field of intensity $2\bar{I}$. If the fields are correlated, the noise in the intensity difference is lower than the standard quantum noise of a field of intensity 2I. Two beams of light having such a correlation are called *twin beams*.

The connection between quadrature squeezed light and twin beams can be shown with the example of a very simple system: the beamsplitter.

3. The beamsplitter and the photon noise

A beamsplitter can be used to perform a homodyne measurement: the incident beam entering in the input channel A is mixed on a 50/50 beamsplitter with the field entering in the other input channel B. Classical optics give the fields going out of the system in channels C and D:

$$E_C = (E_A + E_B)/\sqrt{2} \tag{8a}$$

$$E_D = (-E_A + E_B)/\sqrt{2} \tag{8b}$$

where the minus sign in eq. (8b) comes from the π phase shift between the two reflections on the beamsplitter: one is from air off glass and the other is from glass off air. It can also be viewed as a consequence of the energy conservation $I_C + I_D = I_A + I_B$.

Assuming that field E_A is much more intense than field E_B, the measurement of the intensity difference $I = I_C - I_D$ between the two channels gives:

$$I = I_C - I_D = 2 \langle E_A \rangle E_{B1} \tag{9}$$

where E_{B1} is the quadrature component of E_B in phase with the mean amplitude $\langle E_A \rangle$. The intensity difference I is a signal resulting from the homodyning of the quadrature component E_{B1} by the local oscillator $\langle E_A \rangle$. This property is also true for the fluctuations:

$$\Delta I^2 = 4 \langle E_A \rangle^2 \, \Delta E_{B1}{}^2 . \tag{10}$$

The photon noise reflects the quantum fluctuations of the quadrature component E_{B1}. In particular, when nothing enters in channel B, the vacuum fluctuations

$\Delta E_{B1}{}^2 = \varepsilon^2$ are homodyned by the local oscillator

$$(\Delta I^2)_{st} = 4 <E_A>^2 \varepsilon^2 . \tag{11}$$

Then, the noise in the difference between the intensities in the two output channels is the standard quantum noise. This can also be interpreted as the result of the random distribution of the photons between the two channels in the transmission-reflection process. It is independent of the photon statistics of the input beam. It is however possible to reduce the noise in the intensity difference after the beamsplitter by entering a squeezed field in channel B. Twin photon beams are obtained. We will see below that twin beams can also be generated directly in experiments of nonlinear optics.

4. Direct techniques for quantum noise reduction

Squeezed states have been shown to be generated by several nonlinear optical systems. In most cases the involved phenomenon is parametric generation, which can be achieved in a three-wave mixing process or in a four-wave mixing process. Two photons are created in the signal waves at frequencies ω_1 and ω_2, while one photon of the pump wave is annihilated (in three-wave mixing), or two photons of the pump waves are annihilated (in four-wave mixing). The systems may be used in the degenerate operation in which $\omega_1 = \omega_2$ are identical or in the nondegenerate operation in which the two signal modes have different frequencies or different polarizations.

- Degenerate parametric generation

The first observation of squeezing was achieved in 1985 in an experiment of parametric generation involving four-wave mixing in sodium vapour[1]. The nonlinearity was enhanced by placing the nonlinear medium in an optical cavity. One year later, squeezed light was generated in a similar system, but with three wave-mixing in a nonlinear $\chi^{(2)}$ crystal[2]. In both cases, the systems were operated in the *degenerate* regime.
Let us first briefly discuss the physical process that leads to squeezing. Parametric interaction amplifies input noise, specifically here, quantum noise. The effect can then be pictured in a simple way (at least in the limit of low pumping rates): the system processes the vacuum noise input and squeezes it. The output, having zero mean field, is expected to be squeezed vacuum[3].

In ref.[2], the parametric crystal, pumped with a laser, is placed in an optical cavity in order to enhance the nonlinearity. Below the oscillation threshold, we can neglect the pump depletion, and the change of the signal field in one roundtrip in the cavity can be written as a function of the cavity losses, the parametric gain, and the input field $\alpha_1{}^{in}$. Here, no input field with a finite value is entered in the signal mode. The input field is the vacuum field coming into the cavity through the coupling mirror[4]:

$$\tau \dot{\alpha}_1 = -\gamma \alpha_1 + \eta \alpha_1^* + t \alpha_1^{in} \tag{12}$$

where the α's are the complex components of the fields, normalized to the photon flux, τ is the cavity roundtrip time, γ is the damping coefficient of the field in the cavity; γ is related to the amplitude transmission coefficient t of the output mirror by $\gamma = t^2/2$ (t is assumed to be small); η is the parametric gain, which is proportional to the the pump field, to the $\chi^{(2)}$ coefficient and to the length of the crystal. The parametric gain term in α_1 involves α_1^*, which is characteristic of parametric amplification, and which is the reason for the phase dependence of the process.

Taking the Fourier transform of eq. (12) and of its complex conjugate, we express the solution in terms of the quadratures $q_1 = (\alpha_1 + \alpha_1^*)/\sqrt{2}$ and $p_1 = (\alpha_1 - \alpha_1^*)/\sqrt{2}i$, with similar notations for q_1^{in} and p_1^{in}:

$$\tilde{q}_1(\Omega) = \frac{t\,\tilde{q}_1^{in}(\Omega)}{\gamma - \eta + i\Omega\tau} \qquad \tilde{p}_1(\Omega) = \frac{t\,\tilde{p}_1^{in}(\Omega)}{\gamma + \eta + i\Omega\tau} . \tag{13}$$

On the other hand, the outgoing field α_1^{out} is related to the field inside the cavity and to the incoming vacuum field by the reflection-transmission relation:

$$\alpha_1^{out} = t\,\alpha_1 - \alpha_1^{in} \tag{14}$$

where the reflection coefficient r has been approximated by one. Using eqs. (13) and (14), one gets:

$$\tilde{q}_1^{out}(\Omega) = \frac{\gamma + \eta - i\Omega\tau}{\gamma - \eta + i\Omega\tau}\,\tilde{q}_1^{in}(\Omega) , \tag{15}$$

$$\tilde{p}_1^{out}(\Omega) = \frac{\gamma - \eta - i\Omega\tau}{\gamma + \eta + i\Omega\tau}\,\tilde{p}_1^{in}(\Omega) . \tag{16}$$

Equations (15) and (16) clearly show that the vacuum field is amplified in quadrature q_1, while quadrature p_1 is squeezed and even goes to zero for $\Omega = 0$ and $\gamma = \eta$, that is for zero noise frequency and when the oscillation threshold is approached. The vacuum field is indeed squeezed by the system.

The experiment (fig.1) uses a nonlinear crystal ($LiNbO_3$) which is pumped with a single mode doubled YAG laser. The crystal is placed in an optical cavity which is resonant for both the pump and the subharmonic field. The observation of squeezing in the signal beam is done by homodyne detection as exposed above. The light exiting the parametric amplifier is combined on a beamsplitter with the light at the fundamental frequency from the YAG laser, which acts as a strong

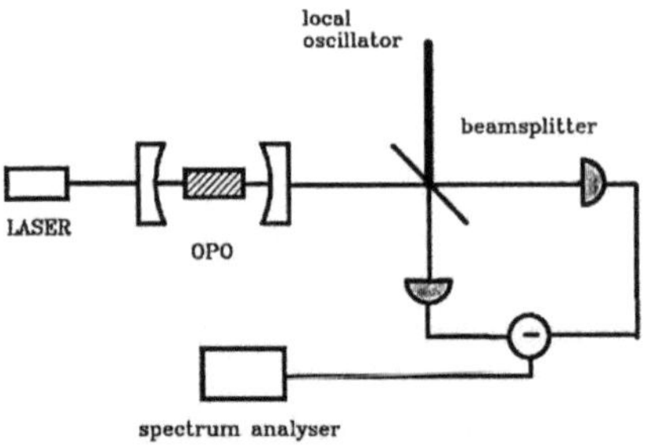

Figure 1: Experimental scheme used by Wu *et al*[2] to generate squeezed vacuum in parametric down conversion.

local oscillator. The photodetection signal is fed into a spectrum analyzer. The signal at a fixed noise frequency oscillates around the shot noise level as a function of the phase of the local oscillator. The noise reduction has been verified to be improved when the incident pump power tends to its threshold value. The optimum noise reduction obtained with this technique is 63%[2]. The best figure obtained with parametric generation in an atomic beam is 25%[1].

Squeezing of about 20% has also been observed using forward four-wave mixing in a long silica optical fiber[5]. However, the experiments have been hampered by index fluctuations due to Brillouin scattering. Short pulses can be squeezed more efficiently with this method because their interaction time is too fast to be sensitive to Brillouin scattering[6]. Other experiments without an optical cavity use pulsed pumping in nonlinear crystals[7-8].

Finally, the systems having $\chi^{(2)}$ nonlinearities seem to yield better squeezing than the $\chi^{(3)}$ systems, mainly for two reasons: the pump frequency is far from the signal frequencies, which allows easy elimination of spurious fields due to pump scattering; in the doubling crystals, the nonlinearity is not resonant, which means that there are no fluctuations due to the nonlinear medium, except for those associated with linear losses.

- Second harmonic generation

Second harmonic generation (SHG) is the reverse process of degenerate parametric generation: the nonlinear $\chi^{(2)}$ medium is excited at ω_1 and a field is generated at $\omega_0=2\omega_1$. In contrast to parametric generation, SHG produces a field with finite mean value. The second harmonic field as well as the pump field at the fundamental frequency can be shown to be squeezed at the output of the nonlinear medium[9]. Two experiments, both using MgO doped $LiNbO_3$ have been performed along those lines. The first one[10] demonstrated 13% squeezing in the fundamental mode, the second one[11] yielded 40% squeezing in the upconverted mode. In the latter case, a monolithic cavity, in which the mirror coatings are deposited directly on the ends of the crystal was used.

- Non-degenerate parametric generation

In parametric generation, signal photons are emitted by pairs, which results in photon correlation and intensity correlations between the two signal beams[12]. In the non-degenerate operation, the two signal fields differ by either their frequencies or their polarizations and can be separated.

Experiments on twin beams have been performed in various conditions: with low intensity cw pump lasers, or with high power pump lasers. In the former case, the crystal emits parametric fluorescence. A maximum correlation of 67% has been obtained in 1990[13]. In the latter case[14], intense twin beams are emitted by the crystal yielding up to 75% noise reduction[15].

To increase the effective parametric interaction length, the nonlinear crystal can be placed in an optical resonant cavity. When pumped with a high enough intensity, the system oscillates like a laser. The optical parametric oscillator (OPO) emits twin laser-like beams, having a high degree of quantum correlation due to the photon pair emission. However, the cavity introduces some decorrelation between the twin photons, since they do not necessarily go out of the cavity after the same number of round trips. The noise in the difference between the intensities of the two signal beams is reduced below the standard quantum limit only for the frequencies within the bandwidth of the cavity.

To theoretically determine the quantum fluctuations in this case, the semiclassical method used for the OPO below threshold can be extended to the derivation of the above threshold case, but the treatment must take into account the depletion of the pump field α_0[16]. As in eq. (12), the equations for the signal fields α_1 and α_2 are driven by the vacuum fluctuations $\alpha_1{}^{in}$ and $\alpha_2{}^{in}$ entering the cavity through the coupling mirror:

$$\tau \dot{\alpha}_1 = -\gamma \alpha_1 - \chi \alpha_2{}^* \alpha_0 + t \alpha_1{}^{in} \qquad (17)$$

$$\tau \dot{\alpha}_2 = -\gamma \alpha_2 - \chi \alpha_1{}^* \alpha_0 + t \alpha_2{}^{in} \qquad (18)$$

where χ is the parametric coupling coefficient and the other notations are the same as in eq. (12). By linearizing the equations around a particular steady state, one can write the evolution of the fluctuations in the various components of the field. We define the quadrature components $q_i = (\alpha_i + \alpha_i{}^*)/\sqrt{2}$ and $p_i = (\alpha_i - \alpha_i{}^*)/\sqrt{2}i$ for $i= 0,1,2$, and we introduce the quantities: $q_- = (q_1 - q_2)/\sqrt{2}$ and $p_- = (p_1 - p_2)/\sqrt{2}$. The mean fields α_i are assumed to be real. The equation for the fluctuations δq_- in the amplitude difference q_- is

$$\tau \delta\dot{q}_- = -\gamma \delta q_- + \chi\, q_0/\sqrt{2}\, \delta q_- + t\, \delta q_-{}^{in} \, . \qquad (19)$$

Taking the Fourier transform and using the input-output relations on the coupling mirror, one gets

$$\delta\tilde{q}_-{}^{out}(\Omega) = \frac{\gamma + \chi\, q_0/\sqrt{2} - i\Omega\tau}{\gamma - \chi\, q_0/\sqrt{2} + i\Omega\tau}\; \delta\tilde{q}_-{}^{in}(\Omega) \, . \qquad (20)$$

Since $\gamma = - \chi\, q_0/\sqrt{2}$ when the OPO operates above threshold, it can be seen that the fluctuations in the difference between the amplitudes of the signal fields go to zero for zero noise frequency. The same is true for the noise in the intensity difference, which is proportional to the noise in the amplitude difference. This property does not depend on the noise of the pump beam, whose contribution cancels out in the difference, nor on the pump power: in contrast to quadrature squeezing below threshold, noise reduction in the twin beams takes place even far from threshold.

The only processes which may degrade the squeezing are the losses, which are not included in the preceding formula. Their effect can be understood by considering that each time a photon is lost inside the cavity, it degrades the correlation between the twin beams. As a result, the predicted squeezing at zero noise frequency is not perfect any more. The remaining noise is equal to the ratio of the internal losses in the cavity to the total cavity damping coefficient[12].

The experimental set-up used in ref[17] comprises a nonlinear KTP crystal pumped by an Ar^+ laser and placed in an optical cavity (fig.2). The cavity has a higher finesse for the signal and idler fields than for the pump. The twin beams emitted by the OPO have orthogonal polarizations and are separated at the output of the optical cavity with a polarizing beamsplitter. The intensities of the two beams are detected with photodiodes and the photocurrents are subtracted. This signal is fed into a spectrum analyser.

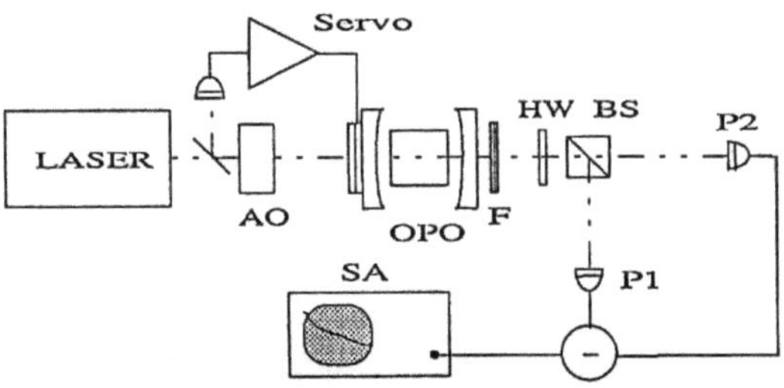

Figure 2. Experimental set-up for twin beams generation (AO: acousto-optic modulator, F: green filter, HW: half-wave plate, Bs: polarizing beamsplitter, P1, P2: Photodiodes, SA: spectrum analyzer)

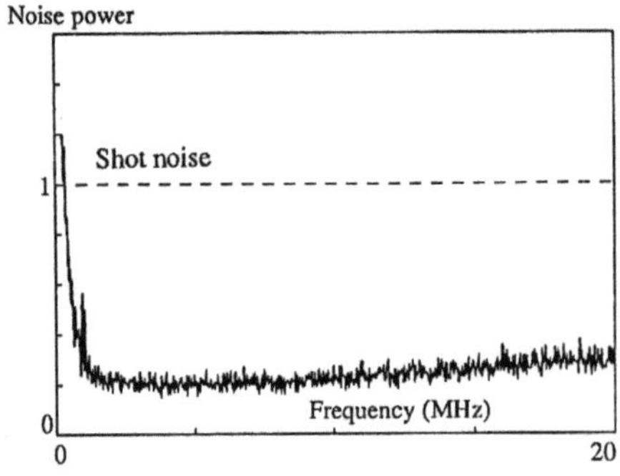

Figure 3. Observed photon noise reduction factor as a function of the frequency.

The noise in the intensity difference is found to be squeezed on a frequency range of the order of the inverse of the cavity storage time (Fig.3). Noise reductions up to 82% have been observed in such conditions in our laboratory. Generation of twin photons has also been observed in a parametric oscillator based on four-wave mixing. The experiment performed in the phase conjugation geometry in sodium vapor[18] has yielded 10% noise reduction.

5. Reduction of quantum fluctuations using servo techniques

A common technique to reduce the intensity fluctuations of a laser beam consists in using an opto-electronic servo loop. A beamsplitter reflects off a part of the beam, which is detected with a photodiode. The photocurrent is then processed in an electronic loop to provide an appropriate correction signal reacting on either the pump beam intensity (feedback) or on the signal beam itself (feedforward). However, this scheme cannot be used for the correction of quantum fluctuations, because the beamsplitter couples the vacuum fluctuations to the beam through its unused port.

It is possible to generate squeezed light in such a system by using a twin beam generator instead of a beamsplitter. The system now allows correction of the quantum fluctuations. The theoretical noise reduction attainable can be determined using the semi-classical approach. The significant effect of a feedforward loop is to correct the intensity fluctuations δI_1 of beam 1 by a term proportional to the intensity fluctuations δI_2 of beam 2 (fig.4). Neglecting the losses of beam 1, the intensity fluctuations δI_{ff} after correction become

$$\delta I_{ff} = \delta I_1 - g\ \delta I_2\ . \tag{21}$$

The intensity noise spectrum $S_{ff}(\Omega)$ normalized to its shot noise reference is then given by:

$$S_{ff}(\Omega) = (1 - g^2)\ S_I(\Omega) + 2g\ S_{I_1-I_2}(\Omega) \tag{22}$$

where $S_I(\Omega)$ is the intensity noise spectrum of a single beam at the OPO output (assumed to be the same for the two beams) and $S_{I_1-I_2}(\Omega)$ is the noise spectrum of the intensity difference. For each frequency Ω, the intensity noise $S_{ff}(\Omega)$ reaches its minimum value when the gain g is set to a value $g = g^{opt}(\Omega)$, which depends on $S_I(\Omega)$ and $S_{I_1-I_2}(\Omega)$.

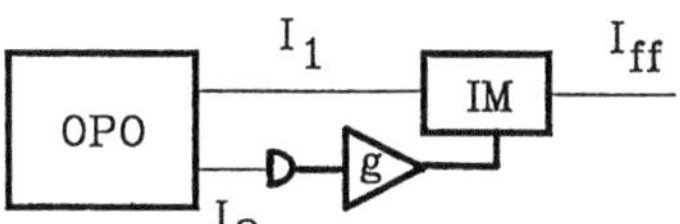

Figure 4. Feedforward control scheme (IM: intensity modulator, g: variable gain amplifier)

When the noise $S_I(\Omega)$ is large compared to one, the optimum gain g^{opt} is equal to one and the optimally corrected intensity noise is simply $S_{ff}(\Omega) = 2\, S_{I_1-I_2}(\Omega)$. In particular, when the two beams at the OPO output are not correlated at the quantum level ($S_{I_1-I_2}(\Omega) \geq 1$), the feedforward loop cannot reduce the intensity noise below the shot noise: only classical excess noise of beam 1 can be corrected. When excess noise becomes smaller ($S_I(\Omega) \approx 1$), noise levels less than $2\, S_{I_1-I_2}(\Omega)$ can be reached for gains smaller than one.

The experimental layout[19] comprises a feedforward correction loop consisting of a variable gain amplifier driven by the photodetector of beam 2 which monitors an electro-optic modulator inserted beam 1. The modulator acts as an analog shutter with a transmission proportional to the applied voltage. Total optical losses of beam 1 (including transmission bias) are about 20%.

The noise reduction is sensitive both to the gain and phase lag of the correction signal, which are properly matched in a frequency range centered here around 5 MHz. The reduction in intensity noise is up to 24% below the shot noise level. Intensity noise reductions of 22 % have been observed in a feedback configuration using the twin beams emitted in parametric fluorescence[20]

A different technique has also been used to squeeze the intensity. It does not use a parametric generator, but a laser, the output noise of which is reduced by modifying the noise of the pumping system[21].

In conclusion, we have seen that in recent years, there has been significant success in squeezing. Applications of quantum noise reduction are now foreseen. They concern ultrasensitive measurements in physics (detection of gravitational waves, parity violation in atomic physics, new spectroscopic techniques) and intrinsicly new phenomena (quantum nondemolition measurements, two-photon interferences, inhibited relaxation), or even technical uses which go beyond the frame of fundamental physics (communication systems, gyrolasers).

The Laboratoire de Spectroscopie Hertzienne de L'Ecole Normale Supérieure et de l'Université Pierre et Marie Curie is associated with the CNRS (URA 0018). This work has been supported in part by the EEC contract ESPRIT BRA 3186.

1- Slusher R.E, Hollberg L.W., Yurke B., Mertz J.C. and Valley J.F. 1985, Phys. Rev.Letters **55**, 2409; Slusher R.E., Yurke B., Grangier P., LaPorta A., Walls D.F.and Reid M. 1987, J. Optical Society Am. **4**, 1453

2- Wu L.A., Kimble H.J., Hall J.L.and Wu H. 1986, Phys. Rev. Let. **57**, 2520; Wu L.A., Min Xiao and Kimble H.J. 1988, J. Opt. Soc. Am. **B4**, 1465

3- Yuen H.P. and Shapiro J.H. 1979, Optics Lett. **4**, 334

4- Reynaud S. and Heidmann A. 1989, Opt. Commun. **71**, 209

5- Shelby R.M., Levenson M.D, Perlmutter S.H., DeVoe R.G. and Walls D.F. 1986, Phys. Rev. Lett. **57**,691; Sshumaker B.L., Perlmutter S.H., Shelby R.M. and Levenson M.D. 1987, Phys. Rev. Lett. **58**, 357

6- Bergman K. and Haus H.A. in OSA Meeting Technical Digest 1990, Vol. 15 (OSA, Washington D.C., 1990) p 265

7- Slusher R.E., Grangier P., LaPorta A., Yurke B. and Potasek M.J. 1987, Phys. Rev. Lett. **59**, 2566
8- Hirano T. and Matsuoka M. 1990, in IQEC Technical Digest Series Vol. 8 (OSA, Washington D.C.) p 391
9- Kennedy T.A.B., Anderson T.B. and Walls D.F. 1989, Phys. Rev. **A40**, 1385
10- Pereira S.F., Min Xiao , Kimble H.J. and Hall J.L. 1988, Phys. Rev. **A38**, 4931
11- Sizmann A., Horowicz R.J., Wagner G. and Leuchs G. 1990, Opt. Commun. (to be published)
12- Reynaud S. Fabre C. and Giacobino E. 1987, J. Optical Society Am. **B4**, 1520
13- Rarity J.G. and Tapster P.R. 1990, in IQEC Technical Digest Series Vol. 8 (OSA, Washington D.C.) p 8
14- Abram I., Raj R.K., Oudar J.L. and Dolique G. 1987, Phys. Rev. Lett. **57**, 2516
15- Aytur O. and Kumar P. 1990, Phys. Rev.Lett. **65**, 1551
16- Fabre C., Giacobino E., Heidmann A. and Reynaud S. 1989, J. de Physique **50**, 1209
17- Heidmann A., Horowicz R.J., Reynaud S., Giacobino E., Fabre C. and Camy G. 1987, Phys. Rev. Lett. **59**, 2555; Debuisschert T., Reynaud S., Heidmann A., Giacobino E. and Fabre C. 1989, Quantum Opt. **1**, 3
18- Vallet M., Pinard M. and Grynberg G. 1989, Europhys. Lett. **11**, 739
19- Mertz J., Heidmann A., Fabre C., Giacobino E. and Reynaud S. 1990, Phys. Rev. Lett. **64**, 2897
20- Tapster P.R., Rarity J.G. and Satchell J.S. 1988, Phys. Rev. **A37**, 2963
21- Machida S. and Yamamoto Y. 1988, Phys. Rev. Lett. **60**, 792; Richardson W.H., Machida S. and Yamamoto Y 1990, in IQEC Technical Digest Series Vol. 8 (OSA, Washington D.C.) p 395

Optical Switching Induced by the Variation of the Squeezing Phase

P. Galatola[1], *L.A. Lugiato*[1], *M.G. Porreca*[2], and *P. Tombesi*[2]

[1]Dipartimento di Fisica, Politecnico di Torino, Corso Duca degli Abruzzi 24, I-10129 Torino, Italy

[2]Dipartimento di Fisica, Università di Roma "La Sapienza", Piazzale Aldo Moro 2, I-00185 Roma, Italy

Abstract. We consider the standard model of single-mode optical bistability and assume that the atoms interact, in addition to the cavity mode which is pumped by an external coherent field, also with a beam in a broadband squeezed vacuum state. We show that the presence of the squeezed field changes the atomic decay rates and therefore it influences the shape of the steady state transmission function of the output versus the input field. Under appropriate conditions the controlled variation of the phase of the squeezed beam can force the system to switch from a low to a high transmission state and vice versa, without changing any other parameter of the bistable system (for example the input field).

1. Introduction

The distinctive feature of the squeezed states of light is the reduction of the quantum fluctuations in one quadrature component below the vacuum limit, at the expense of an increase in the noise in the orthogonal quadrature component. Some recent works [1–3] have pointed out that this phase sensitivity of the squeezed states can induce relevant effects in the interaction with matter. Examples of these phenomena are the inhibition of the atomic phase decay rates [1] or the suppression of one sideband of the resonance fluorescence spectrum [2].

We report another interesting effect produced by the squeezed states phase dependence, namely the capability of inducing the switching of an optical bistable system, driven by a constant coherent input field, by varying the phase of a squeezed vacuum control beam interacting with the two-level atoms contained in the optical cavity.

2. The Model

The system we consider is the standard model of single-mode optical bistability with two-level atoms [4,5], with the addition of a beam in a squeezed vacuum state injected into the cavity with a slight tilt with respect to the cavity axis, such that its interaction with the atoms is basically phase matched with that of the cavity mode pumped by the coherent input field but it does not contribute

Springer Proceedings in Physics, Vol. 55
Nonlinear Dynamics and Quantum Phenomena in Optical Systems
Editors: R. Vilaseca, R. Corbalán

to the cavity mode itself. Our configuration is thus similar to the "bright squeezing" model [6], with the difference that in our case the tilted beam has a zero mean value and is squeezed: such a beam could be obtained, for instance, by a degenerate optical parametric oscillator below threshold [7].

By introducing the following assumptions: a) the spectrum of the control beam can be considered constant over frequency intervals on the order of the relaxation rates of the system, b) the difference between the mean frequency ω_s of the squeezed beam and the cavity mode resonant frequency ω_c is much smaller than the free spectral range of the cavity, and c) the phase mismatch due to the slight tilt of the squeezed beam is negligible [8], one can describe the semiclassical behavior of our model by the following set of time evolution equations, which represents an extension of the standard one [5] to include also the squeezed control beam [1]

$$\frac{dx}{dt} = k[-(1+i\theta)x + y - 2Cv], \tag{1a}$$

$$\frac{dv}{dt} = \gamma_\perp[-(1+2\bar{N}+i\Delta)v - 2\bar{M}v^* + mx], \tag{1b}$$

$$\frac{dm}{dt} = -\gamma_\perp[x^*v + xv^* + 2(1+2\bar{N})m - 2], \tag{1c}$$

where x is the normalised mean value of the output field, k is the cavity line width, $\theta = (\omega_c - \omega_0)/k$ is the normalised detuning between the cavity eigenfrequency ω_c and the frequency ω_0 of the coherent input field y, C is the bistability parameter, v is the normalised mean value of the atomic polarization, $\gamma_\perp$ is the total decay rate of the polarization, $\Delta = (\omega_a - \omega_0)/\gamma_\perp$ is the normalised detuning between the atomic transition frequency ω_a and ω_0, m is the mean value of the atomic population inversion. The real parameter $\bar{N}$ and the complex quantity $\bar{M}$ are related to the correlation functions of the squeezed vacuum field operators b and $b^\dagger$

$$\langle b^\dagger(t)b(t')\rangle = N\delta(t-t'), \quad \langle b(t)b^\dagger(t')\rangle = (N+1)\delta(t-t'),$$

$$\langle b(t)b(t')\rangle = M\mathrm{e}^{-i\omega_s(t+t')}\delta(t-t'), \tag{2}$$

via

$$\bar{N} = sN \qquad \bar{M} = sM \qquad s = \frac{\gamma_s}{\gamma_\perp}, \tag{3}$$

where γ_s is the damping rate due to the squeezed beam. In Eqs. (2) the limit of delta correlation (white spectrum) is considered. Due to the Heisenberg uncertainty relations one must have $|M|^2 \leq N(N+1)$, which implies $|\bar{M}|^2 \leq \bar{N}(\bar{N}+s)$. The percentage of squeezing of the control beam is given by $200(|M|-N)$. For the sake of simplicity in Eqs. (1) we neglected the collisional decays.

We note that the presence of the squeezed vacuum field introduces a phase dependent relaxation process in the polarization—the term proportional to $\bar{M}$ in Eq. (1.b)—. In the following we want to exploit the effects which arise specifically from the possibility of controlling the phase of $\bar{M}$.

3. Steady-State Solution

The steady-state solution $\dot{x} = \dot{v} = \dot{m} = 0$ of Eqs. (1) leads to the steady state equation

$$y = (1+i\theta)x + 2C\frac{(1+2\bar{N}-i\Delta)x - 2\bar{M}x^*}{(1+2\bar{N})[(1+2\bar{N})^2+\Delta^2+|x|^2-4|\bar{M}|^2]-(\bar{M}x^{*2}+\bar{M}^*x^2)}, \tag{4}$$

where we can fix the phase of the input field by assuming that y is real and positive.

For suitable selections of the parameters of the system, the steady-state transmission function (4) of the normalised output amplitude $|x|$ versus the normalised input amplitude y has the characteristic bistable S-shape, with the negative slope portion unstable. In our situation, however, the actual shape of the bistable curve, as well as the location of the turning points, crucially depends on the phase of $\bar{M}$, i.e. on the orientation of the squeezing ellipse of the squeezed vacuum beam with respect to the input field. Thus the variation of the squeezing phase produces a deformation and a shift in the hysteresis cycle, in such a way that the system under appropriate conditions is forced to switch from the lower to the upper branch, and vice versa. This effect is obtained by keeping all the parameters of the system—including the input field—constant in time; only the phase of the squeezed vacuum field is varied in a controlled way.

Figure 1 shows that, when the phase Φ of the squeezed control beam—where Φ is defined through the relation $\bar{M} = |\bar{M}|\exp 2i\Phi$—is adiabatically varied from 0 to π, the output amplitude covers a full hysteresis cycle including both the low and the high transmission branch.

We would like to remark that the phenomenon we describe in this paper can offer the possibility of testing experimentally the fundamental effect of the modification of the atomic damping rates induced by the interaction with squeezed light, first predicted in [1], in a more accessible experimental configuration.

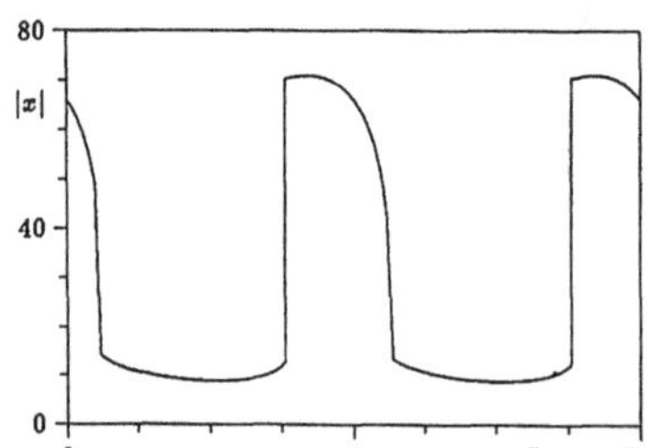

Fig. 1. Steady-state normalised amplitude $|x|$ of the output field as a function of the phase Φ of the squeezed beam. The parameters are $C = 160$, $\Delta = 20$, $\theta = 1$, $y = 75$, $\bar{N} = 0.3$, $|\bar{M}| = (\bar{N}(\bar{N}+s))^{1/2}$, where $s = \gamma_s/\gamma_\perp = 0.5$. With these parameters the control beam is in a minimum uncertainty state with a squeezing of 76%.

Acknowledgments

We are grateful to Dr. R. E. Slusher and Prof. H. J. Kimble for fruitful discussions. This work was completed in the framework of the Esprit Basic Research Action on Noise Reduction in Optical Systems (NOROS).

References

1) C. W. Gardiner, *Phys. Rev. Lett.* **56**, 1917 (1986).
2) H. J. Carmichael, A. S. Lane and D. F. Walls *Phys. Rev. Lett.* **58**, 2539 (1987).
3) J. M. Courty and S. Reynaud, *Europhys. Lett.* **10**, 237 (1988).
4) L. A. Lugiato in *Progress in Optics*, Vol. XXI, ed. by E. Wolf, North-Holland, Amsterdam, 1984; p. 69.
5) F. Castelli, L. A. Lugiato and M. Vadacchino, *Nuovo Cimento* D **10**, 183 (1988).
6) F. A. M. de Olivera and P. L. Knight, *Phys. Rev. Lett.* **61**, 830 (1988)
7) L. A. Wu, H. J. Kimble, J. L. Hall and H. Wu, *Phys. Rev. Lett.* **57**, 2520 (1986).
8) F. A. M. de Olivera and P. L. Knight, *Phys. Rev.* A **39**, 3417 (1989) .

Intensity Squeezing in Lasers with Two-Photon Saturable Absorbers

P. García Fernández[1], *L. Lugiato*[2], *and F.J. Bermejo*[1]

[1]Instituto de Estructura de la Materia, CSIC,
Serrano 119, E-28006 Madrid, Spain
[2]Dipartimento di Fisica, Politecnico di Torino,
Corso Duca degli Abruzzi 24, I-10129 Torino, Italy and
Consorzio INFM, unità di Torino Politecnico

Abstract. It is shown that a laser with two-photon saturable absorber exhibits a phase instability similar to that of two–photon optical bistability. We calculate the spectrum of the intensity fluctuations in the output field, under conditions of exact resonance between the amplifier and the absorber. In the bad–cavity limit the relevant part of the spectrum assumes a simple lorentzian form; as the phase instability threshold is approached, the reduction of quantum noise becomes perfect for small frequencies.

1 Introduction

Despite the extensive literature on the one-photon laser with an intracavity saturable absorber[1,2], to our knowledge only one paper treats the case of a laser containing a saturable absorber which displays a two-photon transition[3]. As in Ref. 3, we focus on the case that the active (i.e. inverted) medium of the laser undergoes a one-photon transition, and we assume exact resonance between the cavity, the frequency of the one-photon transition in the amplifier and one half the frequency of the two-photon transition in the absorber. The configuration of the laser with the two–photon saturable absorber (L2PSA) is closely related to that of two–photon optical bistability[4] (2POB); the basic difference is that in the L2PSA the active and the passive media are contained in the same cavity, whereas in the case of the 2POB the laser is contained in a cavity different from that which includes the passive material.

In this paper we focus on the analysis of the quantum mechanical fluctuations of the output field. Because in the L2PSA, contrary to the case of the 2POB, there is no driving input field, the output field has no reference phase and therefore exhibits phase diffusion. Hence only the fluctuations in the radial direction, i.e. the amplitude fluctuations, can exhibit squeezing at steady state. We calculate, for a single-

Springer Proceedings in Physics, Vol. 55
Nonlinear Dynamics and Quantum Phenomena in Optical Systems
Editors: R. Vilaseca, R. Corbalán

ended cavity, the spectrum of the intensity fluctuations, which, as it is demonstrated in Ref. 5, coincides with the spectrum of squeezing of the amplitude fluctuations.

The most interesting case arises in bad-cavity conditions, i.e. when the ratio of the cavity damping constant, κ, to the atomic relaxation rates, γ, is substantially larger than unity. This situation favours the existence of the phase instability and, as happens in general in the case of two-photon processes[6,7], the best conditions for quantum noise reduction arise when we approach an instability threshold. The analysis of this paper follows closely the same line as described in Ref. 5 for the case of a laser containing an intracavity crystal which produces a second harmonic beam. Therefore we refer the reader to Ref. 5 for the detailed discussion of several technical points that are used also in this paper.

2 The quantum statistical model and its semiclassical equations

We consider a ring cavity, exactly resonant with the frequency ω_o, which contains an active homogeneously broadened two–level medium with transition frequency ω_o and a passive homogeneously broadened medium which undergoes a two–photon transition of frequency $2\omega_o$. The cavity mode of frequency ω_o is associated with creation and annihilation operators a^+ and a which obey the commutation relation

$$[a, a^+] = 1 \tag{2.1}$$

The ith atom of the active medium ($i = 1, 2, ..., N_1$) has raising and lowering operators r_{i1}^+, r_{i1}^-, and an inversion operator $r_{3i1} = (r_{i1}^+ r_{i1}^- - r_{i1}^- r_{i1}^+)/2$, which obey the commutation rules

$$\begin{aligned} [r_{i1}^+, r_{j1}^-] &= 2r_{3i}\delta_{ij}, \\ [r_{3i1}, r_{j1}^\pm] &= \pm r_{i1}^\pm \delta_{ij} \end{aligned} \tag{2.2}$$

The global active medium is described by these collective operators: the macroscopic polarization operators

$$R_1^\pm = \sum_{i=1}^{N1} r_{i1}^\pm e^{\pm i\vec{k}_o\cdot\vec{x}_i}, \tag{2.3}$$

where $|\vec{k}_o| = \omega_o/c$ and $\vec{x}_i$ is the position of the ith atom, and the total population inversion

$$R_{31} = \sum_{i=1}^{N_1} r_{3i1};$$ (2.4)

they obey the angular momentum commutation relations

$$[R_1^+, R_1^-] = 2R_{31}, \ [R_3, R^\pm] = \pm R^\pm.$$ (2.5)

Similarly, the atoms of the absorbing medium are described as effective two-level systems[8]. They are associated with raising and lowering operators $r_{i2}^+, r_{i2}^-, (i = 1, 2, \ldots, N_2)$ and inversion operators $r_{3i2} = (r_{i2}^+ r_{i2}^- - r_{i2}^- r_{i2}^+)/2$ which obey commutation rules identical to (2.2). The global passive medium is described by three collective operators $R_2^\pm$, R_{32} which are defined by Eqs. (2.3), (2.4) after replacement of the index 1 by 2; they obey the same commutation rules (2.5).

In the interaction picture, the density operator of the system obeys a master equation which is obtained, as usual, by combining the conservative Hamiltonian terms with the markoffian dissipative terms which describe the escape of photons from the cavity and the atomic decays and pump. The master equation, which holds in the uniform field limit approximation, reads

$$\frac{d\rho}{dt} = -\frac{i}{\hbar}[H_1 + H_2, \rho] + (\wedge_F + \wedge_{A1} + \wedge_{A2})\rho,$$ (2.6)

where H_1 represents the interaction between the cavity mode and the active system

$$H_1 = i\hbar g_1 (a^+ R_1^- + a R_1^+),$$ (2.7)

while H_2 describes the two-photon transition in the passive medium

$$H_2 = i\hbar g_2 [(a^+)^2 R_2^- + a^2 R_2^+] \, .$$ (2.8)

The Liouvillian $\wedge_F$ describes the losses of the cavity mode:

$$\wedge_F \rho = \kappa \{[a, \rho a^+] + [a\rho, a^+]\},$$ (2.9)

where κ is the damping constant. The Liouvillians $\wedge_{As} (s = 1, 2)$ describe the atomic decay and pump processes

$$\wedge_{As}\rho = \frac{1}{2}\sum_{i=1}^{N_s} \{\gamma_{\uparrow s}([r_{is}^+, \rho r_{is}^-] +$$

$$+[r^+_{is}\rho, r^-_{is}]) + \gamma_{\downarrow s}([r^-_{is}, \rho r^+_{is}] + [r^-_{is}\rho, r^+_{is}])+$$

$$+\eta_s([r_{3is}, \rho r_{3is}] + [r_{3is}\rho, r_{3is}])\} \quad s = 1,2 \tag{2.10}$$

where $\gamma_{\uparrow s}(\gamma_{\downarrow s})$ is the upward (downward) transition-rate from the lower (upper) to the upper (lower) level, and η_s is the dephasing rate due to elastic collisions. In terms of these quantities one defines the transverse and the parallel relaxation rates ($s = 1,2$)

$$\gamma_{\perp s} = \frac{1}{2}(\gamma_{\uparrow s} + \gamma_{\downarrow s} + \eta_s), \; \gamma_{\parallel s} = \gamma_{\uparrow s} + \gamma_{\downarrow s}, \tag{2.11}$$

and the unsaturated inversion per atom

$$\sigma_s = \frac{\gamma_{\uparrow s} - \gamma_{\downarrow s}}{\gamma_{\uparrow s} + \gamma_{\downarrow s}}; \tag{2.12}$$

for the absorber ($s = 2$) one has no pump action, i.e., $\gamma_{\uparrow 2} = 0$, and therefore $\sigma_2 = -1$.

The semiclassical equations obtained from the master equation (2.6) introducing the normalized quantities

$$x = \frac{\alpha}{\sqrt{N_s}}, \bar{v}_1 = \left(\frac{N_1\sigma_1}{2}\sqrt{\frac{\gamma_{\parallel 1}}{\gamma_{\perp 1}}}\right)^{-1} v_1,$$

$$\bar{m}_1 = \left(\frac{N_1\sigma_1}{2}\right)^{-1} m_1,$$

$$\bar{v}_2 = \left(-\frac{N_2}{2}\sqrt{\frac{\gamma_{\parallel 2}}{\gamma_{\perp 2}}}\right)^{-1} v_2, \; \bar{m}_2 = \left(-\frac{N_2}{2}\right)^{-1} m_2, \tag{2.13}$$

where N_s is the saturation photon number

$$N_s = \frac{\gamma_{\perp 1}\gamma_{\parallel 1}}{4g_1^2}, \tag{2.14}$$

read

$$\dot{x} = \kappa[-x + 2c_1\sigma_1\bar{v}_1 - 2c_2\bar{v}_2 x^*]$$

$$\dot{\bar{v}}_1 = \gamma_{\perp 1}[x\bar{m}_1 - \bar{v}_1]$$

$$\dot{\bar{m}}_1 = -\gamma_{\parallel 1}[\frac{1}{2}(x\bar{v}_1^* + x^*\bar{v}_1) + \bar{m}_1 - 1] \tag{2.15}$$

$$\dot{\bar{v}}_2 = \gamma_{\perp 2}[\rho x^2 \bar{m}_2 - \bar{v}_2]$$

$$\dot{\bar{m}}_2 = -\gamma_{\parallel 2}[\frac{\rho}{2}(x^2 \bar{v}_2^* + x^{*2}\bar{v}_2) + \bar{m}_2 - 1],$$

where we introduced the pump parameter of the active medium

$$C_1 = \frac{g_1^2 N_1 \sigma_1}{2\kappa\gamma_{\perp 1}}, \tag{2.16}$$

the bistability parameter of the absorber

$$C_2 = \frac{g_2 N_2}{2\kappa}\sqrt{\frac{\gamma_{\parallel 2}}{\gamma_{\perp 2}}} \tag{2.17}$$

and the parameter

$$\rho = \frac{1}{2}\frac{g_2}{g_1^2}\frac{\gamma_{\perp 1}\gamma_{\parallel 1}}{\sqrt{\gamma_{\perp 2}\gamma_{\parallel 2}}}\,. \tag{2.18}$$

3 Stationary states and linear stability analysis

The stationary–intensity solutions of Eq. (2.15) have been calculated in Ref. 3. There are normal stationary solutions, which obey Eq. (2.15) with vanishing time derivatives; and anomalous solutions in which the variables x, $\bar{v}_1$ and $\bar{v}_2$ display an oscillatory frequency factor. We will consider only the normal solutions. First, there is the trivial solution

$$x = \bar{v}_1 = \bar{v}_2 = 0,\ \bar{m}_1 = \bar{m}_2 = 1, \tag{3.1}$$

which corresponds to no emission. This solution becomes unstable when the pump parameter is increased beyond the threshold $C_1 = 1/2$. On the other hand, the nontrivial stationary solultion of Eq. (2.15) is

$$\bar{m}_{1S} = \frac{1}{1+I},\ \bar{m}_{2S} = \frac{1}{1+\rho^2 I},$$

$$|\bar{v}_{1S}|^2 = \frac{I}{(1+I)^2},\ |\bar{v}_{2S}| = \frac{\rho I}{1+\rho^2 I^2}, \tag{3.2}$$

where the quantity

$$I = |x_S|^2 \tag{3.3}$$

obeys the steady–state equation

$$\frac{2C_1\sigma_1}{1+I} - \frac{2C_2\rho I}{1+\rho^2 I^2} = 1 . \tag{3.4}$$

The phase of x_S is arbitrary, and we select it equal to zero; as a consequence, also the phases of the stationary values $\bar{v}_{1S}$ and $\bar{v}_{2S}$ are fixed equal to zero.

The steady-state curve of I as a function of the pump parameter C_1, governed by Eq. (3.4) is illustrated in Figs. 1,2 for various values of ρ and C_2. Its behaviour is qualitatively very different from the case of the one-photon absorber, studied in Ref. 1, in which under appropriate conditions one can obtain bistability between the trivial and the nontrivial stationary solutions. This phenomenon does not arise in the case of the L2PSA, in which, on the other hand, one can obtain bistability between different nontrivial stationary solutions, (see Fig. 2), an effect which is, conversely, not possible in the case of the one-photon absorber. We observe that for the values of the rate constant specified in the caption of Fig. 2 one has no bistability because almost the entire lower branch of the steady state curve is unstable. However, if we choose the rate constants according to "good-cavity" conditions $\kappa << \alpha_{\perp s}$, $\gamma_{\parallel s} (s = 1, 2)$ both the lower and the upper branch of the steady-state curve are stable and hence one obtains bistability.

Let us now pass to the linear stability analysis of the nontrivial stationary solutions. One introduces the deviations of the variables from their respective stationary values (with the phases equal to zero, as already discussed), and one linearizes the semiclassical equations with respect to the deviations. If one reformulates the linearized system in terms of real and imaginary parts of the deviations, one finds that the two sets of variables obey two separate sets of equations. Precisely, the set of equations for the real parts, which governs the amplitude fluctuations, leads to the following fifth-order eigenvalue equation:

$$\det \begin{bmatrix} -\kappa - 2C_2\bar{v}_{2s}\kappa - \lambda & \kappa 2C_1\sigma_1 & 0 & -2C_2 x_s\kappa & 0 \\ \gamma_{\perp 1}\bar{m}_{1s} & -\gamma_{\perp 1} - \lambda & \gamma_{\perp 1} x_s & 0 & 0 \\ -\gamma_{\parallel 1}\bar{v}_{1s} & -\gamma_{\parallel 1} x_s & -\gamma_{\parallel 1} - \lambda & 0 & 0 \\ \gamma_{\perp 2} 2\rho x_s \bar{m}_{2s} & 0 & 0 & -\gamma_{\perp 2} - \lambda & \gamma_{\perp 2}\rho x_s^2 \\ -\gamma_{\parallel 2} 2 x_s \bar{v}_{2s}\rho & 0 & 0 & -\gamma_{\parallel 2}\rho x_s^2 & -\gamma_{\parallel 2} - \lambda \end{bmatrix}$$

$$= 0; \tag{3.5}$$

the set of equations for the imaginary parts, which governs the phase fluctuations, leads to a cubic eigenvalue equation:

$$\lambda\left[\lambda^2+\lambda(\gamma_{\perp 2}+\gamma_{\perp 1}+\kappa(1-2C_2\overline{v}_{2s}))+\right.$$

$$+\gamma_{\perp 1}\gamma_{\perp 2}+\kappa(\gamma_{\perp 1}+\gamma_{\perp 2})-\kappa(\gamma_{\perp 1}-\gamma_{\perp 2})2C_2\overline{v}_{2s}-$$

$$\left.-\kappa\gamma_{\perp 1}\sigma_1 2C_1\overline{m}_{1s}\right]=0. \tag{3.6}$$

The presence of an eigenvalue $\lambda = 0$ in Eq. (3.6) reflects the arbitrariness of the steady state values of the phases.

The nontrivial stationary solution is stable if and only if all the roots of Eqs. (3.5), (3.6) have a nonpositive real part. Thus, Eq. (3.6) leads to a pair of stability conditions, which read

$$\gamma_{\perp 1}+\gamma_{\perp 2}+\kappa(1-2C_2\frac{y}{1+y^2})>0, \tag{3.7}$$

$$\gamma_{\perp 1}\gamma_{\perp 2}+\kappa\gamma_{\perp 2}-\kappa 2C_2\frac{y}{1+y^2}(2\gamma_{\perp 1}-\gamma_{\perp 2})>0, \tag{3.8}$$

where $y=\rho x_s^2=\rho I$ and we used the steady state relations (3.2), (3.4).

The value of I, i.e., of y, for which (3.8) becomes an equality, corresponds to the threshold of a steady bifurcation, in which an anomalous stationary-intensity solution bifurcates from the nontrivial stationary solution. In the following, we exclude this possibility by assuming that

$$\gamma_{\perp 2}>2\gamma_{\perp 1}, \tag{3.9}$$

so that condition (3.8) is automatically satisfied. On the other hand, the value of y for which (3.7) becomes an equality corresponds to the threshold of a Hopf bifurcation, such that a periodic oscillatory solution bifurcates from the nontrivial stationary solution. By solving Eq. (3.7) with respect to y, we find that the steady-state curve is unstable in the interval

$$y_-=\bar{C}_2-\sqrt{\bar{C}_2^2-1}<y<\bar{C}_2^2-1=y_+, \tag{3.10}$$

when

$$\bar{C}_2=C_2\left[1+\frac{\gamma_{\perp 1}+\gamma_{\perp 2}}{\kappa}\right]^{-1}, \tag{3.11}$$

provided that $\bar{C}_2>1$.

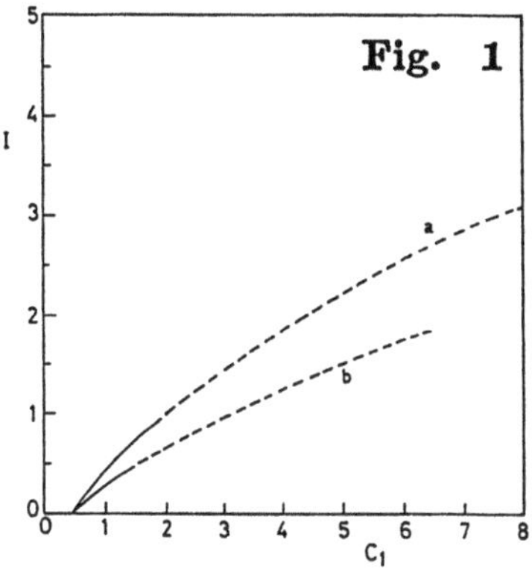

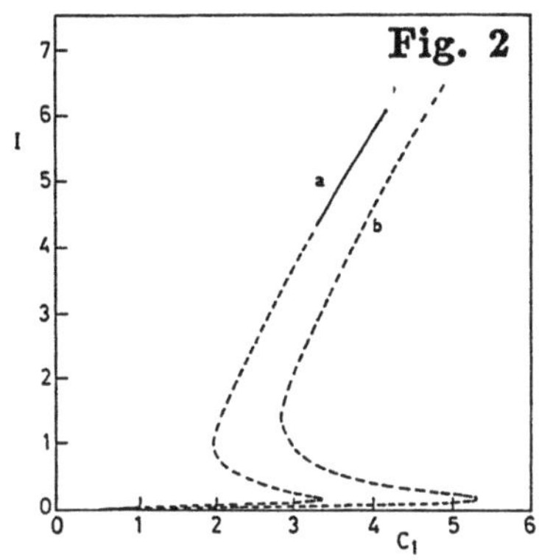

Fig. 1: Steady–state curve of the output intensity as a function of the control parameter C_1. The value of the parameter ρ is 0.1 and (a) corresponds to $C_2 = 5$; (b) corresponds to $C_2 = 10$. The broken line indicates the unstable part of the steady-state curve for $\kappa/\gamma_{\perp 1} = 2000$, $\kappa/\gamma_{\perp 2} = 500$, $\gamma_{\parallel 1} = 2\gamma_{\perp 1}, \gamma_{\parallel 2} = 2\gamma_{\perp 2}$.
Fig. 2: The same as Fig. 1 for $\rho = 10$.

The amplitude characteristic equation (3.5) has been analyzed numerically and the stability of the stationary solutions in the bad–cavity limit is shown in Figs. 1–2. The parameters are $\kappa/\gamma_{\perp 1} = 2000$, $\kappa/\gamma_{\perp 2} = 500$, $\gamma_{\parallel 1} = 2\gamma_{\perp 1}$, $\gamma_{\parallel 2} = 2\gamma_{\perp 2}$, $\sigma_1 = 1$.

Fig. 1 a,b show the case $\rho = 0.1$, $C_2 = 5(a)$, $C_2 = 10(b)$. The phase instability threshold is in $C_1 = 2.01$ for $C_2 = 5$ and in $C_1 = 1.50$ for $C_2 = 10$. It can be observed that the corresponding intensities decrease as C_2 becomes greater.

When $\rho = 10$ (Fig. 2), the steady state curve is S–shaped; the phase–instability domains start when C_1 is extremely close to unity for both $C_2 = 5$ (a) and $C_2 = 10$ (b) and the corresponding intensities are very low. In the case $\rho = 10$, $C_2 = 5$ the phase–instability domain stops at $C_1 = 1.9899$ but the steady state is amplitude unstable until $C_1 = 3.40$ ($I = 4.58$), whereas for $C_2 = 10$ the phase instability domain stops at $C_1 = 2.9950$, but the amplitude instability continues much further.

4 The linearized Fokker-Planck equation in the P–representation

Using the quantum-classical correspondence, one can translate the operator master equation (2.6) into a classical-looking partial differential equation for the quasiprobability distribution $P(\alpha, \alpha*, v_1, v_1^*, m_1, v_2, v_2^*, m_2, t)$ which generalizes to the atomic system the Glauber P-function[10]. The moments of P correspond to the normally ordered expectation values. The partial differential equation for the generalized P–function has derivatives of all orders with respect to the variables m_1 and m_2; however, when the numbers N_1 and N_2 of atoms are very large, the terms with derivatives of order higher than second can be neglected and the equation reduces to a Fokker-Planck equation (FPE)

$$\frac{\partial}{\partial t} P(\alpha, \alpha^*, v_1, v_1^*, m_1, v_2, v_2^*, m_2, t) = \wedge P. \tag{4.1}$$

Normalizing variables as in Eq. (2.14), linearizing around a nontrivial stationary solution and changing variables to real and imaginary parts (compare with Ref. 5 for a detailed description of the procedure), Eq. (4.1) becomes

$$\frac{\partial}{\partial t} P(\delta x_R, \delta x_I, \delta v_{1R}, \delta v_{1I}, \delta m_1, \delta v_{2R}, \delta v_{2I}, \delta m_2, t) = \wedge P. \tag{4.2}$$

From now on, for the sake of simplicity we set $\sigma_1 = 1$. By introducing the 8-component vector:

$$\vec{q} \equiv \begin{bmatrix} \delta x_R \\ \delta v_{1R} \\ \delta m_1 \\ \delta v_{2R} \\ \delta m_2 \\ \delta x_I \\ \delta v_{1I} \\ \delta v_{2I} \end{bmatrix}, \tag{4.3}$$

and the normalized time variable

$$\tau = \kappa t, \tag{4.4}$$

the linearized FPE (4.2) can be cast in a compact form in the following way:

$$\frac{\partial}{\partial \tau} P(\vec{q}, t) = - \sum_{i,k=1}^{8} \frac{\partial}{\partial q_i} M_{ik} q_k P(\vec{q}, t) +$$

$$+ \frac{1}{2} \sum_{i,j=1}^{8} \frac{\partial^2}{\partial q_i \partial q_j} D_{ij} P(\vec{q}, t), \tag{4.5}$$

the drift matrix M_{ik} splits into a 5×5 matrix for the amplitude fluctuations and a 3×3 matrix for the phase fluctuations. The amplitude drift matrix reads:

$$M_R = \begin{bmatrix} -(1+2C_2\bar{v}_{2S}) & 2C_1 & 0 & -2C_2x_S & 0 \\ \tilde{\gamma}_{\perp 1}\bar{m}_{1S} & -\tilde{\gamma}_{\perp 1} & \tilde{\gamma}_{\perp 1}x_S & 0 & 0 \\ -\tilde{\gamma}_{\parallel 1}\bar{v}_{1S} & -\tilde{\gamma}_{\parallel 1}x_S & -\tilde{\gamma}_{\parallel 1} & 0 & 0 \\ \tilde{\gamma}_{\perp 2}\rho 2x_S\bar{m}_{2S} & 0 & 0 & -\tilde{\gamma}_{\perp 2} & \tilde{\gamma}_{\perp 2}\rho x_S^2 \\ -\tilde{\gamma}_{\parallel 2}\rho 2x_S\bar{v}_{2S} & 0 & 0 & -\tilde{\gamma}_{\parallel 2}\rho x_S^2 & -\tilde{\gamma}_{\parallel 2} \end{bmatrix}, \tag{4.6}$$

where we set

$$\tilde{\gamma}_{\perp 1} = \gamma_{\perp 1}/\kappa,\ \tilde{\gamma}_{\perp 2} = \gamma_{\perp 2}/\kappa,$$

$$\tilde{\gamma}_{\parallel 1} = \gamma_{\parallel 1}/\kappa, \tilde{\gamma}_{\parallel 2} = \gamma_{\parallel 2}/\kappa. \tag{4.7}$$

The phase drift matrix reads:

$$M_I = \begin{bmatrix} -(1-2C_2\bar{v}_{2S}) & 2C_1 & -2C_2x_S \\ \tilde{\gamma}_{\perp 1}\bar{m}_{1S} & -\tilde{\gamma}_{\perp 1} & 0 \\ \tilde{\gamma}_{\perp 2}\rho 2x_S\bar{m}_{2S} & 0 & -\tilde{\gamma}_{\perp 2} \end{bmatrix}. \tag{4.8}$$

In Eqs. (4.6), (4.8) the quantities v_{1s}, v_{2s}, m_{1s} and m_{2s} must be replaced by their expressions as a function of x_s, see Eqs. (3.2), (3.3). Let us now consider the normalized diffusion matrix

$$\tilde{D}_{ij} = 8N_sD_{ij}. \tag{4.9}$$

By taking into account the steady-state relations (3.2), (3.3) and the identities

$$\frac{1}{N_1} = \frac{1}{N_s}\frac{\gamma_{\parallel 1}}{8C_1\kappa},$$
$$\frac{1}{N_2} = \frac{\rho}{N_s}\frac{\gamma_{\parallel 2}}{4C_2\kappa}, \tag{4.10}$$

we obtain

$$\tilde{D}_{11} = -8C_2\frac{\rho x_S^2}{1+\rho^2x_S^4}$$

$$\tilde{D}_{22} = \frac{\tilde{\gamma}_{\parallel 1}\tilde{\gamma}_{\perp 1}}{C_1}\left[(2\frac{\tilde{\gamma}_{\perp 1}}{\tilde{\gamma}_{\parallel 1}} - 1)(1+\frac{1}{1+x_S^2}) + 2 + \frac{x_S^2}{1+x_S^2}\right]$$

$$\tilde{D}_{33} = 0$$

$$\tilde{D}_{44} = \frac{2\rho\tilde{\gamma}_{\|2}\tilde{\gamma}_{\perp 2}}{C_2}[-\frac{1}{1+\rho^2 x_S^4}(\rho^2 x_S^4 + 2\frac{\tilde{\gamma}_{\perp 2}}{\tilde{\gamma}_{\|2}} - 1) + 2\frac{\tilde{\gamma}_{\perp 2}}{\tilde{\gamma}_{\|2}} - 1]$$

$$\tilde{D}_{55} = \frac{4\rho\tilde{\gamma}_{\|2}^2}{C_2}\frac{2\rho^2 x_s^4}{1+\rho^2 x_S^4}$$

$$\tilde{D}_{66} = 8C_2\frac{\rho x_S^2}{1+\rho^2 x_S^4}$$

$$\tilde{D}_{77} = \frac{\tilde{\gamma}_{\perp 1}\tilde{\gamma}_{\|1}}{C_1}\left[(2\frac{\tilde{\gamma}_{\perp 1}}{\tilde{\gamma}_{\|1}} - 1)(1 + \frac{1}{1+x_S^2}) + 2 - \frac{x_S^2}{1+x_S^2}\right]$$

$$\tilde{D}_{88} = \frac{2\rho\tilde{\gamma}_{\perp 2}\tilde{\gamma}_{\|2}}{C_2}\left[\frac{1}{1+\rho^2 x_S^4}(\rho^2 x_S^4 - 2\frac{\tilde{\gamma}_{\perp 2}}{\tilde{\gamma}_{\|2}} + 1) + 2\frac{\tilde{\gamma}_{\perp 2}}{\tilde{\gamma}_{\|2}} - 1\right]$$

$$\tilde{D}_{23} = \tilde{D}_{32} = \frac{-2\tilde{\gamma}_{\|1}^2}{C_1}\frac{x_S}{1+x_S^2}, \tag{4.11}$$

the rest of the elements being zero.

5 The spectrum of the intensity fluctuations

As it is shown in Ref. 5, if we use the normalized frequency

$$\bar{\omega} = \omega/\kappa, \tag{5.1}$$

the spectrum of the intensity fluctuations for the output field is given by

$$S(\bar{\omega}) = S_{11}^R(\bar{\omega}), \tag{5.2}$$

where the 5 x 5 matrix S^R is defined as

$$S^R(\bar{\omega}) = (M_R + i\bar{\omega})^{-1}\tilde{D}_R(M_R^T - i\bar{\omega})^{-1}, \tag{5.3}$$

and $\tilde{D}_R$ is the 5x5 matrix given by the elements $\tilde{D}_{ij}$ of the matrix $\tilde{D}$ with $i,j = 1,\ldots,5$. With this definition, one has quantum noise reduction when $S(\bar{\omega}) < 0$ and complete reduction at frequency $\bar{\omega}$ when $S(\bar{\omega}) = -1$.

In Fig. 3 the spectrum is shown for $\rho = 0.1$, $C_2 = 5$; the values of C_1 are 2.0 (a), 1.5 (b) and 1.0 (c) for the same values of the parameters

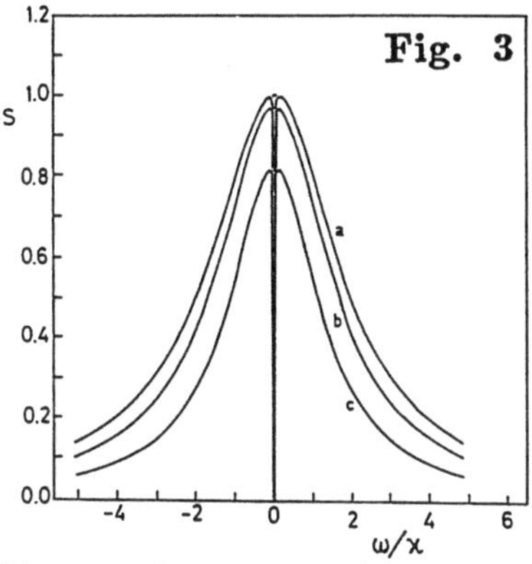

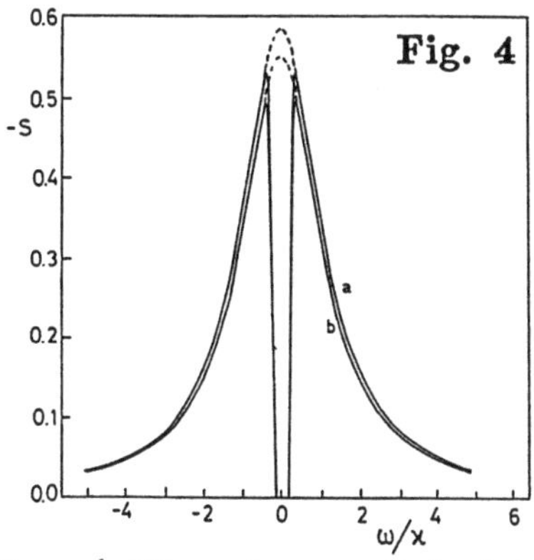

Fig. 3: Spectrum of intensity fluctuations (with a change of sign) in the bad-cavity limit. The values of the parameters are the same as in Fig. 1, and the parameter ρ is 0.1 and $C_2 = 5$. Curve (a) corresponds to $C_1 = 2.0$, (b) $C_1 = 1.5$, (c) $C_1 = 1.0$.
Fig. 4: The same as Fig. 3 for $\rho = 10$, $C_2 = 5$. Curve (a) corresponds to $C_1 = 3.4$, and (b) $C_1 = 3.6$. The broken line is the lorentzian in Eq. (5.4).

$\tilde{\gamma}_{\perp 1}$, $\tilde{\gamma}_{\perp 2}$, $\tilde{\gamma}_{\parallel 1}$, $\tilde{\gamma}_{\parallel 2}$ as in Fig. 1. It can be observed that the degree of squeezing increases as C_1 approaches the phase instability threshold and simultaneously the width of the spectrum becomes greater. In the case of $\rho = 0.1$, $C_2 = 10$, we have found that the squeezing spectra in the neighbourhood of the phase–instability threshold are exactly the same as for $\rho = 0.1$, $C_2 = 5$. The same holds in the cases $\rho = 1.0$, $C_2 = 5$, and $\rho = 10$, $C_2 = 10$. For $\rho = 1.0$, $C_2 = 10$, when C_1 is sufficiently far from the phase–instability threshold, the squeezing becomes much smaller. The same results remain valid for $\rho = 10$, $C_2 = 5$ for intensities below the lower phase–instability threshold y_-; on the other hand, above the upper phase–instability threshold y_+, as C_1 increases, the squeezing decreases and the spectrum becomes narrower (Fig. 4).

These results are in agreement with the general analysis of Ref. 7, which points out that in two–photon processes the squeezing is sizable in the vicinity of the phase–instability threshold and the squeezing spectrum can be divided in two parts, one of width on the order of the cavity linewidth, κ, and the other of width of the other relaxation rates γ, ($\kappa >> \gamma$). The latter part determines only the structure of the spectrum in a neighbourhood of $\omega = 0$. The squeezing spectra obtained in the bad–cavity limit $\kappa >> \gamma_{\perp 1}$, $\gamma_{\perp 2}$ fit the Lorentzian curve predicted in Ref. 7, which is given by

$$S(\bar{\omega}) = \frac{-4D}{(D+1)^2 + \bar{\omega}^2}, \tag{5.4}$$

where

$$D = 2C_2 \frac{\rho x_S^2}{1 + \rho^2 x_S^4}. \tag{5.5}$$

This curve is represented by a broken line in the spectra shown in Figs. 3–4. It should be noted that the two squeezing spectra shown in Fig. 4, which correspond to stationary states in the upper branch of the steady–state curve ($\rho = 10$, $C_2 = 5$) do not present a good agreement with the analytical curve (5.4) for small frequencies. It calls for a closer approach to the bad–cavity limit to obtain a better agreement. A recent paper[11] studies the L2PSA in the good cavity limit, and shows that in this case the level of squeezing in the output field is modest.

References

[1] L.A. Lugiato, P. Mandel, S.T. Dembinski and A. Kossakowski, Phys. Rev. A 18, 238 (1978).

[2] N.B. Abraham, P. Mandel and L.M. Narducci, in Progress in Optics, Vol. XXV, ed. by E. Wolf, North-Holland, Amsterdam 1988, p.1.

[3] P. Mandel and F.K. Fang, Phys. Lett. 83 A, 59 (1981).

[4] P. Galatola, L.A. Lugiato, M. Vadacchino and N.B. Abraham, Opt. Commun. 63, 413 (1989).

[5] P. García–Fernåndez, L.A. Lugiato, F.J. Bermejo and P. Galatola, Quantum Optics 2, 49, 1990.

[6] M.J. Collett and D.F. Walls, Phys. Rev. A 32, 2887 (1985).

[7] L.A. Lugiato, P. Galatola and L.M. Narducci, Opt. Commun. 76, 276 (1990).

[8] L.M. Narducci, W.W. Eidson, P. Furcinitti and D.C. Eteson, Phys. Rev. A 16, 1665 (1977).

[9] H. Haken, Laser Theory, in Encyclopedia der Physik vol XXV/2c (Berlin, Springer-Verlag). p. 153-155.

[10] R.J. Glauber, Phys. Rev. 130, 2529 (1963), 131, 2766 (1963).

[11] M.L. Collett, R.B. Levien and D.F. Walls, to be published.

Quantum Corrections to the Berry Phase for the SO(2,1) Dynamically Invariant Parametric Oscillator

J.M. Cerveró and J.D. Lejarreta

Departamento de Física Téorica, Facultad de Ciencias,
Universidad de Salamanca, E-37008 Salamanca, Spain

Abstract. We prove that an exact expression can be given for the total phase acquired by the exact wavefunction of the vacuum parametric oscillator when a non-adiabatic cyclic change is carried out. The non-dynamical part of this exact phase is shown to be the Berry phase in the adiabatic approximation. The rest of the contributions when a $\hbar$-expansion is made in the exact formula correspond to the quantum corrections which were predicted to exist by Berry and other authors.

1. Introduction

It is now widely known that hamiltonian physical systems possessing a dynamical group of invariance have an exact solution provided by the squeezing operator associated with this dynamical group acting on the ground state of the hamiltonian at $t = 0$. This result has a wide range of applications in statistical models, spin systems and, indeed, quantum optics when we particularize the dynamical group to bosonic groups built up with various kinds of products of photon creation and annihilation operators.

The question to which we address ourselves in this paper is whether we can use these group theoretical properties for constructing the total phase of the wavefunction when the system undergoes temporal closed paths in parameter space. We have made no mention whatsoever of any adiabatic hypothesis so these time excursions do not have to be carried out slowly in order to preserve instantaneous eigenstates. What we are really looking for is a sort of suitably defined generalized Berry phase in such a way that, taking the adiabatic limit, the exact expression can be expanded in powers of $\hbar$. The zeroth order will necessarily be of purely geometrical nature and must coincide with the Berry phase. The rest of the terms in the expansion must be quantum corrections to the above which - to the authors' knowledge - have never been previously calculated. To remind the reader of the basics of the Berry phase we must recall that it is the extra phase that the wavefunction of the system can acquire under adiabatic cycling of the parameters upon which the hamiltonian depends. This

Springer Proceedings in Physics, Vol. 55
Nonlinear Dynamics and Quantum Phenomena in Optical Systems
Editors: R. Vilaseca, R. Corbalán

phase is geometrical in the sense that it only depends on the cycle of the system follows in the space of its parameters and not on the time taken for this adiabatic excursion [1-2].

2. The SO(2,1) dynamically invariant parametric oscillator

Let us consider the following hamiltonian:

$$H(t) = \hbar\omega_0 \, [\, f_0(t)K_0 + f(t)K_+ + f^*(t)K_- \,] \tag{1}$$

where the K-operators are given in terms of the usual products of creation and annihilation operators

$$K_0 = (1/4)[a^+, a]_+ \quad ; \quad K_+ = (1/2)a^{+2} \quad ; \quad K_- = (1/2)a^2 \tag{2}$$

and the f's are functions of time with initial conditions $f_0(0) = 1$ and $f(0) = 0$. The operators close the SO(2,1) Lie Algebra whose commutation rules are given by

$$[K_0 \,, K_\pm] = \pm K_\pm \quad ; \quad [K_- \,, K_+] = 2K_0 \ . \tag{3}$$

Let us now find the time evolution operator U(t), solution of

$$i\hbar \, \partial U(x,t)/\partial t = H(t)U(x,t) \quad ; \quad U(x,0) = 1 \ . \tag{4}$$

As is well known, one can write down U(t) in the form of a squeezing operator [3]:

$$\begin{aligned} U(t) &= \exp\{\beta(t)K_+ - \beta^*(t)\, K_-\} \, \exp\{ih(t)\} \\ &= \exp\{\xi K_+\} \, \exp\{\gamma K_0\} \, \exp\{-\xi^* K_-\} \end{aligned} \tag{5}$$

where $\xi = \tanh|\beta| \, e^{i \arg \beta}$ and $\gamma = -2\ln\cosh|\beta|$. We now solve the Schrödinger equation (4) using (1) and (5). A complete discussion of the solutions can be found in [4]. The final answer for $\xi(t)$, $h(t)$ and $\gamma(t)$ turns out to be

$$\dot{\xi}(t) = -i\omega_0 \, (\, f + f_0\xi + f^*\xi^2 \,) \tag{6.a}$$

$$h(t) = -\omega_0 \int (\, f_0 + f^*\xi + f\xi^* \,)dt' \tag{6.b}$$

$$\gamma(t) = \ln(1 - \xi\xi^*) \ . \tag{6.c}$$

The equation (6.a) is a Ricatti differential equation. Its physical meaning however becomes more transparent when we realize that we are dealing in fact with the Classical Equation of Motion. To see this we perform the change of variables:

$$\xi(t) = -[i\dot{q}(t)/\omega_0 f^*(t)q(t)] - (1/2f^*(t))[f_0(t) + (i/\omega_0)(\dot{f}^*(t)/f^*(t))] \quad (7)$$

with the initial conditions inherited from $f^*(0) = \dot{f}^*(0) = 0$

$$\dot{q}(0) = (i\omega_0 q(0)/2)\ [f_0(0) + (i/\omega_0)(\dot{f}^*(0)/f^*(0))] \quad (8)$$

Using (7) in (6.a) we obtain for the Ricatti equation:

$$d^2q(t)/dt^2 + \Omega^2(t)\, q(t) = 0 \quad (9)$$

where $\Omega^2(t)$ takes the form:

$$\Omega^2(t) = (\omega_0{}^2/4)\ [f_0{}^2 - 4ff^*] - i(\omega_0/2f^*)[\dot{f}_0 f^* - f_0\dot{f}^*] + [\ddot{f}^*/2f^*] - [3\dot{f}^{*2}/4f^{*2}] \quad (10)$$

After some non-trivial changes of variable one can finally recognize in expression (10) the classical frequency of the generalized time dependent harmonic oscillator. A more complete discussion of the relationship among Riccatti equation (6.a) and the equations of motion of the associated classical problem will be reported by the authors in a forthcoming paper to be published elsewhere.
Furthermore, in terms of this new variable q(t), the total phase h(t) takes the following form:

$$h(t) = i\ \ln\{\ [q(t)q^*(0)/q^*(t)q(0)]\ \ [f^*(t)/f(t)]^{1/2}\ \} \quad (11)$$

which we shall find very useful for future discussions to be found below.

3. Cyclic evolution. Quantum phases

Consider now a cyclic evolution of period τ of the hamiltonian H(t). It is clear that:

$$f_0(t+\tau) = f_0(t) \quad ; \quad f(t+\tau) = f(t) \quad \text{and} \quad \Omega(t+\tau) = \Omega(t)\ . \quad (12)$$

In this particular case the Classical Equation (9) is a Bloch equation and using Bloch theorem one can show that solutions exist verifying:

$$q(t+\tau) = \exp(iK\tau)q(t) \quad ; \quad \dot{q}(t+\tau) = \exp(iK\tau)\dot{q}(t) \quad (13.a)$$

$$\xi(t+\tau) = \xi(t) \qquad ; \qquad h(t+\tau) = h(t) - 2K\tau \tag{13.b}$$

$$U(\tau) = \exp \{ ih(\tau)K_0 \} \quad . \tag{13.c}$$

This implies that an initial eigenstate $|k,n>$ of $H(0)$ (or K_0) returns to the initial eigenstate after a time τ UP TO A PHASE:

$$U(\tau) \; |k,n> = \exp \{ ih(\tau) \, \varepsilon_n \} \, |k,n> \tag{14}$$

and indeed $K_0 \; |k,n> = \varepsilon_n \; |k,n>$. What we would like to show now is that without any adiabatic requirements this phase is not only due to the obvious <u>dynamical phase</u> acquired after the time excursion. This dynamical phase is of the form:

$$\Theta_d = - \, (1/\hbar) \int_0^{\tau} <\Psi(t)| \; H(t) \; |\Psi(t)> dt \quad . \tag{15}$$

Since in this hamiltonian we know the exact form of $|\Psi(t)>$ we can calculate explicitly the formal expression (15). The result is:

$$\Theta_d = \varepsilon_n \int_0^{\tau} [\, \omega_0 f_0(t) \; + \; (2\dot{h}(t))/(1 \; - \; \xi\xi^* \;) \;] \; dt \quad . \tag{16}$$

The expression (16) shows an obvious excess of phase compared with $h(t)$ of obvious non-dynamical nature [4-5]:

$$\Theta_g = \varepsilon_n h(t) \; - \; \Theta_d = - \, \varepsilon_n \int_0^{\tau} [\, \omega_0 f_0(t) \; + \; \dot{h}(t) \; \{(1 \; + \; \xi\xi^* \;)/(1 \; - \; \xi\xi^* \;)\} \;] \; dt \quad . \tag{17}$$

Notice that no adiabatic assumption has been made. To investigate the eventual geometrical character of this non-adiabatic phase we have to show first that its adiabatic limit indeed yields Berry's result [2]. This can easily be done by expanding $\xi(t)$ in powers of a small parameter in the form $\xi(t) = \Sigma \, \xi_k \, \varepsilon^k$. Substituting in the Ricatti equation (6.a) and considering only the first order in ε we find for our general expression (17) the following approximate form:

$$\Theta_g = - \, 4 \; \varepsilon_n \int_0^{\tau} [\mathrm{Ref} \; d(\mathrm{Imf}) \; - \; \mathrm{Imf} \; d(\mathrm{Ref})]/[f_0 F \; + \; F^2] \tag{18}$$

where $F^2 = [f_0{}^2 - 4ff^*]$. This is indeed the result of [2]. Also higher orders in ε for the general expression (17) yield the $O(\hbar^n)$ which represent the quantum corrections of the Berry phase [6]. We have omitted these expressions for obvious lack of space due to their outstanding length but nothing in principle can prevent the interested reader from reproducing them by himself following this systematic procedure.

REFERENCES

[1] **Berry M.V.** (1984) Proc. Royal Soc. London A. **392**, 45
[2] **Berry M.V.** (1985) J. Phys. A: Math. Gen. **18**, 15
[3] **Perelomov A.** (1972) Commun. Math. Phys. **26**, 222
[4] **Cerveró J.M. & J.D. Lejarreta** (1990) Quantum Optics **2**, 333
[5] **Cerveró J.M. & J.D. Lejarreta** (1989) J. Phys. A: Math. Gen. **22**, L663
[6] **Berry M.V.** (1987) Proc. Royal Soc. London A. **414**, 31

Nonclassical States of Light and Canonical Transformations

A. Luis and L.L. Sánchez-Soto

Departamento de Optica, Facultad de Ciencias Físicas,
Universidad Complutense, Ciudad Universitaria, E-28040 Madrid, Spain

Representations of nonlinear nonbijective canonical transformations in quantum optics are discussed. The relation of these representations with multiphoton operators is considered.

1 Introduction

Canonical transformations and their unitary representations have a wide range of applications in quantum optics. But while the representations of linear canonical transformations are now clearly established —a relevant example of this is the characterization of squeezed states as generalized coherent states of the real symplectic group—, the general case is not fully understood.

Moshinsky and co-workers [1] have investigated what happens when the transformation relates observables with different spectra. We are faced then with nonbijective canonical transformations.

Here we consider the specific example of a transformation relating an oscillator of unit frequency with another one of frequency κ, where κ is an integer. This example may seem trivial, but it contains all the basic difficulties of the problem. We attempt to relate these nonbijective transformations with the formalism of multiphoton operators, recently introduced in order to produce non-Gaussian squeezing, and that may arise in a k-photon parametric amplifier.

2 Representation of nonbijective canonical transformations

Let us focus firstly on the classical version of the problem we are considering. Taking units in which the mass of the oscillator becomes unity, the implicit definition of this transformation can be expressed as [1]

Springer Proceedings in Physics, Vol. 55
Nonlinear Dynamics and Quantum Phenomena in Optical Systems
Editors: R. Vilaseca, R. Corbalán

$$\frac{1}{2}(p^2+q^2)=\frac{1}{2}\kappa(\overline{p}^2+\overline{q}^2) \quad , \quad \kappa \arctan\frac{p}{q}=\arctan\frac{\overline{p}}{\overline{q}}, \tag{1}$$

which is essentially a dilatation in the action-angle phase space.

It is easily seen that a sector of angle $2\pi/\kappa$ in the (q,p) plane is mapped on the full plane $(\overline{q},\overline{p})$. Thus the canonical transformation is nonbijective and to retrieve bijectivity the phase plane $(\overline{q},\overline{p})$ must have κ sheets connected along the cut in the positive real axis.

Alternatively bijectivity is recovered by noting that the linear canonical transformations

$$\begin{aligned} q_\lambda &= q\cos\left(2\pi\frac{\lambda}{\kappa}\right)-p\sin\left(2\pi\frac{\lambda}{\kappa}\right),\\ p_\lambda &= q\sin\left(2\pi\frac{\lambda}{\kappa}\right)+p\cos\left(2\pi\frac{\lambda}{\kappa}\right), \end{aligned} \tag{2}$$

where $\lambda=0,1,\ldots,\kappa-1$, give us all points (q_λ,p_λ) that map on a single point $(\overline{q},\overline{p})$. The group of linear canonical transformations (2) —which is isomorphic to the cyclic group C_κ— is called the ambiguity group. We can retain the idea of a single sheet, but characterize the functions $f(q,p)$ in it by irreducible representations of the group C_κ. In this way, the functions on the $(\overline{q},\overline{p})$ plane must have κ components, which again can be labeled by the index λ, called the ambiguity spin.

In quantum mechanics a representation of the canonical transformation (1) is a unitary operator U such that it transforms the original *operators* into the new ones through the relation

$$U\ f_\alpha(q,p)\ U^\dagger=\overline{f}_\alpha(\overline{q},\overline{p}). \tag{3}$$

If $\mathcal{H}$ and $\overline{\mathcal{H}}$ are the Hilbert spaces on which the (q,p) and $(\overline{q},\overline{p})$ operators act, the quantum problem associated with (1) is to find a unitary operator $U:\mathcal{H}\longrightarrow\overline{\mathcal{H}}$ such that

$$U\ a^\dagger a\ U^\dagger=\kappa\overline{a}^\dagger\overline{a} \quad , \quad U\ E^\kappa\ U^\dagger=\overline{E}, \tag{4}$$

where, as usual, we have described the oscillator by means of creation and annihilation operators, and E is the phase operator of Susskind and Glogower ($E=1/\sqrt{n+1}\ a$).

If the spectrum of both Hamiltonians were the same we would have the unitary representation [2]

$$U=\int d\epsilon\ \mid\psi_\epsilon\rangle\,\langle\phi_\epsilon\mid, \tag{5}$$

where $\mid\psi_\epsilon\rangle$ and $\mid\phi_\epsilon\rangle$ are the normalized eigenstates of the two Hamiltonians. However, in our case they do not have the same spectrum (in fact, the former has κ levels in the interval in which the later has only one) and then no such unitary operator U exists.

We can split $\mathcal{H}$ in a sum of subspaces in the form

$$\mathcal{H} = \bigoplus_{\lambda=0}^{\kappa-1} \mathcal{H}_\lambda, \tag{6}$$

where $\mathcal{H}_\lambda$ are spanned by $\mid n\kappa + \lambda\rangle$ $(n = 0, \dots, \infty)$ and are invariant under the action of the ambiguity group C_κ, carrying their unitary representations. The spectrum of the original operators restricted to these subspaces is identical with the spectrum of the new ones.

We can now map isometrically each subspace $\mathcal{H}_\lambda$ onto $\overline{\mathcal{H}}$. If we denote by $\mid n)$ the number states in the Hilbert space $\overline{\mathcal{H}}$ and by $\mid n\rangle$ in $\mathcal{H}$, this family of partial isometries is given by

$$U_\lambda = \sum_n \mid n)\,\langle \kappa n + \lambda \mid. \tag{7}$$

Note that they verify $U_\lambda U_\lambda^\dagger = I$ but $U_\lambda^\dagger U_\lambda \neq I$, and transform the original operators in the form

$$U_\lambda\, a^\dagger a\; U_\lambda^\dagger = \kappa \bar{a}^\dagger \bar{a} + \lambda \quad , \quad U_\lambda\, E^\kappa\; U_\lambda^\dagger = \overline{E}. \tag{8}$$

In order to recover bijectivity we shall enlarge the Hilbert space $\overline{\mathcal{H}}$ assuming as final space not $\overline{\mathcal{H}}$, but $\overline{\mathcal{H}} \otimes \mathcal{V}$, where $\mathcal{V}$ is some κ-dimensional space that translates into quantum mechanics the multicomponent structure of the functions in phase space. Taking a basis $\mid e_\lambda\rangle$ in this space $\mathcal{V}$ we can construct a truly unitary operator $U : \mathcal{H} \longrightarrow \overline{\mathcal{H}} \otimes \mathcal{V}$ in the following way:

$$U = \sum_{n,\lambda} \mid e_\lambda\rangle \mid n)\langle n\kappa + \lambda \mid, \tag{9}$$

and we have

$$U a^\dagger a U^\dagger = \kappa \bar{a}^\dagger \bar{a} + \sum_\lambda \lambda \mid e_\lambda\rangle\langle e_\lambda \mid \quad , \quad U E^\kappa U^\dagger = \overline{E}. \tag{10}$$

Note that now any operator in $\mathcal{H}$ is transformed into an operator acting in both $\overline{\mathcal{H}}$ and the new spin variable.

To recover the correspondence between canonical transformations and unitary operators, it has been necessary to add a new degree of freedom that, up to now, has not received a clear physical interpretation. We shall try in the next section to reinterpret these results in terms of multiphoton operators.

3 Multiphoton operators

As it is well known, the abstract commutation relation $[a, a^\dagger] = I$, is usually realized in the familiar Fock space $\mathcal{H}$. In this irreducible representation the action of the operators is

$$a \mid n\rangle = \sqrt{n} \mid n-1\rangle \quad , \quad a^\dagger \mid n\rangle = \sqrt{n+1} \mid n+1\rangle. \tag{11}$$

Any other irreducible representation by closed densely-defined operators in a Hilbert space is unitarily equivalent to it.

For any integer κ we can define a set of operators in $\mathcal{H}$ by

$$a_{(\kappa)} \mid n\kappa + \lambda\rangle = \sqrt{n} \mid (n-1)\kappa + \lambda\rangle, \tag{12}$$

with the commutation relation $[a_{(\kappa)}, a_{(\kappa)}^\dagger] = I$.

It is easily seen that these operators are identical with the multiphoton operators recently introduced [3]. With this definition the Hilbert space $\mathcal{H}$ naturally splits into a sum of subspaces $\mathcal{H}_\lambda$ defined as in (6), in such a way that $(a_{(\kappa)}, \mathcal{H}_\lambda)$ are irreducible representations of the basic commutation relation.

The splitting (6) allows us to make the isomorphism $\mathcal{H} \approx \overline{\mathcal{H}} \otimes \mathcal{V}$, where $\overline{\mathcal{H}}$ is a Hilbert space isomorphic with any of the $\mathcal{H}_\lambda$ and $\mathcal{V}$ is a κ-dimensional Hilbert space. This isomorphism is realized through the unitary operator U in such a way that

$$U a_{(\kappa)} U^\dagger = \bar{a}, \tag{13}$$

and the pair $(\bar{a}, \overline{\mathcal{H}})$ is an irreducible representation of the commutation relation. This operator U is given simply by $U = \sum_{n,\lambda} \mid e_\lambda\rangle \mid n)\langle n\kappa + \lambda \mid$, where $\mid n)$ are the number states associated with $\bar{a}$ and $\mid e_\lambda\rangle$ is a basis of $\mathcal{V}$.

It is clear that this unitary operator and (9) are the same transformation and for this reason we use the same notation. We can see the deep connection between nonbijective canonical transformations and multiphoton operators. The point is that it is not necessary to enlarge our Hilbert space introducing spurious variables , but recognize that $\overline{q}$ and $\overline{p}$, as it shows the previous isomorphism, can not be a complete set of coordinates in phase space. This fact is reflected in that we can find operators that commute with $\overline{q}$ and $\overline{p}$, and are not constants.

We think that the previous identification gives a useful framework to study nonclassical states of light, such as the multiphoton squeezed states[3]. These states could be defined as well through the action on the vacuum of what we might call multiphoton displacement operator

$$| \alpha_{(\kappa)} \rangle \equiv \exp(\alpha a^{\dagger}_{(\kappa)} - \alpha^* a_{(\kappa)}) \mid 0 \rangle. \tag{14}$$

Their probability distribution functions are not Gaussian, presenting squeezing of several orders depending on the parameters α and κ. Translating this definition into the $| n) \mid e_\lambda \rangle$ basis we have

$$U \mid \alpha_{(\kappa)} \rangle = \exp(\alpha \bar{a}^\dagger - \alpha^* \bar{a}) \mid 0 \rangle \mid e_0 \rangle, \tag{15}$$

which shows clearly that these states are nothing but coherent states for the $(\overline{q}, \overline{p})$ variables.

Up to now the space $\mathcal{V}$ has only indexed the subspaces of $\mathcal{H}$. However it can be used to extend the definition of states made in finite-dimensional subspaces of $\mathcal{H}$ to the whole space as happens with the Holstein-Primakof realizations of $SU(2)$ and the phase states of Pegg and Barnett.

References

[1] R. Dirl, P. Kasperkovitz and M. Moshinsky, J. Phys. A: Math. Gen. **21**, 1835 (1988) and references therein

[2] P.A.M. Dirac. *Quantum Mechanics*. Oxford. Clarendon. 1958.

[3] G. D'Ariano, M. Rasetti, M. Vadacchino. Phys. Rev. D **32**, 1034 (1985)

Langevin Equations for Squeezing by Means of Non-linear Optical Devices

F.J. Bermejo[1], *P. García Fernández*[1], *P. Colet*[2], *S. Balle*[2], *R. Toral*[2], *and M. San Miguel*[2]

[1]Instituto de Estructura de la Materia, C.S.I.C., Serrano 119, E-28006 Madrid, Spain

[2]Departament de Física, Universitat de les Illes Balears, E-07071 Palma de Mallorca, Spain

Abstract. A procedure which enables one to obtain Langevin equations from the Heisenberg equations of motion representative of a wide class of non-linear optical devices is applied to the case of two-mode squeezing in a parametric device where allowance is made for different damping constants for the signal and idler modes, and a degenerate parametric amplifier with a fourth-order (Kerr effect) interaction term.

1 Introduction

The customary procedure for the derivation of Langevin equations for multimode optical fields generated by non-linear optical devices involves several steps [1]. First, a master equation for the motion of a density matrix has to be constructed from an effective Hamiltonian which includes the free-field, interaction and damping terms, after adiabatic elimination of the fast (heat-bath) modes, for which some heuristic assumptions regarding time-scale separation are usually considered. A generalized Fokker-Planck equation for an adequate pseudoprobability density is obtained next, after expansion of the density matrix on a suitable basis set. From standard representation theory [2] it is known that the different ways of ordering of the non-commuting operators lead to different equations of motion and that such orderings are equivalent provided that a correct representation exists. Different statistical properties will result as a consequence of the use of different representations since the equations of motion will evolve in time in an ordering-dependent way. However, since most of the observables are related to photo-counting devices, the normally ordered P (or generalized P) representations have been widely used [3], although in principle any other representation could also be employed if the final distribution is converted into normal order before the computation of field-statistical properties.

Springer Proceedings in Physics, Vol. 55
Nonlinear Dynamics and Quantum Phenomena in Optical Systems
Editors: R. Vilaseca, R. Corbalán

The debate regarding the physical soundness of the different representations is still open [4-6], and within it, a recent paper has emphasized the usefulness of the Wigner representation since it allows a direct derivation of the Langevin equations for the optical fields involved. Such a procedure is exact, provided that the effective Hamiltonian contains terms up to second order in the field operators. For higher–order problems, it is well known that the resulting Fokker–Planck equation for the Wigner quasi–probability distribution will, in general contain derivatives higher than second order, and therefore a truncation of those (or a linearization of the operator evolution equations) becomes necessary for obtaining a tractable solution. The available evidence suggests that [4,5] for a certain class of problems such as those related to quantum tunneling between the two stable states of a parametric oscillator operating above threshold, the full quantum problem should be considered. However, for problems not aiming to study the detailed dynamics near threshold and in circumstances where the noise is low in comparison with the mean values of the field amplitudes, the procedure described in [6] seems appropriate. In what follows we will briefly describe the application of the approach given in [6] to two representative cases of non–linear devices currently used for the generation of squeezed light.

2 A non–degenerate parametric amplifier with non–homogeneous linear losses

Apart from its interest as a squeezing device, some recent interest has arisen since it has been suggested that such a setup could be employed for high–accuracy (below shot–noise) absorption measurements [7]. The model herein considered represents an intense (assumed undepleted) laser beam of frequency 2Ω, which upon illumination of a suitable non–linear medium characterized by a second order coupling produces pairs of highly correlated photons with frequencies $\Omega \pm \epsilon$ where $\epsilon < \Omega$ is the modulation frequency.

The effective Hamiltonian written in the Schrödinger picture is written in terms of free–field, interaction and dissipative parts as

$$H = H_0 + H_{int} + H_D \tag{1}$$

$$H_0 = \omega_+ a_+^\dagger a_+ + \omega_- a_-^\dagger a_-$$

$$H_{int} = \frac{1}{2} f(t) a_+^\dagger a_-^\dagger + \frac{1}{2} f^*(t) a_+ a_-$$

$$H_D = \sum_j \omega_j b_j^+ b_j + \sum_j \chi_j^+ (b_j a_+^+ + b_j^- a_+) + \sum_j \chi_j^- (b_j a_-^+ + b_j^+ a_-)$$

where the function $f(t)$ contains the coupling terms $f(t) = i\kappa(t)e^{2i(\phi - \Omega t)}$ and the absorption losses have been included in the usual way in terms of the b_j heat–bath operators.

The Heisenberg equations of motion for the a_+, a_- and b_j modes can be written with ease and the heat–bath modes are then adiabatically eliminated.

By means of the correspondence $a_\pm \longrightarrow \alpha_\pm$ (and $a_\pm^\dagger \longrightarrow \alpha_\pm^*$) the Langevin equations of motion can be obtained for the two complex quantities describing the output modes with the result

$$\dot{\alpha}_+ = -i(\Omega + \epsilon - i\gamma_+)\alpha_+ - if(t)\alpha_-^* + L_+(t)$$

$$\dot{\alpha}_- = -i(\Omega - \epsilon - i\gamma_-)\alpha_- - if(t)\alpha_+^* + L_-(t) \qquad (2)$$

where the damping constants $\gamma_\pm$ denote the linear losses for the two modes and the terms $L_\pm(t)$ represent complex white noise with a mean variance given by

$$< L_\pm(t) >= 0 =< L_\pm(t)L_\pm(t') >$$

$$< L_\pm(t)L_\pm^*(t') >= 2D_\pm\delta(t - t') \qquad (3)$$

in terms of diffusion constants $D_\pm$, which basically represent the spectral densities at the frequencies of the two modes of the heat–bath,

$$2D_\pm = \gamma_\pm(1 + 2\bar{n}) \qquad (4)$$

where $\bar{n}$ is an effective temperature.

The solution of (2) is a Wigner probability density, and a Fokker–Planck equation can be derived after separating the two real parts of the complex stochastic processes

$$\alpha_\pm(t) = [x_\pm(t) + iy_\pm(t)]e^{i[(\Omega \pm \epsilon)t - \phi]} \qquad (5)$$

which transforms the system (2) into

$$\dot{x}_\pm = -\gamma_\pm x_\pm + \kappa x_\mp + L_{\pm x}$$

$$\dot{y}_\pm = -\gamma_\pm y_\pm - \kappa y_\mp + L_{\pm y} \qquad (6)$$

where

$$< L_{\pm x,y}(t) >= 0$$

$$< L_{\pm i}(t)L_{\pm j}(t') >= D_{\pm}\delta_{ij}\delta(t-t'), \quad i,j = x,y \ .$$

It is straightforward to obtain from (6) a Fokker–Planck equation for the Wigner function which contains derivatives up to second order. On the other hand, a stationary solution for (6) exists which, above threshold, can be written as

$$x_+(t) = [A_0 + \alpha(t)]e^{\lambda_+ t} + [B_0 + \beta(t)]e^{-\lambda_- t}$$

$$x_-(t) = \mu_1[A_0 + \alpha(t)]e^{-\lambda_+ t} + \mu_2[B_0 + \beta(t)]e^{-\lambda_- t} \qquad (7)$$

where

$$\lambda_\pm = \frac{\gamma_+ + \gamma_-}{2} \pm \left[\left(\frac{\gamma_+ + \gamma_-}{2}\right)^2 - (\gamma_+\gamma_- - \kappa^2)\right]^{1/2},$$

$$\mu_1 = \frac{\gamma_+ - \lambda_+}{\kappa}, \quad \mu_2 = \frac{\gamma_+ - \lambda_-}{\kappa}$$

$$\alpha(t) = \int_0^t dt' \frac{\mu_2 L_+(t') - L_-(t')}{\mu_2 - \mu_1} e^{\lambda_+ t'}; \ \beta(t) = \int_0^t dt' \frac{\mu_1 L_+(t) - L_-(t')}{\mu_1 - \mu_2} e^{\lambda_- t'}$$

and a solution for the $y_\pm$ variable can be easily found upon the corresponding changes of variable.

The second and fourth–order expectation values can now be computed and the detailed expression for second and fourth–order correlations will be given in a forthcoming paper [8].

The quantity of interest for absorption mesurements would be the spectrum of fluctuations of the intensity difference between the signal I_+ and idler I_- modes

$$S_D(\omega) = \int d\tau e^{-i\omega t} < I_+(\tau) - I_-(\tau), I_+(0) - I_-(0) > \qquad (8)$$

which can be written in terms of the intra- and interbeam fluctuations as

$$S_D(\omega) = S_{++}(\omega) + S_{--}(\omega) - S_{+-}(\omega) - S_{-+}(\omega) \qquad (9)$$

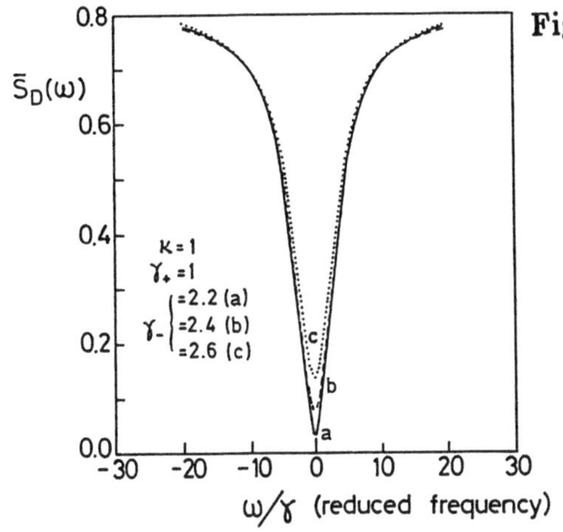

Figure 1

$$S_{++}(\omega)+S_{--}(\omega) = \int d\tau e^{-i\omega\tau}[< I_+(t+\tau), I_+(t) > + < I_-(t+\tau), I_-(t) >]$$

$$S_{+-}(\omega)+S_{-+}(\omega) = \int d\tau e^{-i\omega\tau}[< I_+(t+\tau), I_-(t) > + < I_-(t+\tau), I_+(t) >].$$

The evaluation of the spectral components can be obtained after some calculus and the detailed expressions will be given elsewhere [8].

A normalized spectrum may be defined as

$$\bar{S}_D(\omega) = \frac{S_D(\omega)}{S_{++}(\omega) + S_{--}(\omega)} \tag{10}$$

so that the shot–noise limit is $\bar{S}_D(\omega) = 1$ and perfect noise suppression means $S_D(\omega) = 0$. The shape of the spectrum for several values of the asymmetric losses is shown in Figure 1. As can be seen upon inspection of the figure, the increase in the loss coefficient of one of the modes leads to an increase in the intensity fluctuations and therefore it may be used in order to perform absorption measurements well below the shot–noise limit. A full detailed discussion will be given elsewhere [8].

3 Squeezing from fourth-order interaction in a degenerate parametric amplifier

It has been shown that [9] an intense coherent beam which propagates along an optical Kerr medium, exhibits "self–squeezing" (amplitude squeezing), and that the effect could be remarkably high provided that the incident pump strength is strong enough. The effect of such higher–order interactions coupled to a standard two–photon medium was analyzed by Tombesi [10] some time ago. The interest in the already mentioned study was focused on the possibilities of generating strong squeezed light at short interaction times (i.e. in travelling–wave geometries). Our interest here is to explore the field–statistical properties of a more realistic model of a parametric amplifier where the crystal losses are explicitly accounted for.

The effective Hamiltonian is written as

$$H = \Omega a^+ a + \frac{1}{2} f(t) a^{+^2} + \frac{1}{2} f^*(t) a^2 + \frac{\Gamma}{2} a^{+^2} a^2 + \sum_j \omega_j b_j^+ b_j +$$

$$+ \sum_j \chi_j (b_j a^+ + b_j^+ a) \,, \tag{11}$$

where the first term on the right–hand side is the free–field contribution, the second–order interaction is described by the terms which contain the function $f(t) = i\kappa(t) e^{2i(\phi - \Omega t)}$, where $\kappa(t)$ characterizes the coupling, and the fourth–order term is written in terms of the non–linear coupling coefficient Γ which is taken to be proportional to the third–order susceptibility. The interaction with the heat–bath modes is comprised in the last term.

We will follow an analogous procedure to the one previously used except that a linearization has to be introduced in order to deal with the higher–order term.

The Heisenberg equations for the field operators are readily obtained and upon the elimination of the heat–bath modes one gets

$$\dot{a} = -i(\Omega - i\gamma) a - i f(t) a^* - i\Gamma a^* a^+ + L(t) \tag{12}$$

$$< L(t) L^*(t) >= 2D\delta(t - t'), \quad 2D = \gamma(1 + 2\bar{n})$$

where γ is the damping constant and $L(t)$ are (complex) white noise terms. The linearization of the Heisenberg equation can be done provided that a separation can be made between the c–number part and the part carrying the quantum fluctuactions δa, that for the destruction operator becomes

$$a = (\bar{a} + \delta a) e^{-2i(\Omega t - \phi)} \,.$$

We can now replace the quantum noise term δa by the complex stochastic process $\delta\alpha$

$$\delta\dot{\alpha} = -i(\Omega - i\gamma)\delta\alpha - if(t)\delta\alpha^* - i\Gamma\bar{\alpha}(2\bar{\alpha}^*\delta\alpha + \bar{\alpha}\delta\alpha^*) + L(t) \qquad (13)$$

which has a solution given as a Wigner probability density. The associated Fokker–Planck equation in the W representation will thus correspond to a reduced form where the derivatives higher than second order have been truncated.

The complex $\alpha(t)$ stochastic processes can be converted into real ones as was done previously (Eqn. (5)) , and the Langevin equations for the two field quadratures will be

$$\dot{x} = (-\gamma + \kappa)x + \Gamma y(y^2 + x^2) + L_x$$

$$\dot{y} = (-\gamma - \kappa)y - \Gamma x(y^2 + x^2) + L_y$$

where the noise terms have the same properties as those given in (6).

By introduction of a rescaled time $\tau = t(\gamma + \kappa)$ and defining the dimensionless constants

$$\mu = \frac{\kappa - \gamma}{\kappa + \gamma}, \qquad \nu = \frac{\Gamma}{\gamma + \kappa}$$

the equations can be written as

$$\dot{x} = \mu x + \nu y(x^2 + y^2) + L_x$$

$$\dot{y} = -y - \nu x(x^2 + y^2) + L_y \qquad (14)$$

which have a deterministic stationary solution which verifies

$$\frac{\bar{x}_s^2}{\bar{y}_s^2} = \frac{1}{\mu} \ .$$

In consequence, it can easily be verified that for $\mu < 0$ the trivial solution $(x_s, y_s) = (0,0)$ is obtained and two other stationary solutions are found above threshold (i.e. for $\mu > 0$)

$$(\bar{x}_s, \bar{y}_s) = (0,0) \quad \text{for } \mu < 0$$

$$\bar{x}_s = \mp[\mu^{1/2}/\nu(1+\mu)]^{1/2}; \;\; \bar{y}_s = -\mu^{1/2}\bar{x}_s \;\; \text{for } \mu > 0 \qquad (15)$$

A local stability analysis can be carried out by means of a linearization around the deterministic solution and one finds that below threshold only the trivial solution is stable whereas a saddle point is approached at threshold and the two solutions above threshold become the only stable ones (a 'pitchfork'–type bifurcation is developed as one passes through the threshold).

In the limit of vanishing damping ($\nu = 1$) the system becomes of Hamiltonian nature and evolves according to

$$\dot{x} = x + \nu y(y^2 + x^2) = \frac{\partial H(x,y)}{\partial y}$$

$$\dot{y} = -y - \nu x(y^2 + x^2) = -\frac{\partial H(x,y)}{\partial x}, \quad H(x,y) = xy + \frac{\nu}{4}(x^2 + y^2)^2 \, .$$

From the equations (14) the fluctuations around the steady states as well as the effects due to the presence of the non–linear term can be computed and some results are presented in an accompanying paper [11].

4 Summary

The method of computing the statistical properties of non–linear optical devices described in [6] which allows one to obtain the field–statistical properties from the Heisenberg equations of motion has been applied to two representative classes of model Hamiltonians. In the first case, since the model involved only contains terms up to second order, the obtained results should be considered as exact as the ones obtained by means of the standard P representation. Since a linearization has been introduced on the second case, the results have a range of validity which may exclude the parameter region located near threshold. A more detailed discussion on this topic is contained in an accompanying paper [11].

References

[1] M. Sargent, M.O. Scully, and W.E. Lamb, 'Laser Physics', Addison–Wesley, Reading, MA, 1974, Chapt.16.

[2] M. Hillery, R.F. O'Conell, M.O. Scully, E.P. Wigner, Phys. Rep. 106, 121, (1984).

[3] C.W. Gardiner, 'Handbook of Stochastic Methods', 2nd edition, Springer–Verlag, Berlin, 1985, Chapt. 10.

[4] H.J. Carmichel, M. Wolinsky, Phys. Rev. Lett. 60, 1836, (1988).

[5] P.D. Drummond, P. Kinsler, Phys. Rev. A40, 4813, (1989).

[6] T.W. Marshall, E. Santos, Phys. Rev. A41, 1582, (1990).

[7] A.S. Lane, M.D. Reid, D.F. Walls, Phys. Rev. Lett. 60, 1940, (1988).

[8] P.García-Fernández, S. Bal.le, F.J. Bermejo, (submitted).

[9] R. Tanas, S. Kielich, Quantum Opt. 2, 23, (1990).

[10] P. Tombesi in 'Quantum Optics IV', J.D. Harvey, D.F. Walls (Eds.), Springer, 1986, p. 81.

[11] C. Cabrillo, P. García-Fernández, P. Colet, R. Toral, M. San Miguel, F.J. Bermejo (this volume).

Part II

Processes in Very Strong Fields

Atomic Physics and Nonlinear Optics in Very Strong Laser Fields

J.H. Eberly

Department of Physics and Astronomy, and
The Institute of Optics, University of Rochester,
Rochester, NY 14627, USA

Abstract. The purpose of this paper is to review recent progress in understanding the behavior of electrons and atoms in strong laser fields. Our focus will be on developments during the 1980's, and the different ideas and physical pictures that have been put forward to deal with them. There will not be space for anything approaching a complete set of citations of the original work, but some overviews have already appeared and these may be consulted for further references [1-5].

1. Brief Review of Multiphoton Ionization

The first experimental programs of multiphoton investigations were started around 1965 by the groups of Delone in Moscow [6] and of Mainfray and Manus in Paris (Saclay) [7]. According to the usual theory of multiphoton ionization (MPI), the N-photon ionization rate is proportional to the Nth power of the laser photon flux Φ (intensity divided by photon energy, $\Phi = I/\hbar\omega$)

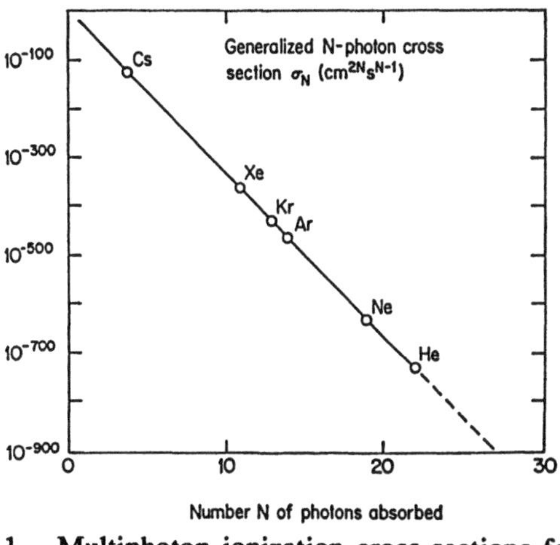

1. Multiphoton ionization cross sections for six atoms. The multiphoton order ranges from N = 4 for cesium to N = 22 for helium at the laser wavelength 1064 nm. [Private communication from G. Mainfray].

$$R_N = \sigma_N \Phi^N, \tag{1.1}$$

where σ_N is a generalized cross section (with units $cm^{2N} s^{N-1}$).

Experimental tests of this law were undertaken for many atomic species and for a variety of laser wavelengths in the period 1965-75. Fig. 1 shows that a kind of "universal" ionization response was gradually revealed. These studies were carried out at 1064 nm with a range of laser flux levels corresponding to intensities up to about 10^{13} W cm^{-2}, typically from temporally and spatially incoherent 30 ns pulses with energy up to about 50 J.

2. Discovery of Above-Threshold Ionization

According to the Einstein theory of the photoelectric effect, trivially generalized to allow for multiphoton absorption, the kinetic energy of a photoelectron is fixed at the value

$$W = N\hbar\omega - E_I, \tag{2.1}$$

where E_I is the atomic ionization energy and N is the number of laser photons of frequency ω necessary to liberate the electron.

However, in 1979 the Saclay group measured photoelectron kinetic energies from 6-photon ionization of xenon atoms [8], and in contrast to the prediction of (2.1), two distinct values of kinetic energy were observed in the photoelectron spectrum. A second, higher energy, satellite peak appeared beside the peak predicted by (2.1), as indicated in Fig. 2. The experiment used 2.34 eV photons from the second harmonic of suitably amplified 12 ns pulses from a Nd:glass laser at an intensity in the range 6-10 x 10^{12} W cm^{-2}.

The Saclay group named this effect Above-Threshold Ionization [9], and the name, as well as its convenient abbreviation ATI, has generally been adopted by other workers. For a period it was not fully accepted that the ionized electron had simply absorbed another photon. However, additional results produced by Saclay and by the FOM group in Amsterdam showed electron spectra with as many as 10 or 12 peaks, all spaced by $\hbar\omega$. A schematic representation of ATI data is shown in Fig. 3.

Another finding came from the FOM data [10], and is also shown in Fig. 3. This is the virtual disappearance of the first photopeak in the energy spectrum as the laser intensity is increased. This tendency of low-order peaks to become relatively smaller is called "peak suppression." It became an important focus of attention and was quickly verified in other experiments. ATI effects are generally more pronounced if the laser wavelength is longer, but observations with wavelengths as short as 193 nm light (from an excimer laser) have been reported.

We can mention again that there are several overviews already published or in preparation that contain material closely related to our subject and where detailed citations can be found [1-5].

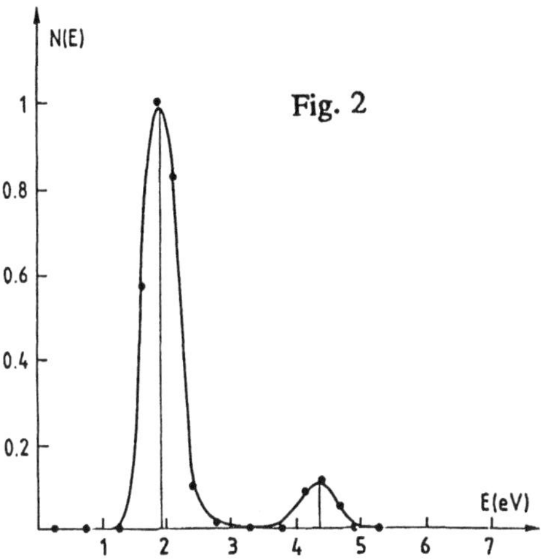

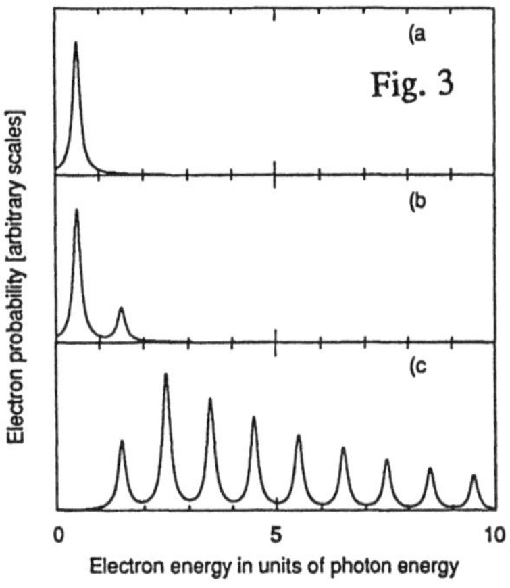

2. First reported photoelectron energy spectra showing ATI. The atom was xenon and circles represent data at 532 nm. [From Agostini et al. [8]].

3. A sketch of the three historical stages of multiphoton ionization experiments, in terms of photoelectron energy spectra: (i) low-intensity ionization leading to a single peak; (ii) the first observation of an ATI peak, with laser intensity approaching 10^{13} W cm^{-2}; and (iii) many ATI peaks and suppression of the first two peaks at intensities above 10^{13} W cm^{-2}.

3. Perturbation Theory and its Limits

Theoretical interest was attracted to ATI almost immediately because the observation of multiple electron peaks indicated that N-photon and (N+1)-photon processes were being induced with roughly the same probability by the same laser field. This is contrary to the nature of perturbation theory and suggests that ATI arises from an atomic system undergoing strong perturbations. This would be remarkable because the intensities employed in the earliest ATI experiments, in the range 10^{12} - 10^{13} W cm^{-2} , are much lower than the value of intensity that is usually taken to mark the limit of perturbation theory, namely the so-called atomic unit of intensity [11], 3.5×10^{16} W cm^{-2}. Thus the unmistakeably non-perturbative appearance of the multiple-peaked ATI electron spectra in Figs. 2 and 3 provoked a fresh examination of the limits of perturbation theory in atomic radiative processes.

As is well-known [11], the cross section in (1.1) is derived from an N-photon matrix element $M^{(N)}$ that is made up out of intermediate-state sums over N-fold products of dipole matrix elements x_{ij} and resonance denominators $1/(E_i$ - $m\hbar\omega$ -

E_k). This very complex quantity can be estimated crudely without knowing the values of the individual matrix elements and denominators. We simply invoke the concept of an average dipole moment μ^* and an average detuning Δ^*. Then we obviously have $M^{(N)}$ = constant $\times (\mu^*/\hbar\Delta^*)^N$. We can apply the same idea to the ionization rate R_N itself. To do this we simply multiply the cross section (essentially the square of $M^{(N)}$) by the factor I^N = constant $\times\, \mathcal{E}^{2N}$. Then the MPI rate (1.1) can be written

$$R_N = \text{constant} \times (\mu^* \mathcal{E}/\hbar\Delta^*)^{2N}. \tag{3.1}$$

This form for the rate is directly relevant to any discussion of the limit of perturbation theory. It is clear that perturbation theory cannot be valid for field strengths larger than the critical value implied by $\mu^* \mathcal{E}_{cr}/\hbar\Delta^* = 1$, because at this value one has

$$(\mu^* \mathcal{E}/\hbar\Delta^*)^{2N} = (\mu^* \mathcal{E}/\hbar\Delta^*)^{2(N+1)} \tag{3.2}$$

which means that higher order processes are equally strong as lower order ones, and the convergence of any perturbation series would be very questionable.

Generally $\hbar\Delta^*$ will be no larger than about one incident photon energy, say $\hbar\Delta^* \approx 1$ eV, and for μ^* a reasonable value is some average of high Rydberg and ground state dipole moments, say $\mu^* \approx 3\, ea_0$. Thus we get

$$\mathcal{E}_{cr} \approx \frac{1\ \text{eV}}{3ea_0} = \frac{1}{3 \times 13.6} \times \frac{13.6\ \text{eV}}{ea_0}. \tag{3.3}$$

Based on this line of reasoning, we get a critical field strength that leads to the corresponding critical intensity $I_{cr} = (c\varepsilon_0/2)\mathcal{E}_{cr}^2$:

$$I_{cr} \approx 6 \times 10^{12}\ \text{W cm}^{-2}. \tag{3.4}$$

This value is not only much lower than the atomic unit of intensity but is also consistent with the intensities that have repeatedly been observed to give multiple ATI peaks such as those shown in Fig. 3. In light of this and other confirmations that one can regard ATI physics as at least partially non-perturbative, several non-perturbative pictures of electrons and atoms in strong fields have been proposed. We will mention some of these next.

4. Classical Free Electron Picture

It is certainly a simplification, but one that may have merit in the case of ATI, to regard ATI electrons as completely free, both because they have energies greater the ionization potential (positive kinetic energies) and also because the laser field is so strong that the residual effect of the coulombic binding potential can be totally ignored. Then a classical picture is reasonable. We may recall that the velocity of a classical free electron in an plane wave monochromatic electric field is given by the formula:

$$v(t) = \frac{e\mathcal{E}_0}{m\omega}(1 - \cos \omega t)\,. \tag{4.1}$$

The cycle-averaged kinetic energy associated with the rapidly oscillating part of the motion (so-called jitter motion) is

$$W_J = \frac{\overline{mv_{osc}^{\;2}}}{2} = \frac{e^2\mathcal{E}_0^{\;2}}{4m\omega^2}. \tag{4.2}$$

One can say that just to put an electron in a strong field requires it be given this much average kinetic energy. This energy can be converted to non-oscillatory drift energy by the action of the ponderomotive force $\mathbf{F} = -\nabla V_P$, where the ponderomotive potential is [12]:

$$V_P = \frac{e^2}{4m\omega^2}|\mathcal{E}(r)|^2 \tag{4.3}$$

where $\mathcal{E}(\mathbf{r})$ is the amplitude of the electric field. Since V_P is greatest at the center of the laser beam where $\mathcal{E}(\mathbf{r})$ is greatest, $\mathbf{F}$ acts to push a newly ionized electron out of the beam, and its potential energy V_P is converted into drift energy in the process [see Fig. 4]. It is probably just as well to think of the jitter energy as potential energy in the first place.

It is generally accepted now that the ponderomotive effect is the origin of "peak suppression" in ATI [recall the absence of the lowest-energy peak in Fig. 3c]. The explanation is as follows. When a free electron is created by ionization it requires its quota of "walking around" kinetic energy in the laser field. Therefore, counting this energy, the minimum energy for ionization is not E_I , but $E_I + W_J$, which is also equal to $E_I + V_P$. In this sense, the ponderomotive potential energy raises the ionization threshold. This is sometimes called the ponderomotive threshold shift. Thus we see that the minimum number of photons required for ionization is not $n = E_I/\hbar\omega$, but rather $n+m = (E_I + V_P)/\hbar\omega$, where m is the number of photons needed to overcome the threshold shift.

How does this affect ATI? The lowest kinetic energy of an electron able to overcome the shifted threshold will be (while it is still in the laser beam)

$$W_{in} = (n+m)\hbar\omega - (E_I + V_P)\,. \tag{4.4}$$

However, as the electron emerges from the laser beam, its potential energy is

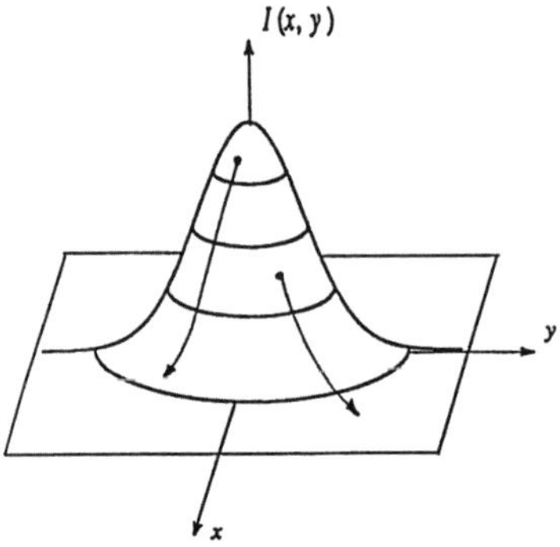

4. Two hypothetical trajectories representing ionized electrons leaving the laser beam under the influence of ponderomotive acceleration

adiabatically converted into translational kinetic energy just equal to V_P . The smallest kinetic energy of an electron outside the field (which is what experiments measure) is therefore $W_{out} = W_{in} + V_P$, or

$$W_{out} = (n+m)\hbar\omega - E_I \, . \tag{4.5}$$

This last formula contains no reference to either ponderomotive or jitter energy, and W_{out} is precisely the energy one would have predicted for the mth ATI peak, not for the lowest peak. Thus, due to the threshold shift, all ATI peaks up to the mth should be missing. This is just what is needed to explain the experimentally quite conspicuous suppression of the lowest peaks in the electron spectrum [recall Figure 3c] with increasing field intensity (increasing threshold shift). The fact that the lowest peaks are usually only relatively suppressed and not absent altogether is sensibly ascribed to the range of intensities in every laser pulse -- some ionization always occurs in the rising and falling edges of the pulse where the intensity (and thus the threshold shift) is less than the maximum.

A quantum mechanical calculation in lowest order perturbation theory shows that the classical "ponderomotive" threshold shift is a just special case of the AC Stark shift of atomic levels.

5. Ionization Pictured as Tunneling

The binding energy of most atoms is about 5-10 times larger than the energy of an optical photon. This means that a ground state electron executes 5-10 orbits during every optical cycle and the optical field is comparatively static. In a static external field of magnitude $\mathcal{E}_0$ in the x direction the total potential energy of a hydrogenic electron is $V(r) = -e^2/r - ex\mathcal{E}_0 \sin\omega t$. The external field has its greatest influence when $\sin\omega t = 1$, in which case (in atomic units $e = \hbar = m = 1$)

$$V(r) = -\frac{Z}{r} - x\mathcal{E}_0 . \tag{5.1}$$

There is a saddle point in the potential surface created by the external field. The value E_0 of the energy there is easily found from $\partial V/\partial x = 0$, which gives $\mathcal{E}_0 = 1/x^2$, or

$$E_0 = -2\sqrt{Z\mathcal{E}_0} . \tag{5.2}$$

With the aid of (5.2) we can determine the field strength needed to free the electron and ionize the atom. We simply require that the external field pushes the saddle point down to the level of the ground state, i.e., $|E_0| = E_I$, where E_I is the ionization potential. On the next revolution in its orbit the electron escapes. If we solve (5.2) for $\mathcal{E}_0$ we can find the corresponding "appearance intensity" $I_{app.}$ at which ionization should first be easily observed, as a function of E_I and Z for any atomic or ionic species:

$$I_{app.} = (c\varepsilon_0/2)\,\mathcal{E}_0^2 = \frac{c\varepsilon_0}{32}\frac{E_I^4}{Z^2} . \tag{5.3}$$

Experiments with intensities in the range of 10^{16}-10^{17} W cm^{-2} now provide strong support [see Fig. 5] for tunneling in the optical domain, and for expression (5.3) as a satisfactory formula for appearance intensity. Earlier data from MPI

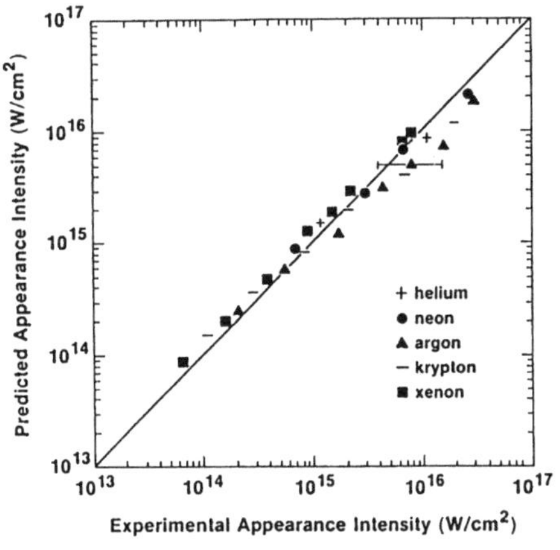

5. Multiphoton ionization data on 27 different charge states of noble gas atoms from Augst, et al. [13] comparing experimental appearance intensity and the theoretical expression for appearance intensity (5.3) obtained from the barrier-suppression argument.

experiments with CO_2 lasers supported the existence of a tunneling regime at lower frequencies where tunneling is more natural. Compared to ATI, tunneling ionization is an opposite case, in the sense that tunneling is not expected to lead to discrete ATI peaks characterized by integer numbers of photons. To date ATI electrons have not been measured at the laser field strengths and wavelengths of the Rochester tunneling experiments [13].

6. The Volkov-Keldysh Free-electron Picture

In 1965 Keldysh [14] used the so-called Volkov wave function of a free electron in a radiation field to make a combined theory of tunneling and multiphoton ionization. Further developments, in the direction of ATI, were made by Faisal [15] and Reiss [16], and the resulting theory is generally known by the initials KFR. The KFR theory can be considered a picture with some quantum improvements on the classical free-electron picture of ATI.

Quantum theory permits the obvious introduction of the quantum energy unit $\hbar\omega$ and the quantum size parameter a_0. However, after initial apparent success in comparison with ATI experiments, it is now not so clear that KFR theory is quantitatively realistic, at least for the most common low-energy ATI photoelectrons. A careful comparison of KFR theory with the predictions of numerically exact electron wave functions, under restricted conditions, does not give encouraging results.

7. The Kramers-Henneberger Picture

At the same time as Keldysh's paper, a different approach to atomic response in a strong field was suggested by Henneberger [17]. His proposal was to view the atom from the time-dependent frame of a free electron in the field. This frame had been used earlier by Kramers [18] and we will refer to it as the Kramers-Henneberger (K-H) frame. In the K-H frame the atomic nucleus moves rapidly back and forth. Because the nucleus spends relatively more time at the turning points of its trajectory, the electron sees the attractive potential of the nucleus preferentially located at the two extremes of its motion, giving rise to what is effectively a double-well potential. That is, the binding potential becomes time-dependent:

$$\frac{-1}{|r|} = \frac{-1}{|r + \alpha(t)|} = \frac{-1}{|r + \alpha_0 \sin\omega t|} \tag{7.1}$$

and the time average of this is a double well, as sketched in Fig. 6. Here α_0 is the amplitude of free-electron response, obtained by integrating the classical velocity (4.1):

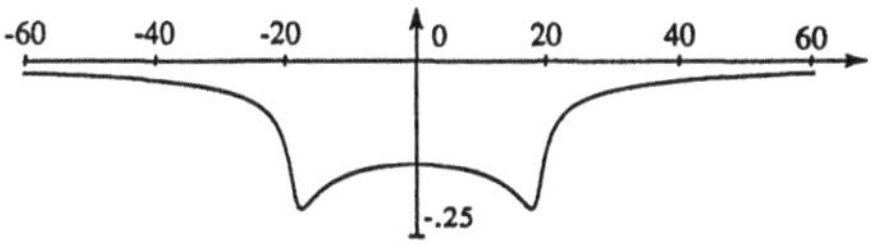

6. Double-well potential obtained by time-averaging the K-H version of the one-dimensional quasi-coulomb potential described in Sec. 10. The two wells indicate the most likely time-averaged positions of the nucleus from the point of view of a classical electron oscillating back and forth in a strong laser field. In this case $\mathcal{E} = 5$ and $\omega = 0.52$, so the K-H parameter $\alpha_0 \approx 18.5$. These numbers, as well as the axis labels are in atomic units.

$$\alpha_0 = -(e\mathcal{E}_0/m\omega^2)\,. \tag{7.2}$$

The K-H Hamiltonian is obviously time-independent when time-averaged. The electron in the laser-dressed double-well potential of Fig. 6 will be stably bound [19]. In other words, if the laser frequency is high enough to justify the time averaging of the nuclear motion, the atom will make no transitions, no matter how strong the field, and in this sense the atom would be stabilized rather than ionized by the external field.

This would be a striking development, and completely contrary to experimental indications to date, which simply show a continuation of the traditional trend in the sense that stronger fields lead to more rapid ionization. Remarkably, however, preliminary computer simulations indicate that it may be possible for atoms to be stabilized in laser fields that are not much more intense than one atomic unit, with electron wave functions taking a localized and strongly two-peaked form reminiscent of the K-H potential. This possibility may indicate a high degree of future utility for the K-H point of view and it is discussed further in Sec. 10.

8. Recent ATI Observations

It is clear from the discussion of the ponderomotive effect in Sec. 4 that very short (100 fsec - 1 psec) pulses would make a difference in ATI observations. If the laser pulse is sufficiently short, then the electron does not have enough time to run down the potential "hill" created by the ponderomotive force [recall Fig. 4]. In this case the electron cannot regain the kinetic energy of its jitter motion and no peak suppression should be observed. The main effect is a shifting toward the red of the ATI electron peaks, since they are all missing the ponderomotive contribution V_P. Such a peak shifting has been reported in ATI experiments and also inferred from *ab initio* calculations of model atomic wave functions. Figure 7 shows an example from Agostini, et al. [20].

An additional ultra-short pulse ATI effect was reported in 1987 by Freeman, et al. [21], namely ATI peak substructure. The structure was found to correlate well with high-lying bound states in xenon [see Fig. 8]. The explanation, as first proposed by the ATT Bell group, is based on intensity-dependent level shifts (AC Stark shifts such as mentioned at the end of Sec. 4) during the laser pulse. At the

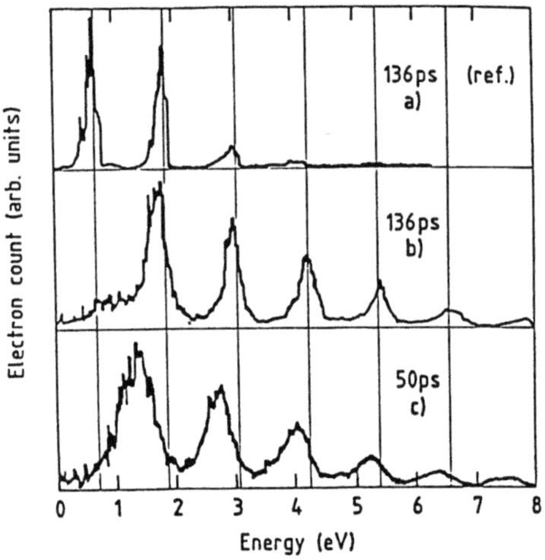

7. The three stages of peak suppression. In (a) the laser pulse is not intense enough to shift the threshold and suppress the first peak. In (b) the laser is almost four times as intense. The s = 0 peak is suppressed, and the s = 1 peak is the first one observed. The s = 1, 2, 3, ... peaks appear with the same kinetic energy as in (a) because they have regained their energy via ponderomotive acceleration. In (c) the laser intensity is the same as in (b) so the s = 0 peak is still suppressed. However, the other peaks are slightly shifted to the red compared with (b). This is because the pulse is shorter and there is not enough time for the photoelectrons to regain all of their ponderomotive energy.

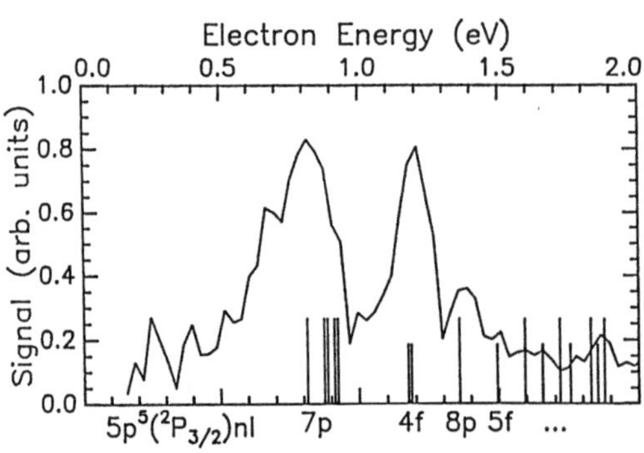

8. Non-free electron behavior: "mapping" of xenon bound levels into ATI peak structure at very short pulse lengths. [From Freeman, et al. [21]].

right laser intensity during the pulse each high-lying level is in exact resonance. At that point it is one-photon ionized and thus gives rise to a replica of itself in the ATI spectrum. This interpretation has been supported by the subsequent observation of giant Stark shifts (up to 1.5 eV) of various excited levels in xenon during 7-photon ionization using 150 fsec pulses by Agostini, et al. [22].

Additional experiments have explored two-laser pump-probe reactions, low frequency multiphoton ionization where ATI peak envelopes but not individual ATI peaks can be observed, experiments with hydrogen, and strong-field effects in molecules. We will not have space here to describe these developments, which are still being actively explored.

9. High-order Harmonic Generation

The excitation of electronic ATI "levels" far above threshold obviously raises the question whether recombination or scattering might lead to interesting nonlinear optical events. Some of these possibilities are sketched in Fig. 9.

Beginning in 1987 reports of MPI light scattering experiments became available from the Illinois-Chicago and Saclay laboratories. These first experiments concentrated mostly but not exclusively on detection of harmonics of the incident laser light [23, 24]. In the observations of Li, et al. [25] odd harmonics of the laser radiation are clearly observed up to the order 33 as well as an extremely long plateau with little variation of the harmonic intensity. Some experimental data is shown compared with calculations in Sec. 10.

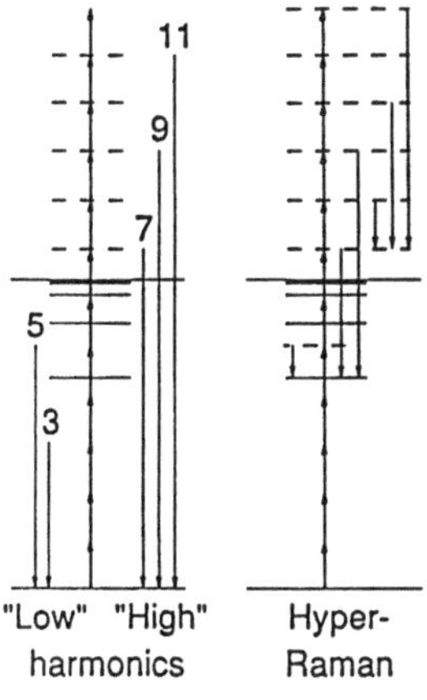

9. Non-linear light scattering and multiphoton ionization. (a) Lower harmonics are generated by transitions from bound states to the ground state, and higher harmonics are generated by transitions from continuum states to the ground state. (b) Other (weaker) nonlinear light scattering processes can also occur in which the final state is not the ground state.

The prediction of traditional nonlinear optics is that the intensity of the Nth harmonic is proportional to the Nth power of the laser intensity and to a field-independent N^{th} order susceptibility,

$$I_{N\omega} \propto (I_\omega)^N \, | \chi^{(N)}(N\omega, -\omega, \ldots, -\omega) |^2. \tag{9.1}$$

The Nth order susceptibility may be expressed in terms of sums over dipole moment matrix elements and resonance denominators. In practice a susceptibility is computed closely analogously to the multiphoton matrix elements discussed in Sec. 3.

The conventional theory of nonlinear optics is based on the supposition that higher order processes are successively smaller and smaller. From this standpoint, the observed long plateaus of nearly equal harmonic peak intensities are highly unexpected. This is the same message carried earlier by the observations of nearly equal ATI peaks of many different orders. Further experiments on high-order harmonics are expected, and difficult theoretical questions, for example concerning phase-matching and the spectral content of side-scattered light, must be addressed.

10. Numerical Simulations

We have already explained that perturbation theory is not adequate as a theoretical tool in the cases of most interest. This creates enormous difficulty in building a useful treatment of ATI. One (pratically impossible) approach would be a direct numerical solution of the correct Schrödinger equation using a realistic description of the atom and the light. More modest attempts have, however, been made with some success. Progress with numerical work is reported in a recent special issue of JOSA B [26].

Despite the wide variety of different approaches and approximations that have characterized numerical work on ATI, the various results largely reinforce each other. In this section we attempt an overview of the kind of results that have been obtained. We do this by concentrating mostly on the results of our own work [27-30], which we hope is reasonably representative. In most cases we start with the Hamiltonian of a model one-electron and one-dimensional atom in a laser field: $H(t) = H_0 + V(t)$, using atomic units:

$$H(t) = -\frac{1}{2}\frac{\partial^2}{\partial x^2} - \frac{1}{\sqrt{1+x^2}} - x\mathcal{E}(t)\sin\omega t\,. \tag{10.1}$$

Our atom is defined on the whole real axis, hence parity is a good quantum number and bound states of the atom do not have permanent dipole moments. The tail of the potential is the same as for the Coulomb potential, and highly excited bound states form a Rydberg series converging to the continuum limit

The "bare" eigenvalues W and eigenvectors |W> of the field-free Hamiltonian H_0 are obtained first, and then we calculate the time-dependent solution $|\Psi(T)\rangle$ for the full Hamiltonian (10.1), where t = T defines the end of the laser pulse. By projecting the wave function $|\Psi(T)\rangle$ onto the energy eigenstates |W> one obtains the ATI photoelectron energy spectrum

$$P(W;T) = |\langle W|\psi(T)\rangle|^2. \tag{10.2}$$

In some cases it is attractive to regard the atom as defined by a set of energy levels, rather than by a binding potential V(x). Schrödinger's equation determines the time dependence of the amplitudes associated with these levels, and with modest assumptions about matrix elements these amplitudes can be determined perturbatively or fully numerically. Roso has shown that such energy-level models are particularly useful for examining separately effects due to bound-free or free-free transitions [30].

We now illustrate with a few examples the way in which various ATI effects described in previous sections manifest themselves in numerical computations of atomic wave functions.

Ponderomotive threshold shift. To observe the threshold shift directly a very short pulse is needed [21], so that the ATI peaks do not have time to convert their jitter energy into drift energy. We plot P(W;T) for a very short 4.25 cycle pulse. An example for nominal five-photon ionization with ω = 0.148 for the three field strengths $\mathcal{E}_0$= 0.05, 0.07071 and 0.085 is shown in Fig. 10. Notice that the spectrum extends down into Rydberg states below W = 0 that are still not resolved after 4.25 cycles. We can see an ionization channel close as the intensity is increased; i.e., the peak labeled 0 moves below W = 0.

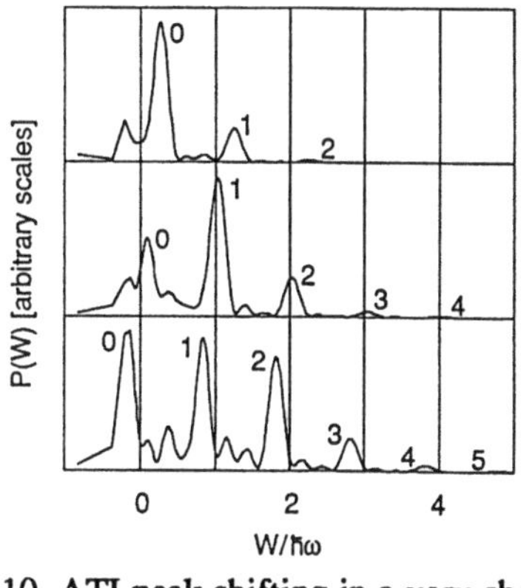

10. ATI peak shifting in a very short laser pulse, as predicted by numerical solution of Schrödinger's equation. From top to bottom the laser intensity increases by approximately a factor of two, and the ATI peaks move to the red since the upward ponderomotive threshold shift (AC Stark shift) increases correspondingly.

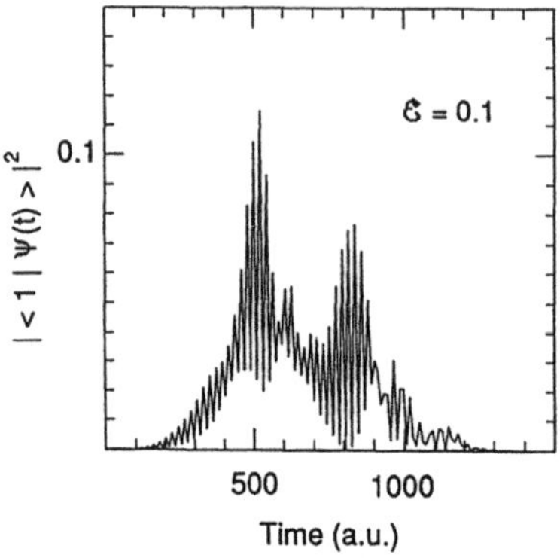

11. Population of the first excited level calculated for a model atom undergoing 5-photon ionization by a smooth 32-cycle pulse. Strong excitation of the level occurs only at $t \approx 500$ a.u. and $t \approx 820$ a.u., and is attributed to AC Stark-shifting of the first level into resonance at those two times during the pulse turn-on and turn-off. [From Javanainen, et al. [27]].

Bound Level Shifts and Population Transients. By projection onto field-free bound states $|n\rangle$, one can compute the populations of these states continuously as a function of time during the pulse: $P_n(t) = |\langle\Psi(t)|n\rangle|^2$. In a smoothly varying laser pulse the atom will be exposed to a continuously changing range of intensities and thus a continuous range of AC Stark shifts. In Fig. 11 we show the population of the first excited level of our model atom as a function of time (in atomic units) during excitation by a smooth pulse. The Stark shifts implied by Fig. 11 are quite large, since in the absence of the laser pulse the first level is out of resonance by $0.05 \approx 0.3\ \omega$. We have obtained similar evidence of shifts of other excited levels as well as the first [27].

Harmonic generation. To compute the intensity $S(\omega_{scatt})$ of light scattered by the atom that is undergoing ionization one needs the expectation value of the atomic dipole moment as a function of time. This is easily obtained from a numerical wave function:

$$d(t) = \langle\Psi(t)|\, ex\, |\Psi(t)\rangle, \tag{10.3}$$

We then carry out a Fourier transform, and finally square the result. Apart from geometric constants, which we ignore, this is essentially the scattered intensity:

$$S(\omega_{scatt}) = |\tilde{d}(\omega_{scatt})|^2. \tag{10.4}$$

Actually, $S(\omega_{scatt})$ is only the coherent component of the light scattering spectrum, but this is the part that contains harmonics propagating in the forward direction.

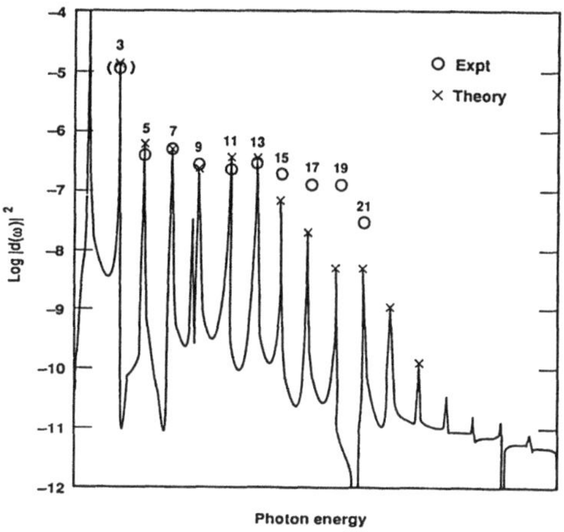

12. Comparison of measured [24] harmonic intensities in xenon at the intensity of 3×10^{13} W cm^{-2} with computations by Kulander and Shore [31]. The vertical scales are chosen such that the computed and measured seventh harmonics agree.

Kulander and Shore have performed three-dimensional simulations of various noble gas atoms using time dependent Hartree-Fock equations. Fig. 12 shows the comparison of the harmonics measured by Ferray, et al., [24] and calculated by Kulander and Shore [31] for xenon at the intensity of 3×10^{13} W cm^{-2}. Their results reproduce the qualitative experimental findings of the initial and final fall-offs and the plateau in the harmonic spectrum.

The computed one-atom response may not be directly comparable with the experiments, because in the experiments the conditions of phase matching are not the same for all harmonics and in formulas such as (10.4) several factors of ω are not included (e.g., the ω^4 Rayleigh factor). Studies of phase-matching under very strong fields have been started, but results to date are incomplete.

We have carried out one-dimensional simulations [32] of the photon spectrum using the same wave functions that we use for ATI. This work presents an opportunity for detailed comparisons between the ATI spectra and the photon spectra obtained under identical conditions. As we predicted [28], the photon spectrum displays an initial drop, a plateau region, and a final fall-off as a function of harmonic order, similar to that reported in the experiments. As a result of calculations with a quite different atom, a two-level atom without continuum levels at all, Sundaram and Milonni [33] have suggested that a harmonic plateau is a general feature of the emission from any strongly driven nonlinear system.

Stabilization and Electron Localization

Atomic stability under strong-field excitation is not easily reconciled with experience in atomic physics. Indeed, in 1985 Lambropoulos [34] predicted that it would not even be possible to subject an atom to very strong fields. However, the K-H picture mentioned in Sec. 7 suggests that the argument may not be compelling if the incident photon frequency is high enough.

Pont, et al. [35], have obtained numerical wave functions for the time-average of the K-H coulomb potential (7.1). One of the striking features of their static eigensolutions is a splitting of the electron probability distribution into a two-peaked or "dichotomous" distribution as soon as $|\alpha_0|$ is greater than about 10 Bohr radii. This is consistent with the picture developed in Sec. 7, but does not really address the issue of stability because the time-averaged K-H potential ignores optical-frequency contributions to the total Hamiltonian that will certainly cause various transitions, including ionization.

We have studied stabilization numerically with the same one-dimensional model atom mentioned above. We find that the exact time-dependent numerical wave functions strongly support the possibility of stabilization [although so far only in a one-dimensional context]. These calculations automatically incorporate all of the time dependences in the full Hamiltonian, including what is ignored by the time-averaged K-H potential. Nevertheless, the time-averaged K-H potential provides a good basis for understanding the results.

One type of results involves the ionization probability

$$P(t) = \int dW \ |\langle W|\Psi(t)\rangle|^2 \tag{10.5}$$

as a function of time. As we have shown [36], plots of P(t) indicate a tendency to saturate before reaching P = 1, and even show that the *amount of ionization becomes less at higher field strength*, at least for $\mathcal{E} \geq 1$. This is what is sensibly meant by stabilization (the suppression of ionization) by a strong-field environment. Uncertainty about the relevance of the bare eigenstates $|W\rangle$ at very high fields, however, makes graphs of electron space-time probabilities more persuasive. In addition to *stabilization*, these plots clearly show electron *localization*.

In Fig. 13 we show the electron space-time probability $|\Psi(x,t)|^2$ as a function of x for a variety of early times near to the turn-on, for a smooth laser pulse with $\mathcal{E} = 5.0$. The graphs show that the electron is well-localized in its ground state at t = 0 before the laser pulse turns on, and then at t = 5 cycles a significant packet of electron probability begins to move toward negative x. At t = 10 and 15 cycles we see this packet continues to move away. The other part of the electron packet, near to x = 0, remains in place without significant further depletion, implying that stabilization has taken effect and ionization has practically ended. Its structure fits K-H concepts fairly well. For example, its confinement boundaries are nearly identical with $\pm\alpha_0$ at all times [36]. We have confirmed this behavior in other circumstances as well, such as longer and shorter pulse turn-on, higher and lower photon frequencies, etc.

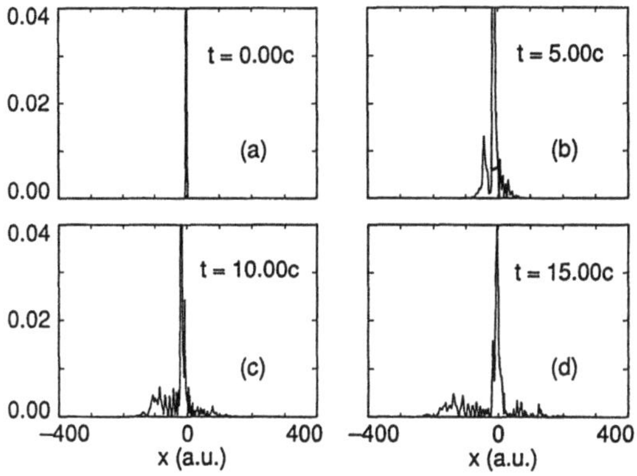

13. Plots of the electron's spatial probability density $|\Psi(x,t)|^2$ during the ionization process. A packet of probability can be seen moving away from x = 0, leaving behind a relatively high density of localized probability. [From Su, et al., [36]].

11. Summary and Outlook

After a series of surprising new observations in the 1980's, the picture of strong-field atomic physics and nonlinear optics is gradually becoming clearer. We understand that one must judge the effect of a given laser intensity in several ways. Of the various ways, the most familiar is the power-law dependence in the historically important relation given in (1.1). But laser intensity plays quite different roles in the electron jitter energy or ponderomotive energy, the K-H displacement α_0, the coulomb barrier suppression (tunneling) threshold, and in bound-level Stark shifts.

It appears that nonlinear optical harmonic generation of very high order and above-threshold electron production are two parallel aspects of an electron's simultaneous interaction with its atomic binding potential and a strong pump field. Although these studies began with ATI, the most intriguing applications may come in nonlinear optics. Indeed, the nature of nonlinear optics with extremely strong pump fields is mostly unknown, and the studies mentioned here have barely inaugurated what may turn into a very active research field.

Apart from the potential for further work on either ATI-type electron processes or on high-order nonlinear optical effects, it would be very interesting to study correlations between the properties of electrons and photons emitted in the same strong field ionization experiment. This is not easy to do, however, since the typical requirements for the two kinds of experiment are almost mutually exclusive. To observe single atom ATI spectra we need a small number of atoms to avoid "space charge" effects, but to detect scattered light we need relatively dense samples with many atoms to produce an observable photon signal. This innate difficulty points up the necessary continuation of theoretical work with atom models, which are fortunately immune to concerns about detectable signal strength.

The behavior of atomic systems in fields near or above the atomic unit of intensity seems likely to become the future focus of attention. The apparent success of the Kramers-Henneberger coordinate frame in helping to understand computer experiments carried out in very strong-field situations, where we are now predicting phenomena such as stabilization and localization, will almost certainly be exploited further. It is already possible to think that the K-H frame may provide a platform for an improved version of perturbation theory [37]. However, other aspects are very cloudy. For example, what advantages can the K-H frame have for describing nonlinear optical effects? This must be worked out.

Connections with classical behavior, including chaos, in the extremely strong-field region have just begun to be explored, and the prediction of wave-function localization will spur these efforts. Of course, it is best to extrapolate cautiously from present knowledge. Multiphoton physics has never been very well-behaved in the sense of obeying the projections made for it. The safest course is to be prepared for more surprises in the 1990's, possibly surprises as interesting as those of the 1980's.

Acknowledgements

The author is pleased to thank Q. Su for assistance with the manuscript, and to acknowledge many conversations during the past 3 years about different aspects of strong-field physics with him as well as with other colleagues, particularly including J. Javanainen, P.W. Milonni, L. Roso-Franco and K. Rzazewski. This research was partially supported by the US National Science Foundation. We are grateful to the Allied-Signal Corporation, University of Rochester Laboratory for Laser Energetics, John von Neumann Supercomputing Center and Pittsburgh Supercomputing Center for support of computational resources.

References

[1] J.H. Eberly, Hyperfine Int. **37** (1987) 33.

[2] S.J. Smith and P.L. Knight ed., Multiphoton Processes, (Cambridge University Press, New York 1988).

[3] N.B. Delone and M.V. Fedorov, Progr. Quant. Electron. **13** (1989) 267.

[4] J.H. Eberly, J. Javanainen and K. Rzazewski, Phys. Reports (1991) .

[5] P.H. Bucksbaum and R.R. Freeman, J. Phys. B (in preparation) .

[6] J.S. Voronov and N.B. Delone, Sov. Phys. JETP **23** (1966) 54.

[7] P. Agostini, G. Barjot, J.F. Bonnal, G. Mainfray and C. Manus, IEEE J. Quant. Electron. **QE-4** (1968) 677.

[8] P. Agostini, F. Fabre, G. Mainfray, G. Petite and N.K. Rahman, Phys. Rev. Lett. **42** (1979) 1127.
[9] Y. Gontier and M. Trahin, J. Phys. B **13** (1980) 4381.
[10] P. Kruit, J. Kimman, H.G. Muller and M.J. van der Wiel, Phys. Rev. A **28** (1983) 248.
[11] F.H.M. Faisal, Theory of Multiphoton Processes (Plenum Press, New York 1987).
[12] T.W.B. Kibble, Phys. Rev. **150** (1966) 1060.
[13] S. Augst, D. Strickland, D.D. Meyerhofer, S.L. Chin and J.H. Eberly, Phys. Rev. Lett. **63** (1989) 2212.
[14] L.V. Keldysh, Sov. Phys. JETP **20** (1965) 1307.
[15] F.H.M. Faisal, J. Phys. B **6** (1973) L312.
[16] H.R. Reiss, Phys. Rev. A **22** (1980) 1786.
[17] W.C. Henneberger, Phys. Rev. Lett. **21** (1968) 838.
[18] H.A. Kramers, Collected Scientific Papers (North-Holland, Amsterdam 1956).
[19] J.I. Gersten and M. Mittleman, J. Phys. B **9** (1976) 2561.
[20] P. Agostini, J. Kupersztych, L.A. Lompré, G. Petite and F. Yergeau, Phys. Rev. A **36** (1987) 4111.
[21] R.R. Freeman, P.H. Bucksbaum, H. Milchberg, S. Darack, D. Schumacher and M.E. Geusic, Phys. Rev. Lett. **59** (1987) 1092.
[22] P. Agostini, P. Breger, A. L'Huillier, H.G. Muller and G. Petite, Phys. Rev. Lett. **63** (1989) 2208; and erratum, ibid. 64 (1990) 815.
[23] C.K. Rhodes, Physica Scripta **T17** (1987) 193.
[24] M. Ferray, A.L. L'Huillier, X.F. Li, L.A. Lompré, G. Mainfray and C. Manus, J. Phys. B **21** (1988) L31.
[25] X.F. Li, A.L. L'Huillier, M. Ferray, L.A. Lompre and G. Mainfray, Phys. Rev. A **39** (1989) 5751.
[26] K.C. Kulander and A. L'Huillier ed., special issue on advances in theoretical understanding of intense-field and multiphoton phenomena, J. Opt. Soc. Am. B **7** 1990) p. no. 4.
[27] J. Javanainen, J.H. Eberly and Q. Su, Phys. Rev. A **38** (1988) 3430.
[28] J.H. Eberly, Q. Su and J. Javanainen, Phys. Rev. Lett. **62** (1989) 881.
[29] J. Javanainen and J.H. Eberly, Phys. Rev. A **39** (1989) 458.
[30] L. Roso-Franco and J.H. Eberly, J. Opt. Soc. Am. B **7** (1990) 407.
[31] K. Kulander and B.W. Shore, Phys. Rev. Lett. **62** (1989) 524.
[32] J.H. Eberly, Q. Su and J. Javanainen, J. Opt. Soc. Am. B **6** (1989) 1289.
[33] B. Sundaram and P.W. Milonni, Phys. Rev. A **41** (1990) 6571.
[34] P. Lambropoulos, Phys. Rev. Lett. **55** (1985) 2141.
[35] M. Pont, N.R. Walet, M. Gavrila and C.W. McCurdy, Phys. Rev. Lett. **61** (1988) 939.
[36] Q. Su, J.H. Eberly and J. Javanainen, Phys. Rev. Lett. **64** (1990) 862.
[37] J. Mostowski and J.H. Eberly, J. Opt. Soc. Am. B (submitted) (1990) .

Multiphoton Absorption Induced by Chaotic Phenomena

C.L. Sigüenza, P.F. Gonzáles-Díaz, and M. Santos
Instituto de Óptica "Daza de Valdés", C.S.I.C.,
Serrano 121, E-28006 Madrid, Spain

Abtract. A fully quantized model for the multiphoton absorption of an N-level system perturbed by $N-1$ simultaneous different wavelengths has been constructed assuming a coherent cooperative interaction mechanism. The Rabi frequency for the center of energies and a fundamental relative frequency are obtained. This model is used to explain the stimulated generation of intramolecular chaos in multiphoton dissociation of rather small molecules induced by two different wavelengths.

1 Introduction

The study of multiphoton excitation and dissociation by an anharmonic level system has received a great deal of attention in recent years [1]. A well established model [2] based on energy stochastization has been widely called for to explain the existence of these processes. Although such a simple model provides us a fairly satisfactory justification for the unimolecular dissociation of large molecules, it presents severe difficulties when one tries to apply it to small or very symmetric molecules. For in these cases, the low density of states makes stochastization difficult to accommodate [3]. One interesting proposal to solve this problem has been given by J.R. Ackerhalt et al. [4] who suggested that deterministic intramolecular generation of instabilities and chaos may be necessary to explain the unimolecular dissociation process, at least for small molecules [5].

Selectivity and efficiency of unimolecular dissociation are sometimes drastically enhanced when the molecules are at the same time irradiated with a second non-resonant laser beam [1].

In this work we have constructed a general simple model which can be accommodated to account for the multiphoton cooperative absorption of a number of different characteristic wavelength photons by a N-level system and applied it to discuss the above-mentioned interpretational problem.

2 The Model

We consider a simple model [6] for the multiphoton absorption of a N-level system perturbed by $N-1$ simultaneous different wavelengths. In constructing this model, two main assumptions have been made: (1) The absorption of any

Springer Proceedings in Physics, Vol. 55
Nonlinear Dynamics and Quantum Phenomena in Optical Systems
Editors: R. Vilaseca, R. Corbalán

number of monochromatic photons is assumed to proceed through a coherent cooperative mechanism, and (2) we assume our system to be in a thermal bath at zero temperature, so that every transition starting from energy levels other than the fundamental one is disregarded. The fully quantized Hamiltonian for the total absorption of $M_1 ..., M_n$ photons with frequency $\omega_{L1} ..., \omega_{Ln}$ by a N level system with transition frequencies ω_i, (where i denotes the ith level from which a one-step transition is taking place upward) in the rotating wave approximation (RWA), can be written

$$H(N) = H(N)_F + H(N)_A + H(N)_I, \tag{1}$$

where $H(N)_F$ is the Hamiltonian for the $N-1$ laser fields, as expressed in terms of creation and annihilation operators, and the characteristic frequencies ω_{Ln}.

In defining the material (atomic or molecular) Hamiltonian, $H(N)_A$, we first of all have to extract the unphysical gauge due to the ambiguity in the origin of energies by absorbing [7] that ambiguity as a coefficient for the identity NxN matrix $\hat{I}$. This distinguishes two different material Hamiltonians according to the parity of the number of levels, N. We obtain

$$H(N)_A = E_G \hat{I} + \frac{1}{2}\hbar\{[\sum_{j=1}^{\alpha}\sum_{i=j}^{N-j} D(N)_j - \sum_{j=1}^{\beta}(\sum_{i=j}^{\beta} - \sum_{i=\alpha+1}^{N-j})(D(N)_j)^2]\omega_i\} \tag{2}$$

where, for N even, $\alpha = \beta + 1 = \frac{N}{2}$, $E_G = \frac{1}{2}(E_\alpha + E_{\alpha+1})$; and, for N odd, $\alpha = \beta = (N-1)/2$, $E_G = E_{\alpha+1}$, E_i is the energy for the ith level, and $D(N)_j$ is a diagonal NxN matrix which is antisymmetric under reflection at the secondary diagonal, the half-upper semidiagonal containing only one $+1$ situated at position jth, with j running from 1 to α, and $[D(N)_j, D(N)_i] = 0$.

Now, the unit matrix $\hat{I}$ can be dropped from (2) by simply shifting the origin of energies [8], gauging off the physically irrelevant degrees of freedom in the original material Hamiltonian.

In the cooperative interaction approximation we are using, the interaction Hamiltonian $H(N)_I$ can be written

$$H(N)_I = \hbar\beta\{[\prod_{n=1}^{N-1} \hat{a}_n^{\dagger M_n}]S_d(N) - [\prod_{n=0}^{N-1} \hat{a}_n^{M_n}]S_u(N)\}, \tag{3}$$

where β is the dipole moment matrix element for the whole cooperative absorption of all photons involved in the process. $S_u(N)$ and $S_d(N)$ are matrices which are defined such that:

$$[D(N)_1, S_u(N)] = 2S_u(N); [S_d(N), D(N)_1] = 2S_d(N)$$

$$\{D(N)_1, S_u(N)\} = \{D(N)_1, S_d(N)\} = 0. \tag{4}$$

From the total Hamiltonian (1), we obtain now the fundamental frequencies that characterize the considered interaction process. Even in a simple model such as that being used here, the calculation of the fundamental frequencies is not an easy task, for nonlinear interaction effects may give rise to the appearance of additional independent radiative phenomena, which can be characterized by nonlinear fundamental frequencies [7]. In our simple interaction approximation, due to the form of the material Hamiltonian (2), one may expect the contribution of only one additional nonlinear fundamental frequency which would come about from the existence of a nonzero term depending on $[D(N)_1]^2$. Counting the number of constants of motion becomes an essential ingredient for characterizing the process. In our case, we have as constant of motion ($[\hat{C}(N)_n, H(N)] = 0$):

$$\hat{C}(N)_n = \hat{a}_n^\dagger \hat{a}_n + \frac{1}{2} M_n D(N)_1, n = 1, ..., N-1. \tag{5}$$

Now, from the Heisenberg equations, we can obtain an expression for the second time derivative of the matrix $D(N)_1$, where the resulting coefficients for $D(N)_1$ and $[D(N)_1]^2$ can be interpreted [9] as general expressions for the two fundamental frequencies:

$$\Omega_{Rc}(N) = -\frac{\epsilon(N)_n}{2}[\sum_{i=1}^{N-1} \omega_i - \sum_{n=1}^{N} M_n \omega_{Ln}] + 2 \mid \beta \mid^2 B(N)_n \tag{6}$$

and

$$\Omega_{Rr}(N) = \frac{\epsilon(N)_n}{2}[\sum_{i=1}^{\beta} \omega_i - \sum_{i=\alpha+1}^{N-1} \omega_i] + 2 \mid \beta \mid^2 A(N)_n \tag{7}$$

where

$$\epsilon(N)_n = 2(\sum_{i=1}^{N-1} \omega_i - \sum_{n=1}^{N} M_n \omega_{Ln})$$

and

$$A(N)_n = \sum_{n=1}^{N} M_n \prod_{p=1, p\neq n}^{N} \hat{a}_p^{\dagger M_p} \hat{a}_p^{M_p} \sum_{q_n=0}^{M_n-1} \hat{a}_n^{\dagger q_n} \hat{a}_n^{M_n-1} \hat{a}_n^{\dagger M_n - q_n - 1}$$

$$B(N)_n = \prod_{n=1}^{N} (\hat{a}_n^{\dagger M_n} \hat{a}_n^{M_n} + \hat{a}_n^{M_n} \hat{a}_n^{\dagger M_n}).$$

These two fundamental frequencies are, respectively, the Rabi frequency for the center of energies when the whole system is considered as an effective three-level system with the zero of energies playing the role of the intermediate energy level for N odd, or the central transition being reduced to such an intermediate level for N even (in both cases, the upper and lower levels of this effective system are N and 1), and a relative frequency. These two frequencies are no longer constants of motion, having a non-linear dependence on the laser intensities.

3 Multiphoton Dissociation Induced by Two Wavelengths

Although for a detailed account of the multiphoton interaction process one should require the lifting of at least some of the approximations contained in our simple model, this still retains some essential physical features which may prove to be important for a suitable explanation of the process of multiphoton dissociation induced by two different wavelengths.

For processes induced by a single wavelength, the model alluded to is rather useless, and corresponds to the case of $N = 2$.

For dissociation processes induced by two wavelengths, we can distinguish two main situations. For sufficiently large and/or very asymmetric molecules, both radiations could effectively simultaneously contribute to the absorption process throughout the energy interval and we cannot define any independent constant of motion covering a restricted energy subinterval. Thus, the material system would behave as an effective two-level system ($N = 2$).

For sufficiently small and/or very symmetric molecules the energy interval where the two wavelengths simultaneously contribute shrinks to a width typically comparable to that of the rotational structure of the involved vibrational levels, and hence, it can be regarded as an effective intermediate level in our model and one can define here two independent constants of motion. For most practical cases, the thus generated intermediate level would be closer to the ground level than to the dissociation threshold. Here, superimposed on the conventional absorption effect driven by Ω_{Rc}, there will be a new force driven by Ω_{Rr} conveying the molecules upward along the level system, weakening the bottleneck effect [3] and increasing the dissociation effectiveness with respect to that in single-wavelength experiments. This enhancement would not be present therefore in large molecules. These predictions have been qualitatively confirmed experimentally [10].

4 The Stimulated Intramolecular Chaos

In our model, a new "radiative" process which is characterized by the new fundamental frequency Ω_{Rr} may appear. Ω_{Rr} actually couples a level of the active mode to a fictitious level which should necessarily be a given component of a reservoir of energy. For polyatomic molecules it is most natural and economic to take the available multimode background formed by all molecular modes other than the active mode to play the role of that reservoir in unimolecular excitation. Clearly, the simultaneous irradiation of a molecule with two different field radiations induces an intramolecular periodic coupling from the active mode to the background. Such a coupling is characterized by a frequency Ω_{Rr} which would operationally manifest as a *recurrence frequency*, having an associated *recurrence time*, Ω_{Rr} which would impose a given density of states, ρ, in the background given by $\Omega_{Rr}^{-1} \propto \rho$. It appears that the larger the anharmonicity the larger Ω_{Rr} and hence, the smaller ρ. In the Ackerhalt et al. model [4] the system spontaneously generates a similar recurrence mechanism coupling also the active mode to the background. That mechanism is taken to be responsible for the generation of chaos, and has the same dependence on the anharmonicity and ρ as the mechanism stimulated by the excitation with two different wavelengths.

For small and/or very asymmetric molecules the Letokhov stochastic model can no longer justify the existence of the unimolecular dissociation process.

In such cases, it is the *spontaneous* generation of intramolecular chaos in single wavelength, with a superimposed *stimulated* generation of chaos in double wavelength excitation, that is to be considered as the ultimate cause for the occurrence of the process. Thus, it is the stimulated generation of chaos that increases the effectiveness of unimolecular dissociation when the molecules are irradiated with two wavelengths.

References

[1] *Multiple-Photon Excitation and Dissociation of Polyatomic Molecules*, edited by C.D. Cantrell (Springer-Verlag, Berlin, 1986).

[2] V.S. Letokhov, *Progress in Optics XVI*, edited by E.Wolf (North Holland, Amsterdam 1978).

[3] T.B. Simpson, J.G. Black, J. Burak, E. Yablonovitch and N. Bloembergen, J. Chem. Phys. **83**, 628 (1985).

[4] J.R. Ackerhalt and P.W. Milonni, Phys. Rev. **A34**, 1211 (1986).

[5] M.V. Kuzmin and A.A. Stuchebrukhov, *Laser spectroscopy of highly vibrational excited molecules*, edited by V.S. Letokhov (Adam Hilger, Bristol and Philadelphia 1989).

[6] A full discussion of this model will appear elsewhere.

[7] M. Santos and P.F. González-Díaz, *Nonlinear Rabi frequency in multiphoton absorption*, Inst. de Optica Preprint, C.S.I.C.-IRM-02-90 (1990).

[8] P.L. Knight and P.W. Milonni, Phys. Rep. **66**, 21 (1980).

[9] N.N.Bogolubov Jr., F.L.Kien and A.S.Shumovski, Phys.Lett. **101A**,201(1984)

[10] M. Santos, C. Sigüenza, J.A. Torresano and P.F. González-Díaz, Spectrochim. Acta **46A**, 455 (1990).

Coherence and Interference in Multiphoton Processes

A. Sanpera and L. Roso-Franco

Departament de Física, Universitat Autònoma de Barcelona,
E-08193 Bellaterra, Spain

Abstract. A test of a recently developed short-time perturbation theory to study photoionization is presented for a one-dimensional model atom with a δ-potential. The theory allows the identification of coherences and interferences between different processes.

1. Introduction

New experimental domains have been appearing during the last few years with the availability of high intensity and very short duration laser pulses. From the theoretical point of view, the interaction of a high intensity field with an atom has to be calculated without introducing the usual perturbation theories; this is a hard simulation problem that needs massive computing [1]. In order to qualitatively describe some experimental features a new kind of perturbation theory that considers the short time influence of the non-resonant terms has been recently proposed [2].

In this paper we report a test of this perturbation theory in a situation where the non-resonant terms are dominant. To do this we consider a simple unidimensional model, we solve it numerically and also we find the solution predicted by perturbation theory, evidencing the degree of agreement between them. The model tries to be as simple as possible and, therefore, the atomic potential is provided by a Dirac δ -function. The reasons for our choice are that this simple potential has a ground state which is similar to the ground state of a real three-dimensional hydrogen atom, also it has only one bound state, with no intermediate resonances, and finally, the δ-potential is very convenient to calculate coupling between states. A number of studies on the ionization of an electron bound to a δ-potential have been published so far [3].

2. One-Dimensional Schrödinger Equation

We solve the Schrödinger time dependent equation for a one-dimensional atom in the dipole approximation

Springer Proceedings in Physics, Vol. 55
Nonlinear Dynamics and Quantum Phenomena in Optical Systems
Editors: R. Vilaseca, R. Corbalán

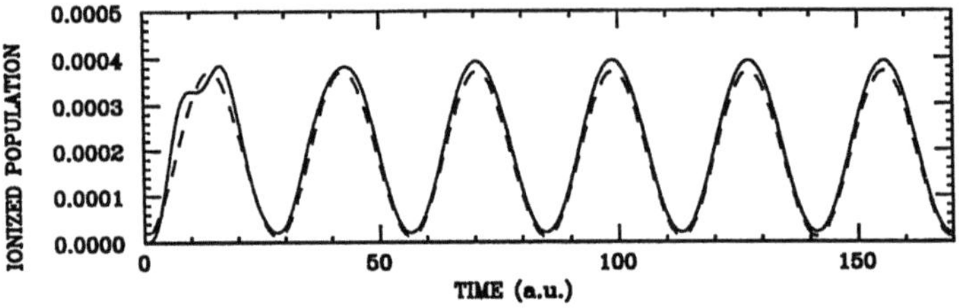

Figure 1.- Time dependence of the ionized population for the first three cycles after turning on of the laser.

$$i\frac{\partial\Psi(x,t)}{\partial t} = \Big(-\frac{1}{2}\frac{\partial^2}{\partial x^2} - g\delta(x) - xE_0(t)\sin\omega_L t\Big)\Psi(x,t) \tag{1}$$

(We take atomic units, $m = \hbar = e = 1$). The field has been treated classically. In the result presented, only a square envelope has been considered, $E_0(t) = E_0$, for $0 \leq t \leq T$, and $E_0(t) = 0$, otherwise.

We consider the one-dimensional atom confined in a box of length $2L$ (with $L >> 1$ a.u.) centered at the origin (where the atom is). The bound state, of even parity, is

$$\phi_0(x) = \frac{\sqrt{g}}{\sqrt{1 - exp(-2gL)}} exp(-g|x|) \tag{2}$$

and its energy is $\omega_0 = -g^2/2$. So the wave function before the field is turned on ($t \leq 0$) is, $|\Psi(x,t)\rangle = |\phi_0(x)\rangle \exp(-i\omega_0 t)$ with the boundary condition $|\Psi(x, t = 0)\rangle = |\phi_0(x)\rangle$. Our model presents only one bound state, therefore the ionized population after a time t is,

$$P_{ion}(t) = 1 - |\langle\phi_0(x)|\Psi(x,t)\rangle|^2, \tag{3}$$

The three free parameters that determine the properties of the atom-field system are g, the strength of the potential well, ω_L, the frequency of the laser, and its intensity, I, proportional to E_0^2. To make all scales in our model as similar as possible to an hydrogen atom, we take $g = 1$, to obtain $\omega_0 = -g^2/2 = -1/2$ [1 a.u.= $27.2eV$]. We have considered $\omega_L = 1/9$ a.u., i.e. five photons are needed to ionize. The last parameter, E_0, has been taken equal to 1.5 x 10^{-3} a.u. (1 a.u. = 3.54 x $10^{16} W/cm^2$). It corresponds to an intensity, where the non-resonant terms dominate. The computed result is represented by the continuous line in Fig. 1, that shows the bound state population for the first three optical cycles.

3. Energy Eigenstates in the Box

Now we want to use the solutions obtained through direct computation of the Schrödinger equation as a test of a new kind of perturbation theory. To do this it is necessary to calculate energy eigenstates in the continuum. This is very easy for a δ-potential in a box. The free states are determined by the values of k allowed by the periodic boundary conditions, $k_n = n\pi/L$ for n integer. The states can be labeled by the integer index n and their energies are, $\omega_n = k_n^2/2$. For each positive energy ω_n the two free eigenstates [4], $\phi_{n,even}$ and $\phi_{n,odd}$, of definite parity are,

$$\phi_{n,odd}(x) = \frac{i}{\sqrt{L}} sin(k_n x) \tag{4}$$

$$\phi_{n,even}(x) = \frac{1}{\sqrt{L}}[exp(-ik_n|x|) - \frac{g - ik_n}{g + ik_n} exp(ik_n|x|)] . \tag{5}$$

On the basis of atomic eigenstates it is not difficult to calculate the dipole matrix elements that account for the coupling with the radiation. The laser field just couples states of different parity. The non-vanishing bound-free dipole matrix elements are,

$$V_{0n} = \langle \phi_{n,odd} | E_0 x | \phi_0 \rangle = \frac{\sqrt{g}}{\sqrt{L}} \frac{4igk_n}{(g^2 + k_n^2)^2} E_0 \tag{6}$$

and the free-free matrix elements are,

$$V_{mn} = \langle \phi_{n,odd} | E_0 x | \phi_{m,even} \rangle = \frac{-4k_n k_m}{(k_n^2 - k_m^2)(g + ik_m)} E_0 . \tag{7}$$

The next step is to write the wave function $\Psi(x,t)$ on this basis,

$$|\Psi(x,t)\rangle = b(t)|\phi_0\rangle + \sum_n c_{n,even}(t)|\phi_{n,even}\rangle + \sum_n c_{n,odd}(t)|\phi_{n,odd}\rangle . \tag{8}$$

The Schrödinger equation becomes a large system of ordinary differential equations for the amplitudes $b(t)$, $c_{n,even}(t)$, $c_{n,odd}(t)$,

$$i\frac{db}{dt} = \omega_0 b(t) + \sum_n c_{n,odd}(t) V_{0n}^* sin\omega_L t \tag{9}$$

$$i\frac{dc_{n,odd}}{dt} = \omega_n c_{n,odd}(t) + \sum_m c_{m,even}(t) V_{mn}^* sin\omega_L t \tag{10}$$

$$i\frac{dc_{n,even}}{dt} = \omega_n c_{n,even}(t) + \sum_m c_{m,odd}(t) V_{nm} sin\omega_L t + b(t) V_{0n} sin\omega_L t . \tag{11}$$

4. Perturbation Theory Result

We are now in condition to apply perturbation theory [2] for short-times to study the evolution of the system. That theory introduces a perturbative Fourier-Floquet series to describe the amplitudes. We use the adiabatic approximation to solve the set of equations for the amplitudes that are far away from the resonance term (in our case it will be a five-photon resonance). If we neglect all other possible orders and keep only the contribution due to the first order, we find that the ionized population is

$$P_{ion}(t) = \sum_n \left[\frac{|V_{0n}|^2/4}{(\omega_n - \omega_0 - \omega_L)^2} + \frac{|V_{0n}|^2/4}{(\omega_n - \omega_0 + \omega_L)^2} - \frac{|V_{0n}|^2 cos(2\omega_L t)/2}{(\omega_n - \omega_0)^2 - \omega_L^2} \right]. \tag{12}$$

It corresponds to equation (14) in Ref. 2. The first sum gives the effect of the rotating process, the second is due to the counter - rotating one and the third is the interference between them. This last term is responsible for the oscillations appearing in Fig. 1.

The sums in (12) can be transformed to energy integrals by introducing the proper density of states,

$$\sum_n \frac{|V_{0n}|^2/4}{(\omega_n - \omega_0 \pm \omega_L)^2} = \frac{4}{\pi} E_0^2 \int_0^\infty \frac{d\omega\sqrt{2\omega}}{(g^2+2\omega)^4(\omega + g^2/2 \pm \omega_L)^2} \tag{13}$$

$$\sum_n \frac{|V_{0n}|^2/2}{(\omega_n - \omega_0)^2 - \omega_L^2} = \frac{8}{\pi} E_0^2 \int_0^\infty \frac{d\omega\sqrt{2\omega}}{(g^2+2\omega)^4((\omega + g^2/2)^2 - \omega_L^2)} \tag{14}$$

and these integrals can be worked out analytically. Substituting this in (12) we find that the first-order contribution gives an ionized population after a time t equal to,

$$P_{ion}(t) = E_0^2(0.445 + 0.450 cos 2\omega_L t)\,, for \omega_L = 1/9 \; a.u. \tag{15}$$

This perturbative expression corresponds to the dashed line in Fig. 1, and presents a good agreement with the computed solutions, evidencing that, although this is a five-photon ionization, the dynamics in this range of intensities and times is governed by the interference of the rotating and counter rotating one-photon processes.

5. Conclusions

We have shown that the δ potential is a good test system to compare real space and energy space calculations. We have used it to check, in the weak field regime, the validity of a new short-time perturbation theory recently presented. We conclude that for short interaction times and low intensities the dynamics is described by the energy non-conserving terms corresponding to the first order time dependent perturbation theory.

Acknowledgements. The authors would like to thank Professor J. H. Eberly for many helpful discussions. This work has been partially supported by the Spanish DGICYT, grant No. PB-86-650-C03-01.

References

1.- J. H. Eberly and J. Javanainen, Eur. J. Phys., **9** , 265 (1988).
2.- L. Roso-Franco and J. H. Eberly, J. Opt. Soc. Am. B **7**, 407 (1990).
3.- S. Geltman, J. Phys. B: Atom. Molec. Phys. **10** ,831 (1977); G. Scharf, K. Sonnenmoser and W. F. Wreszinski, Europhys. Lett., **10** , 19 (1989); K. J. LaGattuta, Phys. Rev. A **40** , 683 (1989); W. Becker, J. K. McIver and M. Confer, Phys. Rev. A **39**, 6068 (1989); A. Sanpera and L. Roso-Franco, Phys. Rev. A **41**, 6515BR (1990).
4.- K. J. LaGattuta, Phys. Rev. A **40**, 683 (1989); L. Schiff, "Quantum Mechanics ", Mc Graw Hill, 1968.

Part III

Basic Quantum Phenomena

Amplification, Lasing and Inversion

*P. Mandel and O. Kocharovskaya**

Université Libre de Bruxelles, Optique Nonlinéaire Théorique,
Campus Plaine C.P. 231, B-1050 Bruxelles, Belgium
*Permanent address: Institute of Applied Physics,
USSR Science Academy, 603600 Nijny-Novgorod, USSR

Abstract. We analyze the Maxwell-Bloch equations for a three-level medium interacting with a bichromatic field in the Λ configuration and show how trap states arise and lead to amplification without inversion.

1. Trap States

It is well known that when a two-level medium interacts with a monochromatic field, amplification will take place when the upper level population, N(2), is larger than the lower level population N(1):

$$N(2) > N(1) .$$

The question that we address in this lecture is: how is the amplification condition generalized when the material medium is a three-level system. As we shall show, the properties of a three-level medium are significantly different from that of a two-level. In particular, amplification is possible when the upper level population is less than the population of each of the lower levels.

Let us consider a three-level medium interacting with a bichromatic field in a Λ configuration. Each atomic level j=1,2 or 3 is characterized by a wave function φ_j and an energy $\hbar\omega_j$ such that $\omega_3 \gg \omega_2 > \omega_1$. A bichromatic electric field E(t), given by

$$E(t) = E_a \cos\omega_a t + E_b \cos\omega_b t , \quad (1)$$

Springer Proceedings in Physics, Vol. 55
Nonlinear Dynamics and Quantum Phenomena in Optical Systems
Editors: R. Vilaseca, R. Corbalán

interacts with the three-level medium. The interaction hamiltonian is $V(t) = -\hat{d}E(t)$, where $\hat{d}$ is the electric dipole. We first assume that the field E_a (E_b) interacts only with levels 1 (2) and 3, with the simplified selection rules:

$$<\varphi_j|\hat{d}E_a|\varphi_3> = \wp E\delta_{j,1}, \quad <\varphi_j|\hat{d}E_b|\varphi_3> = \wp E\delta_{j,2}. \qquad (2)$$

In that case, the transition probability between the upper state and a linear superposition of the two lower states is given, in the rotating wave approximation, by

$$W_1(C_1,C_2) = \left|<C_1 e^{-i\omega_1 t}\varphi_1 + C_2 e^{-i\omega_2 t}\varphi_2|V|e^{-i\omega_3 t}\varphi_3>\right|^2$$

$$= \frac{\wp^2 E^2}{4}\left|C_1 e^{-i(\omega_{31}-\omega_a)t} + C_2 e^{-i(\omega_{32}-\omega_b)t}\right|^2,$$

where $\omega_{ij} \equiv \omega_i - \omega_j$ and $|C_1|^2 + |C_2|^2 = 1$. On resonance, which we define through $\omega_{31} - \omega_a = \omega_{32} - \omega_b$, the transition probability reduces to

$$W_1(C_1,C_2) = \frac{\wp^2 E^2}{4}|C_1 + C_2|^2 . \qquad (3)$$

To fully appreciate this result, let us also consider the situation in which each field E_a and E_b interacts equally with the two levels 1 and 2. In that case, the probability transition between the upper state and a linear superposition of the two lower states is given, on resonance, by

$$W_2(C_1,C_2) = \frac{\wp^2 E^2}{4}\left|C_1(1+e^{-i\omega_{21}t}) + C_2(1+e^{i\omega_{21}t})\right|^2 . \qquad (4)$$

The main difference between (3) and (4) is that there exist a choice of the parameters C_1 and C_2 such that W_1

vanishes for all times:

$$W_1(1/\sqrt{2},-1/\sqrt{2}) \equiv 0 \ .$$

Such a property is impossible for the transition probability W_2. Hence, provided the selection rules (2) hold, there exists a trap state defined by

$$|\Psi_{trap}> = \frac{1}{\sqrt{2}} [|\varphi_1> - |\varphi_2>] \ . \tag{5}$$

This state is an atom trap because atoms in that state do not interact with the upper state via the field E:

$$|<\Psi_{trap}|V|\varphi_3>|^2 = 0 \ .$$

These trap states were found experimentally and theoretically analyzed by the Pisa group [1]. The experiment was performed on Na vapor interacting with a multimode laser in an inhomogeneous magnetic field. The fluorescence emission was recorded and dark lines were observed at space points of the Na cell such that the Zeeman sublevel frequency equals an integer number of the laser intermode spacing. This is tantamount to the resonance condition $\omega_{32}-\omega_b=\omega_{31}-\omega_a$ in the case of the bichromatic field (or two-mode laser interacting with the three-level medium). A slightly different set-up was independently realized by Gray et al.[2] and again dark lines were found which involved the hyperfine structure of the Na D lines. Dark lines in the fluorescence spectrum imply that there is no absorption, hence no population in the upper level and therefore trapping of all atoms in the trap state. Recently, it was shown that population trapping can also be achieved when a three-level medium interacts with a periodic train of ultrashort pulse [3,4]. Here the resonance condition is that the frequency separation between the two low-lying atomic states be equal to an integer number of the pulse repetition frequency.

Population trapping has many applications. The most spectacular applications of trap states are in atom cooling [5] and in the possibility of achieving amplification without population inversion. This effect was predicted theoretically [6] for the coherent amplification of an ultrashort pulse. It was reported by Harris's group [7] that amplification and lasing at 96.9 nm (or 12.8 eV) had been observed without population inversion. A theoretical analysis was proposed by Harris [8]. A different scheme was also proposed by Scully et al.[9]. From then on, an ever larger number of papers had been published on possible mechanisms leading to amplification without inversion. One of the motivations for that research, apart from the fascination of understanding a new phenomenon in the realm of light-matter interaction, is the possibility of producing very short wavelength coherent sources, including X-ray lasers.

The underlying idea in most of the proposed schemes is the following. In a three-level medium, a trap state is induced by the external field. Besides the trap state, there must be a second state, which is also a linear superposition of $|\varphi_1\rangle$ and $|\varphi_2\rangle$ but which can interact with the upper state $|\varphi_3\rangle$. Let $|\Psi_{int}\rangle$ be this state. Since trapped atoms do not interact with the upper level, we are dealing with an effective two-level medium and the amplification condition becomes

$$N_3 > N_{int} .$$

It turns out to be possible to verify that condition without having $N_3 > N_2$ or $N_3 > N_1$. The same conclusion holds for short pulses [6].

2. Maxwell-Bloch Equations

The above discussion was very qualitative. Let us analyze the Λ configuration in a more rigorous way by deriving the semi-classical equations describing the coupling of the classical field (1) with a quantized three-level medium. Introducing the reduced fields

$$\alpha = |\mu_{31}|E_a/2\hbar, \ \beta = |\mu_{32}|E_b/2\hbar \ ,$$

the propagation equations for α and β, derived from the Maxwell equations, are:

$$\partial\alpha/\partial z + (1/c_a)\partial\alpha/\partial t + \kappa_a\alpha = 2\pi i N\omega_a|\mu_{31}|^2\sigma_{31}/c_a\epsilon_a\hbar \ ,$$

$$\partial\beta/\partial z + (1/c_b)\partial\beta/\partial t + \kappa_b\beta = 2\pi i N\omega_b|\mu_{32}|^2\sigma_{32}/c_b\epsilon_b\hbar . \quad (6)$$

In these equations, $c_j=c/\epsilon_j$ (with j=a or b) is the velocity of light at frequency ω_j, ϵ_j the dielectric permitivity, κ_j the phenomenological damping rate of the field E_j and σ_{pq} the slowly varying envelope of the density matrix element ρ_{pq}: $\rho_{31} = \sigma_{31}\exp(-i\omega_a t)$, $\rho_{32} = \sigma_{32}\exp(-i\omega_b t)$. The atomic polarizations σ_{31} and σ_{32} characterizing the optical transitions verify the evolution equation:

$$\partial\sigma_{31}/\partial t = -\sigma_{31}(\gamma_{31}+i\delta_a) + i\alpha n_{13} + i\beta\sigma_{21} \ ,$$

$$\partial\sigma_{32}/\partial t = -\sigma_{32}(\gamma_{32}+i\delta_b) + i\beta n_{23} + i\alpha\sigma_{21}^{*} \ , \quad (7)$$

in terms of the low-frequency coherence (LFC) σ_{21} defined through:

$$\rho_{21}=\sigma_{21}\exp[-i(\omega_a-\omega_b)t] \ .$$

In the material equations (7), we have used the notation

$$n_{ij} = \rho_{ii}-\rho_{jj} \ , \quad \delta_a = \omega_{31}-\omega_a \ , \quad \delta_b = \omega_{32}-\omega_b \ .$$

In a linear approximation, we consider n_{13}, n_{23} and σ_{21} as constants determined by external sources. Let us seek plane wave solutions of the coupled equations (6) and (7):

$$\alpha,\ \beta,\ \sigma_{31},\ \sigma_{32} \sim \exp(-i\omega t+ikz)\ . \tag{8}$$

This leads for the material variables to the relations

$$\sigma_{31} = i(\beta\sigma_{21}+\alpha n_{13})/(\gamma_{31}+i\delta_a-i\omega),$$

$$\sigma_{32} = i(\beta n_{23}+\alpha\sigma_{21}^{*})/(\gamma_{32}+i\delta_b-i\omega)\ . \tag{9}$$

The field equations (6) lead to a pair of homogeneous equations for α and β. The compatibility condition for the field equations is the dispersion relation which has the solutions:

$$k - \omega/c = (i/2)\left\{\Gamma_a+\Gamma_b\pm\left[(\Gamma_a-\Gamma_b)^2+4g_a g_b|\sigma_{21}|^2\right]^{1/2}\right\}.$$

In this expression, we have introduced the notation

$$\Gamma_a = \kappa_a + g_a\, n_{13}\ ,\quad \Gamma_b = \kappa_b + g_b\, n_{23}\ ,$$

$$g_a = \frac{2\pi\omega_a|\mu_{31}|^2 N}{\gamma_{31}\epsilon_a c_a \hbar[1+i(\delta_a-\omega)/\gamma_{31}]}\ ,$$

$$g_b = \frac{2\pi\omega_b|\mu_{32}|^2 N}{\gamma_{32}\epsilon_b c_b \hbar[1+i(\delta_b-\omega)/\gamma_{32}]}\ .$$

Let us fix ω to be real. Then the instability condition is, from (8):

$$\mathrm{Im}(k) < 0\ ,$$

corresponding to an exponentially growing field amplitude as the field progresses in the medium. Hence the instability conditions of this linearized analysis is equivalent to the amplification or gain condition. When the losses are negligible, $\kappa_a=\kappa_b=0$, the gain condition is

$$|\sigma_{21}|^2 > n_{13}n_{23} .$$

Thus, provided the LFC exceeds a critical threshold, a parametric instability develops.

Let us analyze in more detail the threshold condition

$$|\sigma_{21}|^2 = n_{13}n_{23} . \tag{10}$$

We notice first that when (10) is verified, it follows directly from (9) that $\sigma_{31}=\sigma_{32}=0$. Therefore the density matrix at threshold is of the form

$$\rho = \begin{pmatrix} \rho_{11} & \rho_{12} & 0 \\ \rho_{21} & \rho_{22} & 0 \\ 0 & 0 & \rho_{33} \end{pmatrix} ,$$

indicating the absence of coherence at the optical transition but only LFC. The eigenvalues of ρ are the population of the three atomic levels. A simple calculation yields the following eigenvalues at threshold:

$$N_1 = N_3 , \; N_2 = 1-2N_3 , \; N_3 = \rho_{33} .$$

The corresponding eigenvectors are

$$\Psi_1 = \begin{pmatrix} -\sqrt{\rho_{22}-\rho_{11}} \\ \sqrt{\rho_{11}-\rho_{33}} \\ 0 \end{pmatrix} , \; \Psi_2 = \begin{pmatrix} \sqrt{\rho_{11}-\rho_{33}} \\ \sqrt{\rho_{22}-\rho_{11}} \\ 0 \end{pmatrix} , \; \Psi_3 = \begin{pmatrix} 0 \\ 0 \\ 1 \end{pmatrix} .$$

These eigenvectors are not normalized. In terms of the atomic states,

$$\varphi_1 = \begin{pmatrix} 1 \\ 0 \\ 0 \end{pmatrix} , \; \varphi_2 = \begin{pmatrix} 0 \\ 1 \\ 0 \end{pmatrix} , \; \varphi_3 = \begin{pmatrix} 0 \\ 0 \\ 1 \end{pmatrix} ,$$

we have the relations

$$\Psi_1 = -\sqrt{\rho_{22}-\rho_{33}}\,\varphi_1 + \sqrt{\rho_{11}-\rho_{33}}\,\varphi_2 \ ,$$

$$\Psi_2 = \sqrt{\rho_{11}-\rho_{33}}\,\varphi_1 + \sqrt{\rho_{22}-\rho_{33}}\,\varphi_2 \ .$$

These relations generalize the trap and interacting states introduced in Section 1. Indeed, using the relation (9) at threshold, we obtain

$$\alpha/\beta = -\ \sigma_{21}/n_{13} = -\ n_{23}/\sigma^*_{21} = -\ (n_{23}/n_{13})^{1/2} \ .$$

Hence, the eigenvectors Ψ_1 and Ψ_2 can also be written as

$$\Psi_1 = \beta\,\varphi_2 + \alpha\,\varphi_1 \ , \ \Psi_2 = \beta\,\varphi_2 - \alpha\,\varphi_1,$$

up to a normalisation constant. The states Ψ_1 and Ψ_2 are the interacting and the trap states, respectively. Indeed, it is easy to verify that the transition probability

$$W(j,3) = \left|\langle\Psi_j|\hat{d}(\alpha e^{-i\omega_a t} + \beta e^{-i\omega_b t})|\Psi_3\rangle\right|^2 ,$$

vanishes on resonance if j=2 and remains non-zero for j=1. Thus, when there is no population inversion ($\rho_{33}<1/3$), the trap state Ψ_2 is the most populated state. Another interesting physical effect arises from the property that, at threshold, the other two levels are equally populated ($N_1=N_3$). This implies that absorption and emission contribute equally and that there will be no field absorption: the two-level subsystem (Ψ_3,Ψ_1) is transparent. But since the pair of states (Ψ_3,Ψ_2) does not interact, this two-level sub-system is also transparent, though for a completely different reason. Hence, at threshold, the whole system is transparent to the optical radiation E. These results are obtained in a linear approximation, assuming that n_{13}, n_{23} and σ_{12} are given constants. In a nonlinear analysis, which takes into account the Bloch equations for n_{13}, n_{23} and σ_{12}, it turns out that the threshold condition (10) cannot be reached. At best, a partial bleaching of the medium can be induced by the field E. The interesting feature of this bleaching is that it occurs for intensities that are small compared to the coherence intensity I_c (i.e., the saturation intensity of the low frequency transition 2⇔1). This coherence intensity can be significantly smaller than the saturation intensity I_s of the optical transition 3⇔1. Indeed, we have the relation $I_c = (\gamma_2/\gamma_3)I_s$, where γ_2 (γ_3) is the decay rate of the atomic polarization between level 2 (3) and level 1; hence, when the ratio γ_2/γ_3 is much smaller than unity, bleaching occurs well below the optical saturation intensity.

3. The double-Λ Scheme

Since the Λ configuration leads to bleaching but not to amplification without inversion, we have analyzed the double-Λ scheme. We consider a four-level medium with atomic frequencies given by either $\omega_1 < \omega_2 \ll \omega_3 < \omega_4$ or $\omega_1 < \omega_2 \ll \omega_4 < \omega_3$. A strong bichromatic pump field

$$E' = E'_a \cos\omega'_a t + E'_b \cos\omega'_b t, \quad \omega'_a = \omega_{41}, \; \omega'_b = \omega_{42}, \tag{11}$$

is used to excite the LFC. We determine the conditions of amplification of a weak bichromatic probe field

$$E = E_a \cos\omega_a t + E_b \cos\omega_b t, \; \omega_a = \omega_{31}, \; \omega_b = \omega_{32}. \tag{12}$$

Our main results for this configuration are the following [10]. Let N(3)-N(4) be the population difference between level 3 and level 4 before the two bichromatic fields interact with the medium, but including the decay and incoherent pumping processes. A necessary condition for the amplification of the field E is that N(3)-N(4) exceeds a positive critical value Nc which depends in a complicated way on all decay rates and on the populations of the four levels before the fields E and E′ interact with the medium. When this necessary condition is fullfilled, a sufficient condition gives a lower bound that the pump field E′ must exceed to ensure amplification of the probe field E. This lower bound is finite and therefore the process of amplification without inversion in the double-Λ scheme is proved to be feasible. A result of the necessary condition N(3)-N(4)>Nc>0 is that the probe field E will have higher frequencies than the pump field E′ only if $\omega_3 > \omega_4$. This in turn requires that population inversion be realized between the two upper levels 3 and 4, though no population inversion is necessary at the optical transitions, i.e., between levels {3,4} and levels {1,2}.

Having identified a configuration in which amplification can be achieved without inversion at the optical transitions, we have extended that analysis to the laser situation. We consider that the 4-level medium is in a resonant cavity driven by a strong bichromatic field (11). Our main approximations in that analysis are that (i)the external field is sufficiently intense to neglect pump depletion and (ii)the atomic polarizations at optical transitions can be adiabatically eliminated. This study leads to the following conclusions. Lasing of the field (12) occurs provided N(3)-N(4)>N′>0, in which case the lasing threshold condition becomes

$$|\sigma_{12}| = (\kappa/g) + n_{13} \qquad (13)$$

with the simplifying assumption that N(1)=N(2). In (13), κ is the field E damping rate while g is the modulus of the atom-field E coupling constant. These two conditions are obvious generalizations of the amplification conditions and imply that lasing without inversion at the optical transitions is indeed possible. Furthermore, lasing with frequency up-conversion requires only that a population inversion be created between the two upper levels 3 and 4.

References

1. G. Alzetta, A. Gozzini, L. Moi and G. Orriols, Nuovo Cimento B **36** (1976) 5; E. Arimondo and G. Orriols, Nuovo Cimento Lett. **17** (1976) 333; G. Alzetta, L. Moi and G. Orriols, Nuovo Cimento B **52** (1979) 209; G. Orriols, Nuovo Cimento B **53** (1979) 1.
2. H.M. Gray, R.M. Whitly and C.R. Strand Jr., Opt. Lett. **3** (1978) 218.
3. H. Harde and M. Burgraf, in Coherence and quantum Optics, L. Mandel and E. Wolf, eds, (Plenum, London 1984), Vol. V, p. 993.
4. O. Kocharovskaya and Ya. I. Khanin, Zh. Eksp. Teor. Fiz. **90** (1986) 1610 [Sov. Phys.-JETP **63** (1986) 945].
5. A. Aspect, E. Arimondo, R. Kaiser, N. Vansteenkiste and C. Cohen-Tannoudji, Phys. Rev. Lett. **61** (1988) 826.
6. O. Kocharovskaya and Ya. I. Khanin, Pis'ma Zh. Eksp. Teor. Fiz. **48** (1988) 581 [JETP Lett. **48** (1988) 630].
7. C.P.J. Barty, D.A. King, G.Y. Yin, K.M. Hahn, J.E. Field, J.F. Young and S.E. Harris, "12.8eV Laser in Neutral Cesium", in OSA Proceedings on Short Wavelength Coherent Radiation: Generation and Applications, R.W. Falcone and J. Kirz, eds. (Optical Society of America, Washington DC, 1988) vol.2, p. 13-20.
8. S.E. Harris, Phys. Rev. Lett. **62** (1989) 1033.
9. M.O. Scully, S.-Y. Zhu and A. Gavridiles, Phys. Rev. Lett. **62** (1989) 2813.
10. O. Kocharovskaya and P. Mandel, Phys. Rev. A. **42** (1990) 523.
11. O. Kocharovskaya, R.-D. Li and P. Mandel, Opt. Commun. **77** (1990) 215.

Trapping in Some Model Hamiltonians

J.I. Cirac and L.L. Sánchez-Soto

Departamento de Optica, Facultad de Ciencias Físicas,
Universidad Complutense, E-28040 Madrid, Spain

Population trapping in a general Hamiltonian which generalizes the Jaynes-Cummings model is considered. It is shown that when trapping occurs some system correlation functions become stationary.

1 Introduction

The Jaynes-Cummings model (JCM) and its extensions[1] are some of the simplest solvable models that describe the essential physics of the interaction of a single-mode radiation field with a two-level atom. The JCM exhibits many features without classical analogy, perhaps the most interesting would be the collapses and revivals of Rabi oscillations. Recently it has been found that these oscillations may be reduced when a certain condition between the atomic dipole and the field occurs[2]. Furthermore, if the atom and the field are as well in thermal equilibrium, these oscillations disappear and therefore coherent trapping occurs[3].

In this work we report the trapping conditions for a general Hamiltonian which generalizes the JCM. We consider as well the consequences of these trapping states on some two-time correlation functions.

2 Hamiltonian model and dressed states

We consider a two-level atom with transition frequency ω_0 coupled to a single mode of the radiation field of frequency ω. The Hamiltonian reads

$$\mathrm{H} = \frac{1}{2}\omega_0\sigma_3 + \omega a^\dagger a + \lambda(\mathrm{R}^\dagger\sigma^- + \sigma^+\mathrm{R}), \tag{1}$$

where the σ's are the usual pseudospin operators acting in the space of atomic states, a and $a^\dagger$ are annihilation and creation operators for the mode, λ is the coupling constant and R, $\mathrm{R}^\dagger$ mode operators. We assume the atomic transitions to be mediated by m photons, i. e.,

$$\mathrm{R} \mid n\rangle = \begin{cases} \mathrm{R}_n \mid n-m\rangle, & \text{if } n \geq m; \\ 0, & \text{otherwise,} \end{cases}$$

Springer Proceedings in Physics, Vol. 55
Nonlinear Dynamics and Quantum Phenomena in Optical Systems
Editors: R. Vilaseca, R. Corbalán

where $| n\rangle$ is a field number state. This Hamiltonian accounts for the JCM and its extensions, including the multiphoton JCM (R=a^m) and the intensity-dependent coupling model (R=$a\sqrt{a^\dagger a}$). In the basis of the whole system $\{| n,g\rangle, | n,e\rangle,\ n = 0,1,2,...\}$ where g and e represent the ground and the excited states of the atom respectively, the Hilbert space splits into subspaces

$$\mathcal{H}_n = \begin{cases} | n,g\rangle, | n-m,e\rangle & \text{if } n \geq m; \\ | n,g\rangle & \text{otherwise,} \end{cases}$$

which are invariant under the action of the Hamiltonian. Thus the Hamiltonian can be easily diagonalized with the results

$$\mathrm{H} \,| n,g\rangle = (n\omega - \frac{1}{2}\omega_0) \,| n,g\rangle, \quad , \quad n = 0,1,\ldots,m-1;$$

$$\mathrm{H} \,| \Psi_n^{\pm}\rangle = \omega_n^{\pm} \,| \Psi_n^{\pm}\rangle, \quad , \quad n = m, m+1,... \tag{2}$$

$$\omega_n^{\pm} = \omega(n - \frac{1}{2}m) \pm \Omega_{n\Delta} \quad , \quad \Omega_{n\Delta}^2 = (\Delta/2)^2 + (\lambda \mathrm{R}_n)^2;$$

the detuning being $\Delta = \omega_0 - m\omega$. The eigenstates $| \Psi_n^{\pm}\rangle$ are the well known *dressed states* defined as

$$| \Psi_n^{\pm}\rangle = \cos\theta_n^{\pm} \,| n,g\rangle + \sin\theta_n^{\pm} \,| n-m,e\rangle, \tag{3}$$

with $\tan\theta_n^{\pm} = \lambda \mathrm{R}_n(\Delta/2 \pm \Omega_{n\Delta})^{-1}$.

Here, the subindex n stands for the invariant subspaces where these states belong. In this basis the evolution for the general initial state

$$| \Psi(0)\rangle = \sum_{n=0}^{m-1} A_n \,| n,g\rangle + \sum_{n=m}^{\infty} (B_n^+ \,| \Psi_n^+\rangle + B_n^- \,| \Psi_n^-\rangle), \tag{4}$$

is easily found, which allows us to study all the dynamical features of the evolution of the system.

3 Population trapping

In this section we discuss the existence of population trapping, i.e. , a persistent probability to find the atom in its ground or excited states. Since the σ_3 operator leaves invariant the subspaces $\mathcal{H}_n$ which are decoupled under the evolution and, in each of them, the dressed states evolve with different frequencies, the trapping condition will be that each of these subspaces must evolve with only one frequency, i. e., only one of the dressed states of each subspace must be populated initially. For the initial state (4), this condition may be expressed as

$$B_n^+ = 0 \quad \text{or} \quad B_n^- = 0 \qquad \forall n \geq m. \tag{5}$$

We consider now the case in which initially the two-level atom and the field are decoupled. The initial state may be written as

$$| \Psi(0)\rangle = \sum_{n=0}^{\infty} q_n \left(\cos\phi \mid n,g\rangle + \sin\phi \; e^{i\psi} \mid n,e\rangle\right). \tag{6}$$

Here, the coefficients q_n, ψ and ϕ are related to those appearing in (4) through

$$A_n = q_n,$$

$$B_n^{\pm} = \pm\frac{1}{\sin(\theta^- - \theta^+)}(q_n \cos\phi \sin\theta_n^{\mp} - q_{n-m}\sin\phi \; e^{i\psi}\cos\theta_n^{\mp}).$$

Hence, the trapping condition (5) may be written

$$\frac{q_n}{q_{n-m}}\tan\theta_n^{\pm} = \tan\phi \; e^{i\psi}. \tag{7}$$

It is worth mentioning that when this condition is fulfilled not only the population inversion remains constant but also all linear combinations of products of the operators σ_3, $a^\dagger a$, $\mathrm{R}\sigma^+$ and $\sigma^-\mathrm{R}^\dagger$, since they leave invariant the subspaces $\mathcal{H}_n$ too.

Let us examine now, for the sake of simplicity, the case in which the atom and the field are on resonance ($\Delta = 0$). In this case the condition (7) simplifies to

$$\frac{q_n}{q_{n-m}} = \pm\tan\phi \; e^{i\psi}. \tag{8}$$

Here, it is easily shown that a possible explanation of trapping may be that on one side, the atom and the field must be in thermal equilibrium, and that, on the other side, as in the semiclassical case, the atomic dipole interferes destructively with the cavity eigenmode inhibiting the transitions from one level to the other [3].

Let us particularize now the treatment to the multiphoton JCM. The states of the field which keep constant the population inversion, in the case of zero detuning, may be written as [4]

$$| \Psi\rangle_F \propto \sum_{n=0}^{\infty}(\tan\phi)^n e^{in\psi + i\alpha(n)\pi}\sum_{l=0}^{m-1} q_l \mid n\,m+l\rangle, \tag{9}$$

with $\alpha(n)$ an arbitrary function which takes on the values 0 and 1.

In the case $m = 1$, the usual JCM, the states (9) are the eigenstates of the Susskind-Glogower phase operator. They may as well be regarded as generalized coherent states of the $SU(1,1)$ group, and hence they may be obtained from the vacuum through a unitary transformation.

On the other hand the trapping states for the intensity dependent coupling model coincide with those for the JCM in the case of zero detuning. Thus all the properties studied in that case are still valid.

4 Two-time correlation functions

In the previous section we have found the set of initial states of the system which maintain constant several observables, among them, the population inversion. Under the same condition as when trapping occurs it is easily shown that some two-time correlation functions become stationary. This is also due to the existence of only one frequency on each subspace $\mathcal{H}_n$ and to the invariance of these subspaces under the action of the whole operator. In particular, for the first and second atomic correlation functions we find

$$\langle \sigma^+(t)\sigma^-(t+\tau)\rangle = \sum_{n=m}^{\infty} \mid B_n^{\pm} \mid^2 \frac{\sin^2\theta_n^{\pm}}{\sin^2(\theta_n^- - \theta_n^+)} \exp(-i\omega_n^{\pm}\tau)$$

$$\times \Big(\sin^2\theta_{n-m}^- \exp(i\omega_{n-m}^+\tau) + \sin^2\theta_{n-m}^+ \exp(i\omega_{n-m}^-\tau) \Big); \qquad (10)$$

$$\langle \sigma^+(t)\sigma^+(t+\tau)\sigma^-(t+\tau)\sigma^-(t)\rangle = \sum_{n=2m}^{\infty} \mid B_n^{\pm} \mid^2 \sin^2\theta_n^{\pm}$$

$$\times \left(\frac{\sin\theta_{n-m}^- \sin\theta_{n-m}^+}{\sin(\theta_n^- - \theta_n^+)} \right)^2 4\sin^2(\Omega_{n\Delta}\tau), \qquad (11)$$

where the sign + or − depends on which of the B's coefficients is not zero. The second function is related to the photon bunching and antibunching phenomenon. The first function is used to obtain the spectrum of the light emitted by the atom. In the non-dissipative models this correlation function is not stationary and in order to obtain the emission spectrum the *transient physical spectrum* [5] must be used. However, in the case under study, due to the condition (7), we can obtain it as the Fourier transform of the two-time correlation function. The spectrum is given by a summation of δ functions which, in a real experiment, would become Lorentzian-like of width equal to the width of the detector used to measure [4].

In conclusion, we have shown that in a two-level atom coherent trapping may occur. We have found the atomic and field states which keep constant the evolution of the atomic inversion for some model Hamiltonians. When initially the system is in any of these states, some two-time correlation functions are stationary.

References

[1] E.T. Jaynes and F.W. Cummings, Proc. IEEE **51**, 89 (1963); B. Buck and C.V. Sukumar, Phys. Lett. **81A**, 132 (1981); C.V. Sukumar and B.Buck, J.Phys. A, **17**, 885 (1984).

[2] K. Zaheer and M.S. Zubairy, Phys. Rev. A, **39**, 2000 (1989)

[3] J.I. Cirac and L.L. Sánchez-Soto, Phys. Rev. A, **42**, 2851 (1990)

[4] J.I. Cirac and L.L. Sánchez-Soto, Optics Comm. (To appear)

[5] J.H. Eberly and K. Wódkiewicz, J. Opt. Soc. Am. **67**, 1252 (1977)

Optical Bistability in a Ridge Waveguide Resonator Including a Multiple Quantum Well

G.P. Bava[1], P. Debernardi[2], F. Castelli[3], and L.A. Lugiato[4]

[1]Dipartimento di Elettronica, Politecnico di Torino, I-10129 Torino, Italy
[2]CESPA/CNR, c/o Politecnico di Torino
[3]Dipartimento di Fisica, Politecnico di Torino, Corso Duca degli Abruzzi 24, I-10129 Torino, Italy
[4]Dipartimento di Fisica, Università di Milano, Via Celoria 16, I-20133 Milano, Italy

Abstract. We present a model for optical bistability in the ridge waveguid Fabry–Perot resonator. The non–linearity is due to a Multiple Quantum Wel structure inserted into the resonator. Near the excitonic resonance a bistabl behaviour has been found, as shown by examples of steady–state response.

1. Introduction

Multiple Quantum Well (MQW) structures present a noteworthy interest du to their capability of producing large optical nonlinearities at room tempera ture [1,2]. Remarkable effects of bistability have been recently reported [3].

The nonlinear interaction between the electromagnetic field and the mate rial can be conveniently studied by using models elaborated by the Frankfur school [4]; these models allow for the evaluation of the nonlinear contributio $\Delta\epsilon$ to the dielectric constant.

In this paper we report on the dynamical equations which describe the in teraction between the radiation and the free carriers in the case of the Fabry–Pero resonator realized using a ridge waveguide containing the MQW structure. W propose a method to solve these equations at steady state, and give some ex amples of curves which describe the input power as a function of the outpu power; these results show clearly the possibility of a bistable or multistabl behaviour of the device.

2. Model of the system

The equations which govern the interaction between the electromagnetic fiel and the material for the configuration we consider (Fig. 1) are obtained fron the formalism of modal coupling and by assuming for simplicity that the fre carrier density per unit area $N(y,z)$ is equal in all the wells. If we indicate b C_F and C_B the slowly varying amplitudes of the forward and backward trav elling waves (normalized in such a way that their squared moduli correspon to the power), the equations have the form :

Nonlinear Dynamics and Quantum Phenomena in Optical Systems
Editors: R. Vilaseca, R. Corbalán

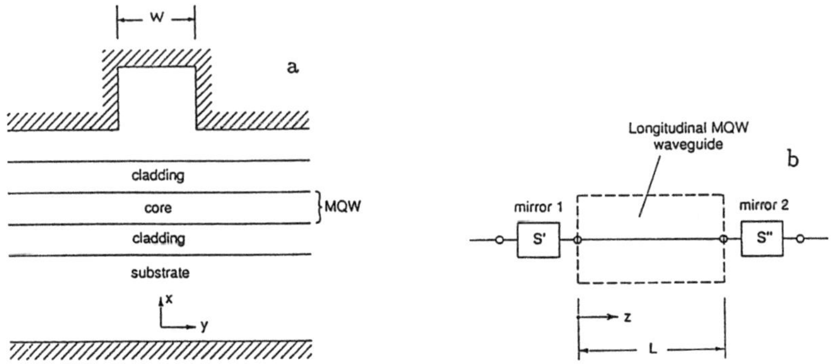

Fig. 1 a) schematic section of the ridge optical waveguide in the transverse $x-y$ plane; b) schematic description of the resonator along the longitudinal direction z.

$$\begin{aligned}
&\frac{\partial C_F}{\partial z}+\frac{1}{v_g}\frac{\partial C_F}{\partial t}=+hC_F \\
&\frac{\partial C_B}{\partial z}-\frac{1}{v_g}\frac{\partial C_B}{\partial t}=-hC_B \\
&\frac{\partial N}{\partial t}=D\frac{\partial^2 N}{\partial y^2}-R(N)+\mathcal{I}m\left\{\frac{\Delta\epsilon}{\epsilon}(N)\right\}\frac{\epsilon d}{\hbar}|E_\mu|^2\left(|C_F|^2+|C_B|^2\right)
\end{aligned} \tag{1}$$

with :

$$h=\frac{i\omega m}{2v_g}\frac{\int_{S_w}\Delta\epsilon|E_\mu|^2\mathrm{d}s}{\int_S\epsilon|E_\mu|^2\mathrm{d}s} \tag{2}$$

where ϵ is the background linear refractive index, v_g is the group velocity of the cavity mode, m is the number of wells, d denotes their thickness, $R(N)=AN+BN^2$ is the recombination term, D is the diffusion coefficient, the function E_μ describes the transverse field distribution, S_w is the well section, S is the sample section and ω is the frequency of the input field.

In the stationary state the input–output relation, which satisfies the boundary conditions of the problem, can be written as follows :

$$\begin{aligned}
|C_{in}|^2 &= \left|\frac{C_{out}}{S'_{21}S''_{21}}\right|^2 e^{2\int_0^L(\alpha+\alpha_i)\mathrm{d}z}\left|1-S'_{22}S''_{11}e^{2L(i\beta_0-\alpha_i)+2\int_0^L h(z)\mathrm{d}z}\right|^2 \\
D\frac{\mathrm{d}^2N}{\mathrm{d}y^2} &= R(N)-\frac{|C_{o}ut|^2}{\hbar v_g|S''_{21}|^2}\eta(y)\times \\
&\left[e^{2\int_z^L(\alpha+\alpha_i)\mathrm{d}z}+|S''_{11}|^2e^{-2\int_z^L(\alpha+\alpha_i)\mathrm{d}z}\right]\mathcal{I}m\left\{\frac{\Delta\epsilon}{\epsilon}(N)\right\}
\end{aligned} \tag{3}$$

where we define :

$$\eta(y)=\frac{\int_d\epsilon|E_\mu|^2\mathrm{d}x}{\int_S\epsilon|E_\mu|^2\mathrm{d}s} \tag{4}$$

and $\alpha = -\mathcal{R}e\{h\}$, while α_i and β_0 are respectively the intrinsic linear attenuation constant and the unperturbed propagation constant of the guided mode. The matrices S'_{ij} and S''_{ij} are related to the two mirrors (Fig. 1).

The solutions of Eqs. (3) can be obtained by using an iterative procedure; one solves first the diffusion equation, for example by the finite difference method and then one consider the input–output relation; the steps are iterated until $N(y,z)$ converges.

3. Numerical results

The MQW structure is assumed to be formed by GaAs/GaAlAs with wells of thickness $d = 100$ Å; for the numerical values relative to the parameters of these materials and for the modelization of the optical nonlinearity we refer to [4]. In particular, Fig. 2 and , 3 show the absorption coefficient and the refractive index, respectively, as a function of the photon energy, for several values of the carrier density. The range of variation for the operating frequency that we consider lies at the feet of the excitonic peak which is evident in Fig. 2. For the recombination coefficients we selected the values $A = 0.2 \times 10^9 \mathrm{s}^{-1}$, $B = 5 \times 10^{-4} \mathrm{cm}^2\mathrm{s}^{-1}$; the diffusion coefficient has been assumed equal to $16 \mathrm{cm}^2\mathrm{s}^{-1}$.

The structure parameters have been chosen as follows : $m = 6$, $\alpha_i = 2.5 \mathrm{cm}^{-1}$, $W = 3\mu\mathrm{m}$; the frequency ω_0 of the cavity mode nearest to resonance with the input field has been taken equal to $(E_{g2D} - 3E_{02D})/\hbar$, where

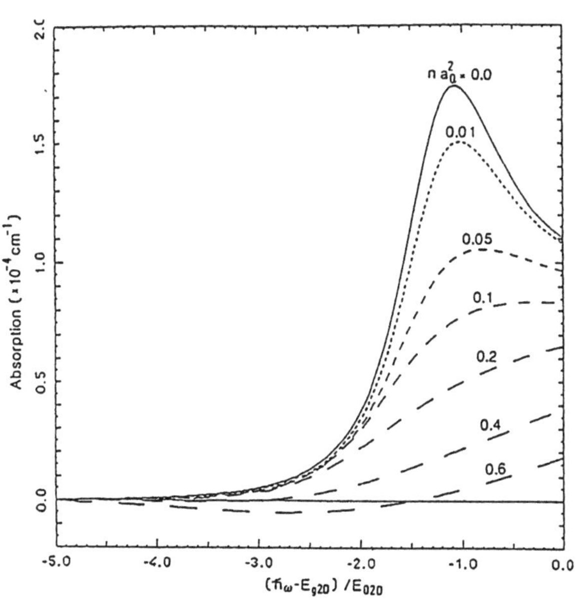

Fig. 2 Absorption coefficient as a function of the operating frequency for several values of Na_0^2, where N is the free carrier density and a_0 is the excitonic Bohr radius.

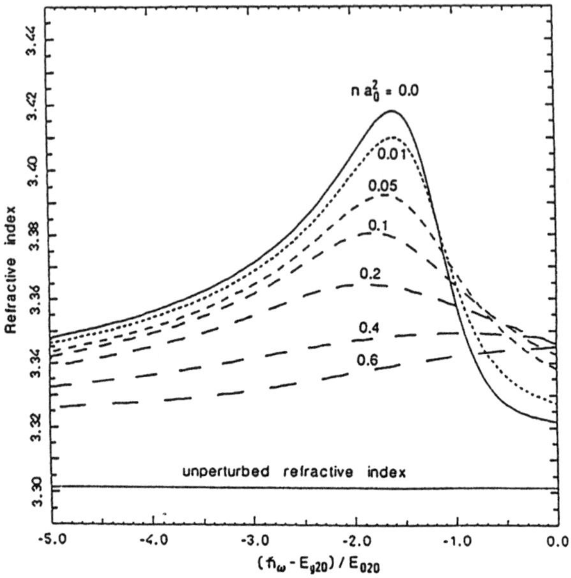

Fig. 3 Refractive index as a function of the operating frequency for several values of Na_0^2.

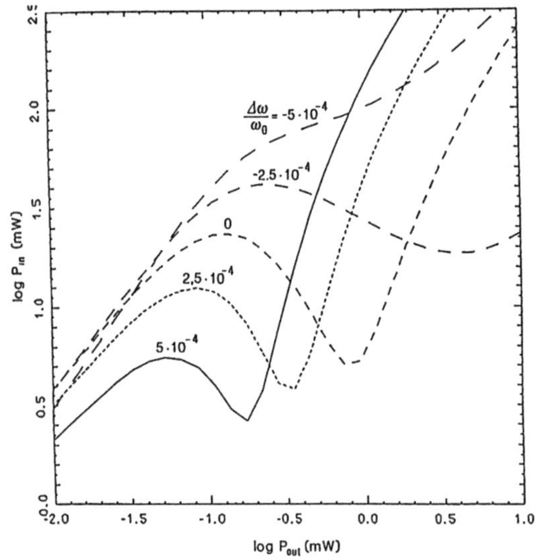

Fig. 4 Input power as a function of output power under variation of the detuning parameter δ.

E_{02D} is the excitonic binding energy (16.8 meV) and E_{g2D} is the bidimensional gap (1.543 meV).

In Fig. 4 we show the curves of the input power as a function of the output power for a device length $L = 50\mu$m, an intensity mirror reflectivity coefficient $R = 0.9$, when the detuning $\delta = (\omega_0 - \omega)/\omega_0$ is varied. Fig. 5 exhibits the

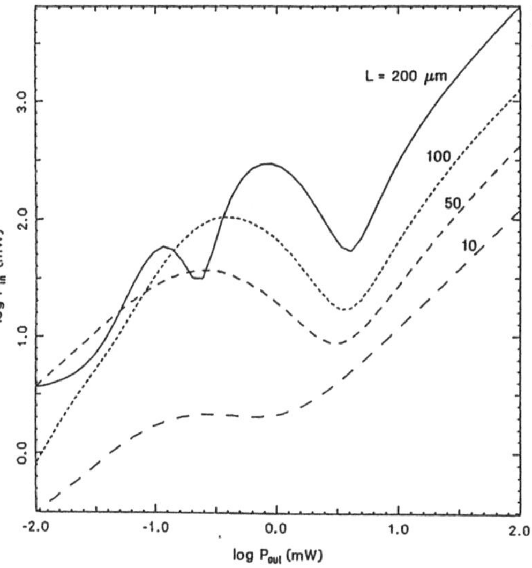

Fig. 5 Same as Fig. 4, under variation of the cavity length.

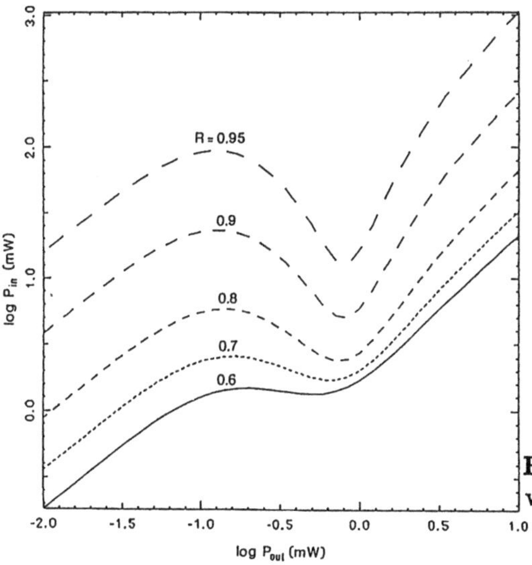

Fig. 6 Same as Fig. 4, under variation of the mirror reflectivity.

input–output curves for $R = 0.9$ and $\delta = 0.0025$ when the length L is varied. In Fig. 6 the control parameter is R, with $L = 50\mu$m and $\delta = 0$. From examination of the figures it emerges that for the same parameters (L and δ) there is an optimum value with respect to the size of the hysteresis cycle. On the other hand, there is obviously a minimum value for R to obtain a bistable response; where R is increased, the phenomenon is enhanced but requires a larger input power. On average the ratio between the values of the power in correspondence to the boundaries of the bistability region is $\simeq 10$ for the output and is somewhat smaller for the input. In Fig. 5 one sees that when L is increased one meets a phenomenon of multistability, due to the possibility of matching between the input frequency and different longitudinal modes of the resonator when the optical power is increased.

References

[1] D.S.Chemla, D.A.B.Miller et al., IEEE J. of Quantum Electr., Vol. QE-20, pp. 265-275, March 1984.

[2] H.M.Gibbs, "Optical bistability: controlling light by light", Academic Press, 1985.

[3] J.L.Oudar, B.Sfez, R.Kuszelewicz, J.C.Michel, R.Azoulay, Phys, Stat. Sol. B, Vol 159, pp. 181-189, 1990.

[4] C.Ell, R.Blank, S.Benner and H.Haug, JOSA B, Vol. 6, pp. 2006-2012, November 1989.

Squeezing Behaviour of a Degenerate Paramp with Fourth-Order Interaction: Stationary and Transient Statistics

C. Cabrillo[1], *P. García Fernández*[1], *F.J. Bermejo*[1], *P. Colet*[2], *R. Toral*[2], *and M. San Miguel*[2]

[1]Instituto de Estructura de la Materia, C.S.I.C., Serrano 119, E-28006 Madrid, Spain

[2]Departament de Física, Universitat de les Illes Balears, E-07071 Palma de Mallorca, Spain

Abstract. A model system representing a degenerate parametric amplifier plus a fourth-order term (Kerr effect) is analyzed. Such higher-order interaction leads to the appearance of a pitchfork bifurcation showing bright-squeezing in the stationary states. A numerical simulation of the transient processes is presented and the transient-squeezing characteristics are finally discussed.

1 Introduction

It has been shown that [1] an intense coherent beam which propagates along an optical Kerr medium, exhibits "self- squeezing" (amplitude squeezing), and that the effect could be remarkably high provided that the incident pump strength is high enough. The effect of such higher-order interactions coupled to a standard two-photon medium was analyzed by Tombesi [2] some time ago. The interest in the already mentioned study was focused on the possibilities of generating strong squeezed light at short interaction times (i.e., in travelling-wave geometries). Our interest here is to explore the field- statistical properties of a more realistic model of a parametric amplifier where the crystal losses are explicitly accounted for. The effective Hamiltonian, the obtention of the semiclassical equations and the stability analysis have been presented in an accompanying paper [3].

2 Fluctuations around steady-states

As it has been shown [3], the stochastic equations for the quadrature fluctuations around the stable solutions are

Nonlinear Dynamics and Quantum Phenomena in Optical Systems
Editors: R. Vilaseca, R. Corbalán

$$\begin{aligned} \delta\dot{x} &= (\mu + 2\nu\bar{y}_S\bar{x}_S)\delta x + \nu(\bar{x}_S^2 + 3\bar{y}_S^2)\delta y + L_x \\ \delta\dot{y} &= -\nu(3\bar{x}_S^2 + \bar{y}_S^2)\delta x - (1 + 2\nu\bar{x}_S\bar{y}_S)\delta y + L_y \,. \end{aligned} \tag{1}$$

The stationary covariance matrix

$$\sigma_{ij} \equiv < x_i x_j > - < x_i >< x_j > \tag{2}$$

can be obtained from the solution of [4]

$$A\sigma + \sigma A^T = -\bar{D}I, \tag{3}$$

where A is the drift matrix (expressed in terms of the coefficients given in eq. (1)) and I is the unit matrix. In the case below threshold, the mean square fluctuations in the two quadrature components are

$$< x^2 >= \frac{\gamma(1+2\bar{n})}{4(\gamma-\kappa)} \,, < y^2 >= \frac{\gamma(1+2\bar{n})}{4(\gamma+\kappa)} \,. \tag{4}$$

The zero-point fluctuations are $< x^2 >=< y^2 >= 1/4$, so that, for sufficiently low temperature ($\bar{n} \rightarrow 0$), the y-component is squeezed. Since the fluctuations are about the steady-state value $(\bar{x}_S, \bar{y}_S) = (0,0)$, this represents a squeezed vacuum state. The product of the mean squares is in this case

$$< x^2 >< y^2 >= \frac{1}{16} \frac{\gamma^2}{\gamma^2 - \kappa^2}(1 + 2\bar{n}^2) \,, \tag{5}$$

that is greater than 1/16. Thus, the state is not a minimum uncertainty state.

Above threshold a pitchfork bifurcation appears leading to two steady-state solutions (bistability, [3]) which are

$$\begin{aligned} \bar{x}_S &= \mp[\mu^{\frac{1}{2}}/\nu(1+\mu)]^{\frac{1}{2}} \\ \bar{y}_S &= -\mu^{\frac{1}{2}}\bar{x}_S \,. \end{aligned} \tag{6}$$

Making use of eq. (2) and diagonalizing the covariance matrix, the stationary variances become

$$\begin{aligned} \sigma_{11} &= \frac{\bar{D}}{16(1-\mu)\mu(1+\mu)^2}[B + (\mu^2 - 1)B^{1/2}] \\ \sigma_{22} &= \frac{\bar{D}}{16(1-\mu)\mu(1+\mu)^2}[B - (\mu^2 - 1)B^{1/2}] \\ B &= \mu^4 + 16\mu^3 + 30\mu^2 + 16\mu + 1. \end{aligned} \tag{7}$$

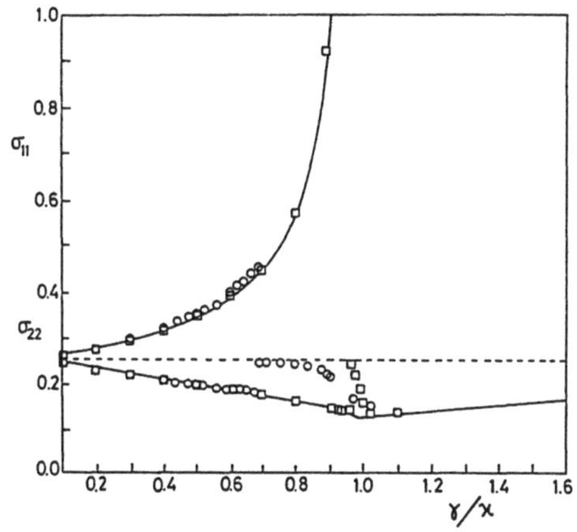

Fig. 1: Comparison between the stationary variances numerically determined (○ correspond to $\Gamma/\kappa = 2\times10^{-2}$, □ to $\Gamma/\kappa = 2\times10^{-3}$) and the variances determined from the linearized theory (solid line).

As it is shown in fig. 1 the 22-component is squeezed. Note that since $(\bar{x}_S, \bar{y}_S)$ are real and different from zero, "bright squeezing" appears.

3 Analysis and discussion of the results

3.1 Stationary behaviour

Besides the linearized analysis of the Langevin equations a numerical approach has been carried out. Both kinds of results are shown in fig. 1. It can be easily seen that for regions far away from threshold, the linearized theory is in good agreement with the simulation results. However, for some values the linearized theory predicts maximum squeezing whereas a substantial increase in σ_{22} is found in the simulation. Such a fact arises from the bistable behaviour of this system. As can be seen [3] the nontrivial solutions are located at points $(\bar{x}_S, \bar{y}_S)$ depending upon the values of Γ/κ and γ/κ. Therefore their relative separation can be controlled by a suitable choice of Γ/κ. For values of $\Gamma/\kappa \leq 2\times10^{-3}$ and $\gamma/\kappa = 0.9$, no jumps between the stable states occur in observable time. For larger values of Γ/κ the probability of hopping attains larger values thus leading to an increase in σ_{22}.

3.2 Transient behaviour

We have analyzed the transient behaviour by means of a numerical procedure. All the obtained curves are for $\bar{n} = 0$ and $\kappa = 0.5$. The case below threshold is shown in figure 2. The figures 3 and 4 display the results for different parameters and initial condition values. It is worth noting that the stationary values are greater than those of the previous analysis. This is due to the fact that the numerical analysis is performed as an average over trajectories and bistability forces,that about half of trajectories go to one of the stationary states while the rest go to the other. The main result common to all curves is the existence of "oversqueezing" with respect to the stationary value. Figure 3 shows the curves which correspond to a coherent initial state and to a squeezed initial state. No dependence upon the initial state in σ_{22} value during "oversqueezing" is observed, i.e., after a certain time the transient evolution loses the memory about the initial conditions. The time where "oversqueezing" occurs is larger, as could be expected, for a initial coherent vacuum.

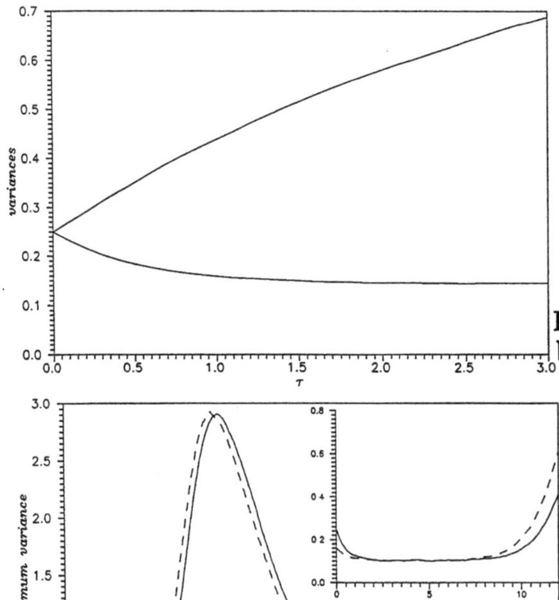

Fig. 2: Transient variances below threshold.

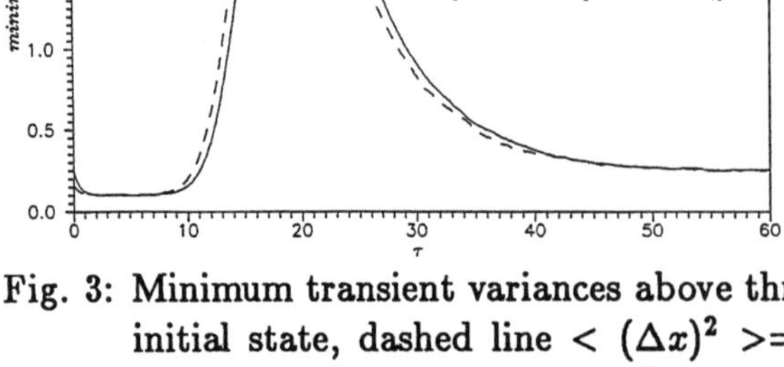

Fig. 3: Minimum transient variances above threshold (solid line coherent initial state, dashed line $<(\Delta x)^2>=1/2$, $<(\Delta y)^2>=1/6$ squeezed initial state. Both $\Gamma/\kappa = 2\times10^{-3}$, $\gamma/\kappa = 0.75$).

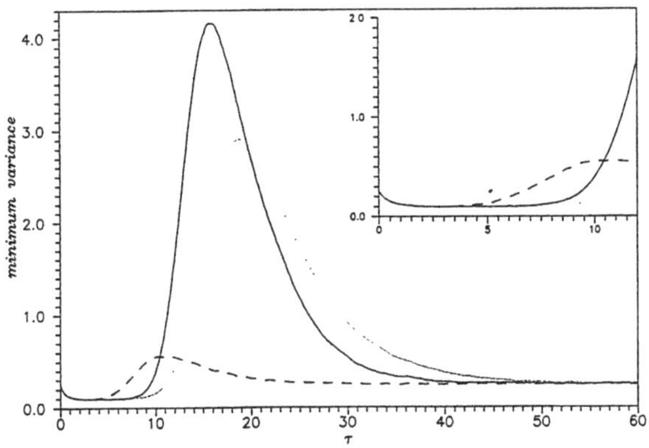

Fig. 4: Minimum transient variances above threshold(solid line $\Gamma/\kappa = 2\times 10^{-3}$, $\gamma/\kappa = 0.64$, dashed line $\Gamma/\kappa = 2\times 10^{-2}$, $\gamma/\kappa = 0.64$, pointed line $\Gamma/\kappa = 2\times 10^{-3}$, $\gamma/\kappa = 0.75$. All with coherent initial states).

Figure 3 shows the dependence with respect Γ/κ and γ/κ. The time-lag where "oversqueezing" occurs is larger for $\Gamma/\gamma = 2\times 10^{-3}$, as it could be expected since the nontrivial stationary states are now far from the origin. On the other hand, as is shown in the figure, such a time-lag becomes larger as one approaches the threshold (i.e., larger for $\gamma/\kappa = 0.64$ than for 0.75).

Finally it is worth remarking that the σ_{22} value in the "oversqueezing" region does not show any noticeable dependence on Γ/κ or γ/κ.

4 Conclusions

The effect of a fourth-order (Kerr-effect) non-linearity in a degenerate parametric amplifier has been analyzed. From the analysis of the semi-classical equations it has been found that maximum squeezing occurs in the vicinity of the threshold. On the other hand, the bistable behaviour encountered above threshold may provide a means for the generation of bright-squeezed light. Some care should be taken with respect to the results obtained near threshold since: a) the model herein analyzed constitutes an oversimplification of a real device (undepleted pump), and b) the full quantum-dynamics has not been explored, and therefore, the estimated jump rates between the two stable states should be taken as semiquantitative [5].

A substantial "oversqueezing" has been found when analyzing the transient dynamics. The time duration where "oversqueezing" occurs is enlarged near threshold. On the other hand, the best results are obtained for small values of Γ/κ.

The obtained results may therefore indicate a way for the generation of bright-squeezed light in travelling-wave devices.

References

[1] R. Tanas, S. Kielich, Quantum Opt. 2, 23, (1990).

[2] P. Tombesi in "Quantum Optics IV", J.D. Harvey, D.F. Walls (Eds.), Springer, 1986, p.81.

[3] F.J. Bermejo, P. García–Fernández, P. Colet, S. Bal.le, R. Toral, M. San Miguel (this volume).

[4] C.W. Gardiner, "Handbook of Stochastic Methods", Springer, Berlin (1985), Chap 4.

[5] P.D. Drummond, P. Kinsler, Phys. Rev. A40, 4813, (1989).

Completely Integrable Systems in Nonlinear Optics of Elliptic Gaussian Beams

A.M. Goncharenko, Yu.A. Logvin, A.M. Samson, P.S. Shapovalov, and S.I. Turovets

Institute of Physics, BSSR Academy of Sciences,
SU-220602 Minsk, USSR

Abstract. It is shown that the systems of equations describing the elliptic gaussian beams propagation in a nonlinear media are Ermakov Hamiltonian systems. It is demonstrated that these systems are completely integrable. The condition under which any Ermakov system is also a Hamiltonian one is determined.

1. Introduction

The cases of completely integrable systems [1] are quite rare in nonlinear physics. The integrals of motion are most often absent from nonlinear systems, which points to the presence of complex (and chaotic) regimes in the dynamics of systems. In many cases the finding of at least several integrals of motion makes it possible to reveal the main properties of the systems. The present paper considers the propagation of elliptic gaussian beams (EGB) [2-7] from the viewpoint of theory of dynamical systems. It is shown that the systems of equations describing the EGB propagation in nonlinear media are completely integrable. Firstly, these systems belong among the Ermakov systems which have recently aroused great interest [8-12]. There exists an Ermakov-Lewis-Ray-Reid (ELRR) invariant for them. Secondly, these systems are Hamiltonian ones; there is another conservative invariant - the integral of energy. We have determined the condition under which any Ermakov system is also a Hamiltonian one. The systems' symmetry permits not only writing the integrals of motion, but obtaining explicit solution, as well [6,7].

Some concrete physical cases are considered in this paper. In these cases, systems of equations and integrals of motion can be written in a compact, generalized form. The optical properties of EGB with general astigmatism are described in more detail in [7].

Springer Proceedings in Physics, Vol. 55
Nonlinear Dynamics and Quantum Phenomena in Optical Systems
Editors: R. Vilaseca, R. Corbalán

2. The Basic Equations

The elliptic gaussian beam propagation in nonlinear inhomogeneous media is describing by the following system of equations [2-7]:

$$\ddot{a} + \alpha a = \left(4(b/a - \varkappa) + \delta\, b/a \frac{1 + 3b/a}{(1-(b/a)^2)^3} \right)/a^2 b \,,$$
$$\ddot{b} + \alpha b = \left(4(a/b - \varkappa) + \delta\, a/b \frac{1 + 3a/b}{(1-(a/b)^2)^3} \right)/b^2 a \,, \quad (1)$$

where the overdot indicates differentiation with respect to longitudinal coordinate of the beam. The parameter $\varkappa$ is the normalized beam power, δ is the beam rotation parameter , α describes the inhomogeneity of the medium [7].

The system (1) is a system of equations for two coupled nonlinear oscillators, it is written in the standard Ermakov form [8-12]:

$$\ddot{a} + \omega^2 a = f(b/a)/a^2 b \,,$$
$$\ddot{b} + \omega^2 b = g(a/b)/b^2 a \,. \quad (2)$$

It is seen that in our case the function g coincides with f and parameter α plays the role of ω^2.

The systems of two coupled nonlinear oscillators have been sufficiently studied in the dynamical chaos theory [1]. The Ermakov systems occupy a special place from the viewpoint of dynamical chaos. They may be linearized [12] and, therefore, there are only regular trajectories for any Ermakov system.

3. Integrals of the System

It can easily be seen that the system (1) is a Ermakov Hamiltonian one. It can be obtained in the following form:

$$\ddot{a} = -\partial H/\partial a \,, \qquad \ddot{b} = -\partial H/\partial b \quad (3)$$

from the Hamiltonian

$$H = (\dot{a})^2/2 + (\dot{b})^2/2 + V(a,b)$$

where V(a,b) is the potential function:

$$V(a,b) = \alpha(a^2+b^2)/2 + 2(a^2+b^2)/a^2b^2 - 4\varkappa/ab + \delta^2/2\ (a^2+b^2)/(a^2-b^2)^2. \quad (4)$$

For the Ermakov systems there exists the ELRR-invariant [9]:

$$I = (\dot{a}b - \dot{b}a)^2/2 + 2(a^2+b^2)^2/a^2b^2 - 4\varkappa(a^2+b^2)/ab + \delta^2/2\ (a^2+b^2)^2/(a^2-b^2)^2. \quad (5)$$

The ELRR-invariant is independent of the inhomogeneity parameter α, since α in (1) plays the same role as ω^2 in (2) and disappears when ELRR-invariant is constructed. It can easily be shown that the integrals of motion H and I are in involution:

$$\{ H , I \} = 0, \quad (6)$$

i.e. the Poission bracket for H and I is zero. The existence of two integrals of motion in involution for the Hamiltonian system (1) proves its complete integrability. It is well known from the classical mechanics that for the motion in a central force field (i.e. when the potential function depends only on $r = (a^2+b^2)^{1/2}$) the angular momentum of the particle $J = a\dot{b}-b\dot{a}$ is conservative. Figure 1 shows the isolines for the potential function (4). It is seen that the picture remains unaltered at inversion of the coordinate system ab relative to the origin and at mirror reflection relative to the bisectrices of the coordinate angles. It can be stated that for the given symmetry the ELRR-invariant is the analogue of angular momentum for the case of central forces. Comparing (4) and (5), we obtain

$$I = J^2/2 + V(a,b)\ r^2 - \alpha\ r^2 . \quad (7)$$

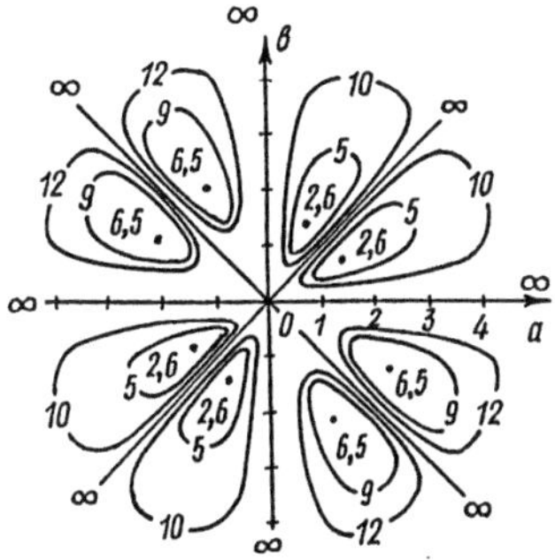

Fig. 1 The isolines for the potential function (4), $\alpha = 1$, $\varkappa = 1$, $\delta = 1$.

In the case of central forces the ELRR-invariant transforms to the square of the angular momentum.

4. Ermakov Hamiltonian Systems

Let us show that the Ermakov system (2) becomes a Hamiltonian one if the functions f and g have some limitations. The Ermakov system is Hamiltonian if the right-hand sides of Eqs.(2) are partial derivatives of the function V(a,b), i.e.

$$\begin{aligned} f(b/a)/a^2 b &= \partial V/\partial a \,, \\ g(a/b)/b^2 a &= \partial V/\partial b \,. \end{aligned} \tag{8}$$

This condition is equivalent to the expression

$$\frac{\partial}{\partial b}\,(f(b/a)/a^2 b) = \frac{\partial}{\partial a}\,(g(a/b)/b^2 a), \tag{9}$$

which can be written in the form of the differential equation

$$f(u) - 1/u\, df(u)/du = g(1/u) + u\, dg(1/u)/du \,. \tag{10}$$

Thus, the functions f and g related by (10) lead to the Ermakov Hamiltonian system. Formulas (9),(10) are quite similar to expressions (8)-(12) of [11], where the existence of a second integral of motion for a restricted class of Ermakov systems has been established. In our case, however, the second integral of motion is a Hamiltonian, which is fundamentally important for the analysis of nonlinear dynamical systems. The integrability of any nonlinear dynamical system with N-dimensional phase space requires N independent integrals of motion. At the same time, in the case of a Hamiltonian system with N/2 degrees of freedom (N-dimensional phase space) completely integrability is guaranteed by the existence of N/2 integrals of motion which are in involution.

References

1. A.J.Lichtenberg and M.A.Lieberman, Regular and Stochastic Motion (Springer-Verlag, 1983).
2. V.Vorob'ev, Izv. vys. uch. zav. Radiofizika 13, 1905 (1970).
3. C.R.Giuliano, J.H.Marburger and A.Yariv, Appl.Phys.Lett. 21, 58 (1972).

4. F.Cornolti, M.Lucchesi and B.Zambon, Opt. Communs. 75, 129 (1990).
5. A.M.Goncharenko, Gaussian Light Beams (Nauka i tehnika, Minsk, 1977).
6. A.M.Goncharenko, Yu.A.Logvin, A.M.Samson, P.S.Shapovalov, S.I.Turovets, submitted to Phys.Lett.A.
7. A.M.Goncharenko, Yu.A.Logvin, A.M.Samson, P.S.Shapovalov, submitted to Opt.Communs..
8. V.P.Ermakov, Univ. Izvestija. Kiev 20, 1 (1880).
9. H.R.Lewis Jr., J. Math. Phys. 9, 1976 (1968).
10. J.R.Ray and J.L.Reid, Phys.Lett.A 71, 317 (1979).
11. J.Goedert, Phys.Lett.A 136, 319 (1989).
12. C.Athorne, C.Rogers, U.Ramgulam and A.Osbaldestin, Phys. Lett.A 143, 207 (1990).

Nonlinearities in Polar Molecules: Optical Bistability and Phase-Conjugation

M.A. Antón, I. Gonzalo, and J.L. Escudero

Departamento de Optica, Facultad C. Físicas, Universidad Complutense, E-28040 Madrid, Spain

Abstract. From analysis of optical bistability and degenerate four-wave mixing, it is shown that changes in permanent dipole moments under electronic excitation may contribute significantly to the third-order susceptibilities when two-photon transitions are considered. A relevant feature is the coupled-mode oscillation predicted by a two-level model in two-photon phase conjugation by degenerate four-wave mixing in an absorbing medium.

1. Introduction

Polar molecules whose permanent dipole moments change significantly under electronic excitation can generate strong third-order optical nonlinearities in two-photon transitions. By contrast, the influence of permanent dipole moments is negligible in single-photon transitions.

The non-linear materials must possess high damage resistance. Radiation damage thresholds as high as 1 GW/cm^2 have been observed in some molecules [1,2]. The appropriate molecules are in general organic compounds, derivatives of aromatic systems. The origin of their large nonlinearities lies in the high polarization produced by the π-electronic excited states [3]. Some molecules of this type also have such advantages as short switching, of about 35 ps [4], and easy preparation of thin films.

Concerning previous work on optical bistability (OB) in this type of material, it has been shown theoretically that two-photon OB can be obtained in μm size cavities if the difference in permanent dipole moments between the excited and the ground states, $|\mu_{22} - \mu_{11}|$, is greater enough than the transition dipole moment $|\mu_{12}|$ [5]. Also, theoretical analysis has been performed with a double beam [6] and taking into account phonon effects [7].

2. Two-photon phase-conjugation by degenerate four wave mixing (DFWM) in polar molecules

DFWM has been treated according to a two-level model for two-photon resonant transitions between the two levels, |1> and |2>, the difference between the permanent dipole moments, $\mu_o \equiv |\mu_{22} - \mu_{11}|/2$, being greater or of the order of the transition dipole moment. The standard geometric configuration for the two counterpropagating intense pump amplitudes A_1, A_2, the small

Springer Proceedings in Physics, Vol. 55
Nonlinear Dynamics and Quantum Phenomena in Optical Systems
Editors: R. Vilaseca, R. Corbalán

probe A_4 and the conjugate amplitude A_3, is considered. In order to calculate the reflectivity of the conjugate wave, a well-known method is followed [8]. Here, the presence of $\mu_o \neq 0$ makes it necessary to develop each density matrix element in powers of the field amplitude [5,6,9] which implies a non-saturable absorber material for the intensity range considered. The standard coupled equations for the probe and the conjugate amplitudes are

$$\frac{dA_4^*}{dz} = -\alpha^* A_4^* + i\beta A_3 \; ; \qquad \frac{dA_3}{dz} = \alpha A_3 + i\,\beta^* A_4^*$$

and the reflectivity is given by

$$R = \left| \frac{A_3(0)}{A_4(0)} \right|^2 = \frac{|\beta \sin \gamma L|^2}{|\gamma \cos \gamma L + \alpha_R \sin \gamma L|^2} , \qquad (1)$$

where $\gamma = (|\beta|^2 - \alpha_R^2)^{1/2}$. In the present case, and in Gaussian units

$$\alpha_R = \mathrm{Re}\left[i \frac{2\pi\omega}{cn} (A + 2BI) \right] ; \quad |\beta| = \frac{2\pi\omega}{cn} |B| I ,$$

where $I = |A_1|^2 = |A_2|^2$ (the corresponding power: $W = (1/2) \times 3\times10^{10} \times I \times 10^{-13}$ MW/cm^2) and

$$A = -N \frac{|\mu_{12}|^2 \rho_D^{(o)}}{2i\hbar D_1} \; ; \qquad B = B_1 + B_2 + \ldots .$$

$$B_1 = -N \frac{|\mu_{12}|^2 \mu_o^2 \rho_D^{(o)}}{2i\hbar^3} \left(\frac{1}{D_1 D_2 D_3^*} - \frac{1}{D_1^2 D_2} \right)$$

$$B_2 = \frac{N |\mu_{12}|^4 T_1 \rho_D^{(o)}}{i\hbar^3 T_2 D_1 |D_1|^2}$$

$$D_1 = T^{-1} + i(\omega_o - \omega) ; \quad D_2 = T^{-1} + i(\omega_o - 2\omega)$$

$$D_3 = T_1^{-1} + i\omega .$$

$\rho_D^{(o)} \equiv \rho_{22}^{(o)} - \rho_{11}^{(o)} = -1,\ 1$ in the absorbing and inverted medium respectively. The next terms in B are negligible when $\mu_o \geq |\mu_{12}|$. In the numerical example, the molecular density $N = 10^{19}$ cm^{-3} in the absorbing medium, and $N = 10^{17}$ cm^{-3} in the

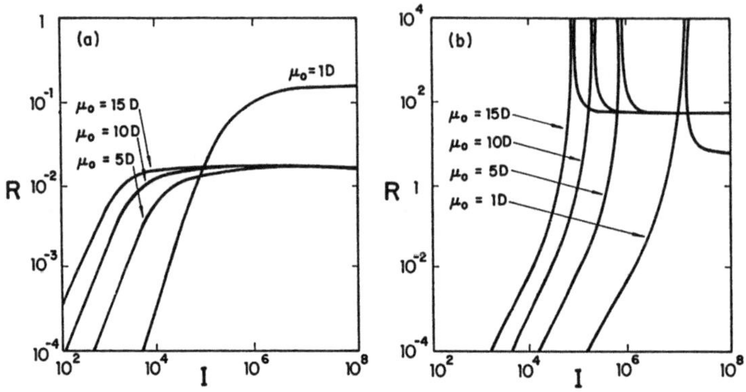

Fig. 1: Reflectivity R versus pump intensity, I, on decimal logarithmic scales, for different μ_o (Debyes) values in an absorbing (a) and inverted (b) medium.

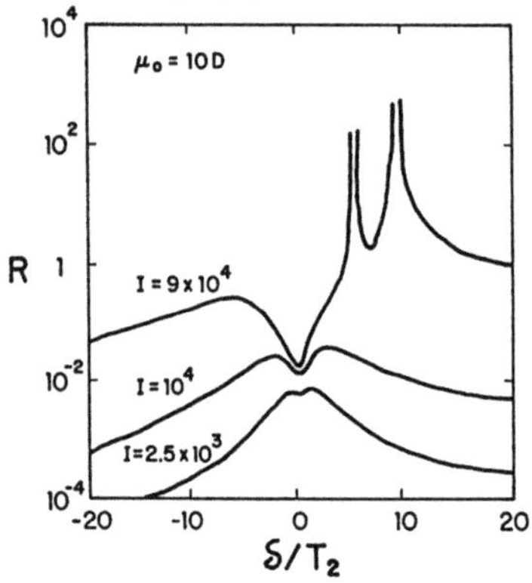

Fig. 2: Reflectivity versus frequency detuning normalized to T_2, for μ_o = 10 D and for different I values, in the absorbing medium.

inverted medium. The energy gap between the levels is $\hbar\omega_o$; the frequency of the field, $\omega = 10^{14}$ s^{-1}; the linear refraction index, n = 1.6; the length of the medium, L = 10 cm and $T_1 = 10^{-11}$ s, $T_2 = 10^{-12}$ s are the longitudinal and transversal relaxation times respectively. Taking μ_{12} = 5 D, the reflectivity R obtained for various μ_o values is shown in Fig. 1. The analysis of the reflectivity as a function of the detuning $\delta = 2\omega - \omega_o$ shows that coupled-mode oscillations can be obtained in the absorbing medium for high enough detuning values at high pump intensities (Fig. 2). For such a case, μ_o = 10 D, $\delta = 6\ T_2$, the reflectivity versus pump intensity is shown in Fig. 3 where a detail with I in linear scale is included.

The origin of this behaviour lies in the competition between B_1 and B_2 terms so that, for certain values of the physical parameters, $|\beta| > \alpha_R$. Then, the functions involved in (1) are

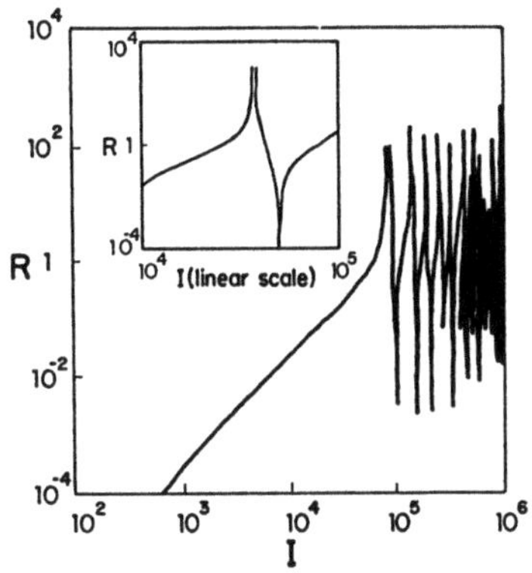

Fig. 3: Reflectivity versus pump intensity for μ_o = 10 D and δ/T_2 = 6 in absorbing medium. A detail is shown with a linear scale for I.

sinusoidal and coupled-mode oscillations are obtained periodically for I values obeying tan $\gamma L = -\gamma/\alpha_R$.

3. Conclusion

Changes in permanent dipole moments can be relevant in two-photon resonant electronic transitions.

Concerning OB, a high enough μ_o value allows small size ($\simeq$ 1 μm) cavities to obtain OB. When OB is considered in impurity centers inserted in a solid matrix, phonon effects become negligible if $\mu_o^2 >> |\mu_{12}|^2$ [7].

With respect to DFWM, as μ_o increases, the reflectivity of the phase-conjugate wave also increases, except at very high pump fields where the limits attained depend on the particular μ_o, μ_{12} values. A complicated structure of the reflectivity versus detuning appears at high pump intensities showing coupled-mode oscillation in the absorbing medium at moderate detuning. Some qualitative similarities can be noticed in the case of DFWM for two-photon transitions in multilevel atoms [10]. In our case a relevant aspect is the competition between the term dependent on $\mu_o^2|\mu_{12}|^2$ and that dependent on $|\mu_{12}|^4$, giving rise to coupled-mode oscillations in the absorbing medium for certain values of the physical parameters; for instance detuning above a threshold value.

4. Acknowledgement

Work partially supported by CICYT proyect n° ROB89-0664.

5. References

[1] A.F. Garito, K.D. Singer, K. Hayes, G.F. Lipscomb, S.J. Lalama, K.N. Desai: J. Opt. Soc. Am., 70, 1399 (1980)

[2] R. Twieg, A. Azema, K. Tain, Y.Y. Cheng: Chem. Phys. Lett., 92, 208 (1982)

[3] P. Kerkoc, M. Zgonik, K. Sutter, Ch. Bosshard, P. Günter: J. Opt. Soc. Am. B, 7, 313 (1990)

[4] S.Y. Hou, W.M. Hetherington III, G.M. Korenowski, K.B. Eisenthal: Chem. Phys. Lett., 68, 282 (1979)

[5] N.C. Kothari, T. Kobayashi: IEEE J. Quantum Elect.: QE-20, 418 (1984)

[6] M.A. Antón, J. Gonzalo, J.L. Escudero: Opt. Quant. Electron., 21, 415 (1989)

[7] I. Gonzalo, M.A. Antón: Opt. Quant. Electron., 21, 471 (1989)

[8] R.A. Fisher. "Optical Phase Conjugation". Acad. Press 1983

[9] M.A. Antón, I. Gonzalo: To be published in J. Opt. Soc. Am., B (1991)

[10] Tao-Yi Fu, Sargent III: Opt. Lett., 5, 433 (1980)

Phase Space Methods in Quantum Optics: Quantization Maps in the (q,p) Representation

J. Martinez

Instituto de Estructura de la Materia, C.S.I.C.,
Serrano 123, E-28006 Madrid, Spain

The paper analyzes the existence of quantization maps associated with the ordering of operators in the coordinate–momentum representation. It is demonstrated that, for hamiltonian systems, the only well–defined quantization map of this class is the one corresponding to symmetric order that leads to the Wigner function.

1 Introduction

It has been shown [1] that the problem of finding an ordered expansion of functions of noncommuting observables is associated with the problem of mapping them to a phase space representation. In quantum optics, phase space methods have been extensively used and formalised from different points of view [1-6].

The debate regarding the physical soundness of the different quasi–probability distributions (q.p.d.) is still open. In particular, the questions regarding the most relevant q.p.d. and the meaning of the correlations computed with different ordering schemes are still to be settled [6].

The purpose of this work is to shed some light onto these problems. We will study the existence of quantization maps associated with the ordering of operators, and particularly the $\nu = -1, 0, 1$ orderings that correspond to the Q,W and P distributions.

2 S-quantization maps, definitions

Let E be the phase space of dimension $2n, n \in N$ (for the moment we will consider $n = 1$). The Weyl quantization map [7] is an application that associates functions on E with operators on a Hilbert space carrying an irreducible representation of the Weyl commutation relations. In coordinate-momentum representation, the Weyl displacement operator is

$$\widehat{M}(\theta,\tau) = e^{i\theta\hat{q}+i\tau\hat{p}} \quad .$$

Nonlinear Dynamics and Quantum Phenomena in Optical Systems
Editors: R. Vilaseca, R. Corbalán

We can choose another set of representatives of the ray representation of this group just by multiplying the generators by a unimodular function of the parameters of the group [8]. In this way, the abelian structure of the group is retained when we replace the generators for the s–ordered Weyl displacement operators defined by

$$\widehat{M}(\theta,\tau;s)=\widehat{M}(\theta,\tau)\,\Omega_s(\theta,\tau)$$

where Ω_s is the filter function of the quantization map

$$\Omega_s(\theta,\tau)=e^{-i\hbar\theta\tau s/2}\quad . \tag{1}$$

This filter function modifies the Baker-Haussdorf factor that appears when "breaking" the operator $\widehat{M}_S \equiv \widehat{M}(\theta,\tau;s)$ so that $s=-1,0,1$ correspond to the standard, symmetric and antistandard expansions for the Weyl displacement operator, namely

$$\begin{aligned}\widehat{M}_1(\theta,\tau) &= e^{i\theta\hat{q}}e^{i\tau\hat{p}}\\ \widehat{M}_0(\theta,\tau) &= e^{i\theta\hat{q}+i\tau\hat{p}}\\ \widehat{M}_{-1}(\theta,\tau) &= e^{i\tau\hat{p}}e^{i\theta\hat{q}}\quad .\end{aligned}$$

The operator $\widehat{M}_S$ "displaces" a state on Hilbert space in the form

$$\widehat{M}(\theta,\tau;s)|\psi>=e^{i\hbar\theta\tau(1-s)/2}\int dq\,e^{i\theta q}|q><q+\hbar\tau|\psi>\quad . \tag{2}$$

The mean value of $\widehat{M}_S$ plays the role of a characteristic function [3]. In fact, the natural expression for the phase space probability distribution is

$$F(q,p;s)=\frac{1}{2\pi}\int\!\!\int d\theta d\tau e^{-i\theta q-i\tau p}Tr[\hat{\rho}\,\widehat{M}_s]\quad . \tag{3}$$

Through equation 3 we see that $\widehat{M}(\theta,\tau;s)$ defines the inverse of the s-quantization map through the expression

$$G(q,p)=2\pi\hbar\,Tr[\widehat{G}(\hat{q},\hat{p})\,\widehat{\Delta}^{\tilde{\Omega}_S}(q-\hat{q};p-\hat{p})] \tag{4}$$

where $\widehat{\Delta}^{\Omega_S}$ is the Fano operator, with the structure of a Dirac Delta operator :

$$\widehat{\Delta}^{\Omega_S}(q-\hat{q};p-\hat{p})=\frac{1}{(2\pi)^2}\int\!\!\int d\theta d\tau\,\Omega_s(\theta,\tau)e^{-i\theta[q-\hat{q}]-i\tau[p-\hat{p}]}$$

and $\tilde{\Omega}_S$ is the reciprocal filter function, defined by the expression $\tilde{\Omega}_S\,(\theta,\tau)=\Omega_S(-\theta,-\tau)^{-1}$.

The s–distributions defined in equation 3 are the generalization in the $(\hat{q},\hat{p})$ representation of the usual quasi–probability distributions of quantum optics associated with an ordering of operators. The present work is devoted to studying the existence of these s–distributions.

In the $(\hat{a},\hat{a}^\dagger)$ representation the Weyl operator adopts the form

$$\widehat{M}|_{(\hat{a},\hat{a}^\dagger)} = e^{\xi\hat{a}^\dagger-\xi^*\hat{a}} e^{\frac{is}{2}|\xi|^2\sin 2\chi} = \widehat{D}(\xi)\, e^{\frac{s}{4}(\xi^2-\xi^{*2})}$$

where we have defined the complex number $\xi = |\xi|e^{i\chi} = i\sqrt{\frac{\hbar}{2}}(\theta+i\tau)$. It is seen from the above equation that there is no way to compensate the Baker-Haussdorf factor with the s filter function for the Glauber $\widehat{D}$ operator [3] as we have done for $\widehat{M}$ in the $(\hat{q},\hat{p})$ representation. Thus, in order to get an ordered expansion for $\widehat{D}$ we need another independent complex parameter λ. Thus the generalized filter function is defined as

$$\Omega_{s,\lambda}(\xi) = e^{\frac{s}{4}(\xi^2-\xi^{*2})} e^{\frac{-\lambda}{2}|\xi|^2} \ . \tag{5}$$

In order to collect these results, let us write the expressions of the generalised Weyl displacement operators in both representations:

$$\begin{aligned} \widehat{M}_{s,\lambda}|_{(\hat{q},\hat{p})} &= \widehat{M}(\theta,\tau)\, e^{\frac{\lambda\hbar}{4}(\theta^2+\tau^2)} e^{-i\hbar\theta\tau s/2} \\ \widehat{M}_{s,\lambda}|_{(\hat{a},\hat{a}^\dagger)} &= \widehat{D}(\xi)\, e^{\frac{s}{4}(\xi^2-\xi^{*2})} e^{\frac{-\lambda}{2}|\xi|^2} \ . \end{aligned}$$

Quantization maps associated with an ordering of operators are obtained for the set $(s,\lambda=0)$ in the $(\hat{q},\hat{p})$ representation and for the set $(s=0,\lambda)$ in the $(\hat{a},\hat{a}^\dagger)$ representation. We will call the former ones s–quantization maps, the later λ–quantization maps. The term ν–quantization maps will be used for both of them.

2.1 Uniqueness condition

Let us now consider the quantum mean value of an operator $\widehat{G}$, i.e. $Tr[\hat{\rho}\widehat{G}]$. If we map both operators through equation 4 into the phase space $F(q,p;s)$ and $G(q,p;s')$ functions and consider the completeness relation for the displacement operator it is easily seen that the necessary condition to express the quantum mean value as a phase space average is

$$\Omega_{S'}(\theta,\tau) = \widetilde{\Omega_S}\,(\theta,\tau)$$

and this implies, with the filter function given by equation 1, that $S'=-S$. The same result is obtained for λ-quantization maps since no conjugation properties of the filter function on equation 5 are involved. In this way ν-quantization maps provide ν mathematically equivalent forms for computing any quantum mean value as a phase space average in the form

$$\begin{aligned} Tr[\hat{\rho}\widehat{G}(\hat{q},\hat{p})] &= \int\int \frac{dq\,dp}{2\pi} F(q,p;-s)G(q,p;s) \propto \overline{G} \\ Tr[\hat{\rho}\widehat{G}(\hat{a}^{\dagger},\hat{a})] &= \int\int \frac{d^2\alpha}{\pi} F(\alpha;-\lambda)G(\alpha;\lambda) \propto \overline{G} \end{aligned} \quad (6)$$

where as we have just seen $\nu' = -\nu$ is the uniqueness condition: the integrals of equations 6 do not depend on ν. It is worth emphasizing that no matter what ζ ordered expansion we adopt for the generic quantum operator $\widehat{G}$ we may use any ν-quantization map although the mapping of the operators is straightforward when $\zeta = \nu$ is chosen.

3 Wigner function in the $(\hat{q},\hat{p})$ representation

As we have seen in the last section, ν–quantization maps are mathematically equivalent to computing quantum mean values as phase space averages, provided that they all exist. We can find analogies of this equivalence in the Theory of Deformations of Symplectic Structures [9], and in Prime Quantization Theory [10].

Nevertheless, in this section we will show that in coordinate–momentum representation the Wigner function stands out. Let us study the behaviour of q.p.d. $F(q,p;s)$ with generic complex s ordering parameter $s = s_1 + is_2$. First let us point out that only purely imaginary values of s lead to real q.p.d. In contrast to this the different behaviour of the $\widehat{D}_\lambda$ operator under conjugation ensures real q.p.d for $\lambda \in \Re$ in complex phase space. Indeed, from the behaviour of $\widehat{M}_s$ under adjoint operation we see that for the q.p.d.

$$F^*(q,p;s) = F(q,p;-s^*) \ .$$

In this way we must restrict the domain of s such that $s = -s^*$, i.e. $s \in \Im(s)$, in order to get real valued q.p.d. In fact we get

$$\begin{aligned} W^*(q,p) &= W(q,p) \\ F^*_{-1}(q,p) &= F_1(q,p) \end{aligned}$$

as noticed first by Agarwal and Wolf [1].

On the other hand, inserting equation 2 in equation 3, the q.p.d. adopts the form

$$F(q,p;s) = \frac{1}{2\pi}\int\int\int d\tau dq' d\theta \, e^{\hbar\theta\tau s_2/2} e^{i\theta[q'-q+\frac{\hbar\tau}{2}(s_1-1)]}$$
$$\times < \Psi|q' > e^{-i\tau p} < q' + \hbar\tau|\Psi > \quad .$$

All the integrals are taken between $-\infty$ and $+\infty$. Thus, whatever the sign of s_2 may be, the exponential factor $e^{\hbar\theta\tau s_2/2}$ will cause a divergence. The $F(q,p;s)$ q.p.d. in coordinate–momentum representation only exist for $s_2 = 0$. We conclude that only the $s = 0$ intersection of both axes $\Re(s), \Im(s)$ provide a real valued, well-defined quasiprobability distribution: the Wigner function.

References

[1] G. S. Agarwal and E. Wolf, Phys. Rev. D2, 2161, 2187, 2206 (1970).

[2] M. Lax, Phys. Rev. 172, 350 (1968).

[3] K. E. Cahill and R. J. Glauber, Phys. Rev. 177, 1857, 1883 (1969).

[4] M. Hillery, R. F. O'Conell, M. O. Scully, E. P. Wigner, Phys. Reports 106, n° 3 (1984), 121-167.

[5] J. E. Moyal, Proc. Cambridge Phil. Soc. 45, 99 (1949).

[6] T. W. Marshall, E. Santos, Phys. Rev. A41, 1582 (1990).

[7] H. Weyl, *The Theory of Groups and Quantum Mechanics*, (Dover, New York, 1931), pg. 274.

[8] V. Bargmann, Ann. Math. 59, 1 (1954).

[9] F. Bayen, M. Flato, C. Fronsdel, A. Lichnerowicz, D. Sternheimer, Ann. Phys. 110, 11 (1978).

[10] S. Twareque Ali and H. D. Doebner, Phys. Rev. A41, 1199 (1990).

Part IV

Nonlinear Dynamics

Chaos and Pulsating Instabilities in Stimulated Brillouin Scattering

R.G. Harrison, J.S. Uppal, Weiping Lu, and A. Johnstone

Physics Department, Heriot-Watt University,
Riccarton, Edinburgh, EH14 4AS, UK

In this paper we investigate experimentally and theoretically the nonlinear dynamical features of stimulated Brillouin scattering processes in single-mode fibre both with and without the presence of external feedback. Good qualitative agreement between theory and experiment establishes the dynamics to be deterministic in origin and indicates the influence of noise to be small. Our results suggest the possible existence of similar dynamical behaviour in other stimulated scattering and three-wave mixing processes. The implications of our findings for various applications of nonlinear optics, notably optical fibre communications and technology, are recognised.

1. Introduction

Nonlinear dynamics in general and in particular deterministic chaos with the recent emergence of structural complexity is in a period of spectacular growth and progress [1,2]. Its implications are being strongly felt in a broad and ever-growing range of physical sciences. Of these, nonlinear optical systems are central in providing relatively simple theoretically tractable systems which operate at high speed. Furthermore the technological importance of these, both lasers and nonlinear optical devices, demands investigation and control of instability mechanisms. From the prolific activity of this field particularly in the area of lasers, attention is now being focussed on other important areas of nonlinear optics.

This work is primarily concerned with exploring the generic nature of nonlinear dynamical behaviour in fundamental nonlinear optical phenomena. Accordingly we identify nonlinear interactions in which the a priori requirement of feedback for the emergence of instability, hitherto commonly imposed by an external cavity, is an implicit and integral feature of the fundamental nonlinear phenomena. Significantly a broad range, if not the majority, of nonlinear optical interactions satisfy this criteria of which general three-wave mixing and parametric processes merit special attention as cornerstones of the field. Through recent major advances in optical fibre technology we are now able to interrogate these phenomena under ideal cw and for low power pulsed optical excitation by utilizing the long interaction length of this medium to readily induce nonlinearity under these conditions. Furthermore plane wave analysis is an excellent approximation in describing beam propagation in such media. In addition, however, fibre systems are critical components to an entire future telecommunications technology and any qualitative insight into the nonlinear

Springer Proceedings in Physics, Vol. 55
Nonlinear Dynamics and Quantum Phenomena in Optical Systems
Editors: R. Vilaseca, R. Corbalán

dynamical behaviour of light propagation in such media has potentially important applications.

In this paper we consider in particular stimulated scattering phenomena and provide, to our knowledge, first evidence of chaotic dynamics in one of these basic processes, namely stimulated Brillouin scattering (SBS). Single mode optical fibre is used to generate SBS under cw single mode pump conditions, resulting in first order Stokes emission only [3]. We find both the transmitted pump and back scattered SBS to exhibit chaotic behaviour under all operating conditions investigated, including those close to the threshold for SBS; the SBS exhibiting massive instabilities with modulation depths ~ 100%. This fundamental and conceptually simple nonlinear wave interaction contrasts with those reported in which feedback either by a reflecting surface or involving counterpropagating pump beams is used as a necessary condition for generating dynamical instabilities. Indeed we find that the inclusion of external optical feedback modifies but does not suppress the chaotic dynamics and also gives rise to a wide range of quasi-oscillatory and bursting instabilities. The generality of these observations has been established for fibre lengths ranging from 38 to 300 metres and for launch powers down to ~ 100 mW (threshold for SBS in 200 m fibre).

Subsequent mathematical analysis, based on coupled nonlinear field (pump and Stokes) and material equations, has established the central role of nonlinear refraction and its interplay with the nonlinear gain of the SBS interaction as being responsible for the dynamical behaviour for both cavityless and cavity operation, omission of coexisting dispersive effects, as in conventional treatments, resulting in stable solutions except for the case of low ratio of SBS intensity to input intensity where stable oscillations appear. Further the generality of this treatment points to the likelihood of similar dynamical behaviour in other stimulated scattering phenomena and 3-wave mixing interactions. Significantly the good agreement of our experimental findings with these theoretical results shows the dynamics to be deterministic in origin and not to be due to, or unduly influenced by, the noise structure of the initial spontaneous Brillouin scattering.

2. Experiment

Our experimental arrangement is schematically shown in Fig. 1. The cw emission of a single-mode argon-ion laser at 514.5 nm, with an instantaneous (≤ 1 msec) linewidth of ~ 15 kHz (Coherent Innova 100) was used as a pump source providing variable output power stabilized to ± 2%. Two 10 x microscope objectives L_1 and L_2, were used to couple the light into and out of the optical fibre, respectively. The fibre comprised a pure SiO_2 core of diameter 4.8 μm with a B_2O_3-doped SiO_2 cladding. It was optically isolated from the argon laser using a Faraday isolator (OFR Model IO-5-532) giving an isolation factor of 35 dB between them. The pump signal and backscattered signal, comprising the SBS signal together with residual scatter, were sampled via the beam splitter in Fig. 1. These signals together with the signal transmitted through the fibre were detected using photodiodes (HP-type BPX65) D_1, D_2 and D_3 which have a rise time of ≤ 0.5 nsec. A transient digitizer (LeCroy TR

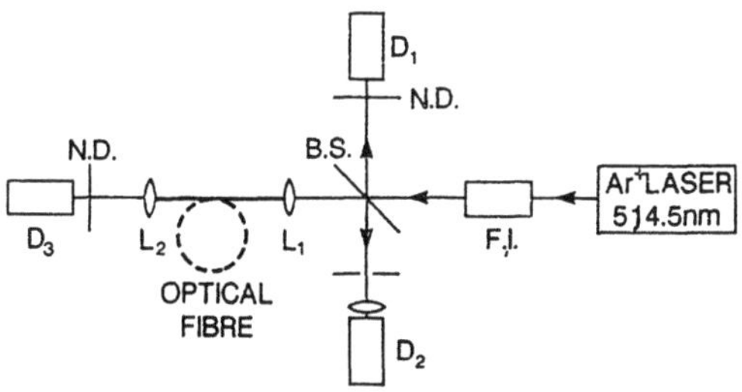

Figure 1 Schematic set-up of experimental arrangement. F.I. is Faraday isolator, N.D. neutral density filters and D_1, D_2, D_3 are photodetectors.

8828C) having a data sampling rate of 5 nsec interfaced with a Masscomp Computer (MC5600) was used for temporal recording and subsequent signal processing to provide power spectra and phase-portrait reconstruction. Results were cross checked from independent recordings on a Tektronix 7104 oscilloscope for both temporal measurements and direct phase-portrait reconstruction of the SBS versus transmitted pump signals (oscilloscope in the x-y mode) and on a Hewlett-Packard spectrum analyzer (HP8590A) for corresponding spectral analysis. Feedback from the end faces of the fibre was eliminated using an index-matching liquid for the cavityless system. Single-mode propagation in the fibre was ensured by looping a small section of the front end of the fibre, thereby suppressing higher-order modes and was confirmed by observation of the spatial form of the transmitted signal. A Fabry-Perot interferometer with variable spacer confirmed throughout only first-order Stokes SBS emission.

At low input power the backscattered signal increased linearly with input power due to Fresnel reflection up to the threshold launch power for SBS (~ 100 mW for a 200-m fibre). The signal thereafter increased rapidly and nonlinearly with launch power, the conversion efficiency to SBS being ~ 40% for a launch power of ~ 300 mW, accounting for losses arising from looping the fibre. Corresponding fibre transmission measurements showed a nonlinear response for input power ≥ 100 mW and for pump powers ≥ 200 mW gave a saturated maximum of 20 mW. These general features are in agreement with the findings of others [4] and are shown in Fig.2. The threshold power for SBS can be estimated from small signal theory of stimulated scattering and is given as $GL_e \sim 21$, where G is the SBS gain factor and, for a pump-laser linewidth $\Delta\nu_L$ much less than the Brillouin scattering linewidth $\Delta\nu_B$ theoretical value of ($\Delta\nu_B \sim 145$ MHz at 514 nm), is given as $KP_L/\lambda^2\Delta\nu_B A$, where K is a constant depending upon the physical and optical properties of the fibre and A is the effective cross-sectional area of the guided mode. The effective interaction length L_e is given as $L_e = \alpha^{-1} \,.\, [1-\exp(-\alpha L)]$, where α is the absorption coefficient experimentally determined to be 4.6×10^{-3} m^{-1} (20 dB/km), and L the length of the fibre. Using physical properties data for fused silica and an estimated value of $A = 0.9 \times 10^{-11}$ m^2, the above relation gives a threshold pump power $P_L \sim 82$ mW. Accounting for losses arising from looping the fibre, the launched power for obtaining SBS is increased by

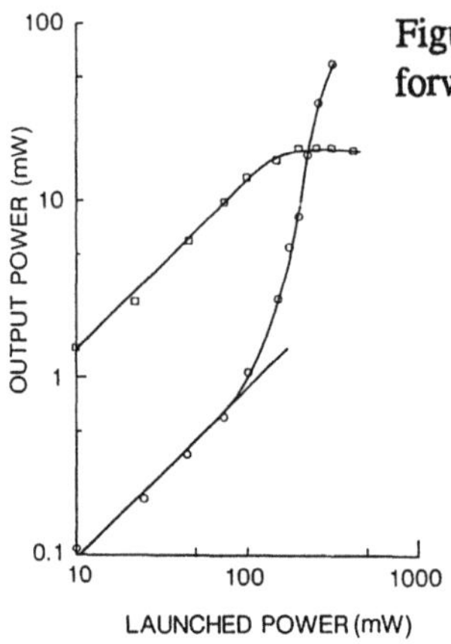

Figure 2 Output power of backward Brillouin wave and forward transmitted laser wave as a function of input power.

an experimentally measured factor of 1.6-1.8, giving a threshold power for SBS~130 mW in reasonable agreement with experimental findings.

Representative recordings of the chopped SBS signal and transmitted pump signal are shown in the traces (a) and (b) of Fig. 3 which demonstrate the magnitude of the instabilities. Trace (a) shows sustained instabilities with modulation depth of ~100%. Similar results were obtained for all input powers, limited to a maximum value of ~700 mW by the laser output power. From trace (b) of the transmitted pump the ratio of the unstable to the underlying stable portion of the transmitted signal provides a measure of the SBS conversion efficiency of 40%, consistent with the measurement above. These general features were confirmed for various fibre lengths from 38 to 300 m with an expected increase in the SBS threshold for reduced fibre length; ~500 mW for the 38-m fibre. Representative forms of the dynamical features from the SBS and the transmitted pump signals, for a 100 m length fibre taken without the chopper and for a launch power of 330 mW, are displayed in the digitized recording of the trace (c) - (h) of Fig. 3 showing the time series, power spectrum and reconstructed phase portraits. The latter were obtained by a two-dimensional plot of data points x_i = 1,2...16,000) versus x_{i-K} for appropriate value of delay K. The time recording of the SBS (trace (c)) shows erratic and aperiodic emission, over a broad time scale down to ~50 nsec, the corresponding power spectrum (trace (e)) exhibiting pure broad band features consistent with fully developed chaos with a spectral bandwidth ~30 MHz which may be compared with the Brillouin gain bandwidth of ~145 MHz. The reconstructed phase portrait (trace(g)) exhibits motion on an outward spiralling and folding trajectory particularly evident from observing the temporal evolution of the trajectory. The corresponding recordings of the pump signal transmitted through the fibre (traces (f) to (h)) exhibit similar chaotic features though over a reduced spectral bandwidth. The phase portrait essentially mirrors that of the SBS due to the parametric nature of the nonlinear interaction providing further evidence of the deterministic rather than stochastic nature of this process, also corroborated by the continuous form of the trajectories spiralling and folding motion within the digitization resolution.

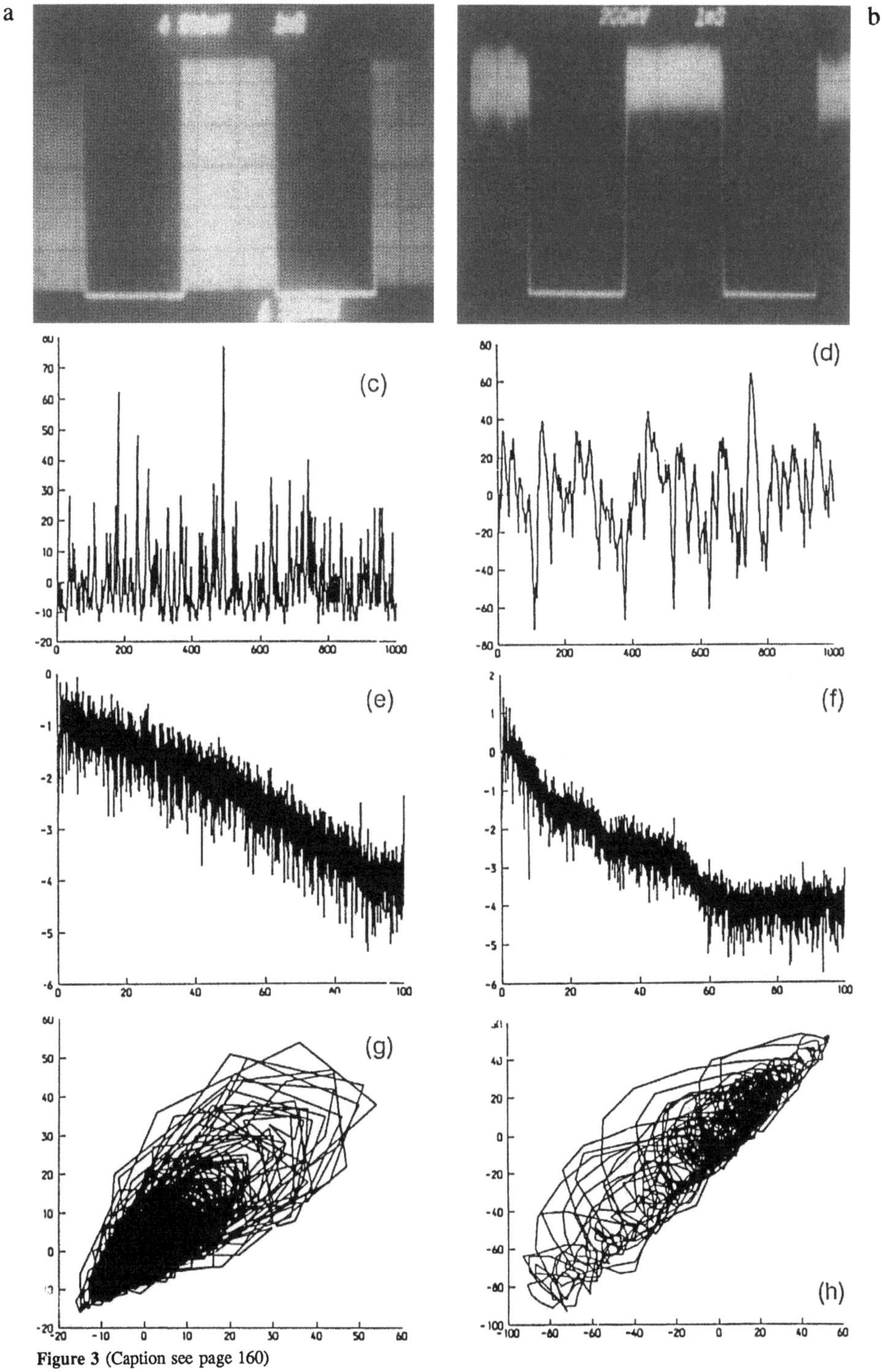

Figure 3 (Caption see page 160)

More generally the spectral broadening was found to depend upon the input pump power, varying from ~10 MHz close to threshold to ~20 MHz at twice the input power. The general features of traces (c)-(d) are typical of those obtained for a wide range of pump powers and were found to be highly reproducible. Interestingly no evidence of precursor routes to chaos were found.

In further investigations of these phenomena we also considered [5] the influence of external feedback on the dynamics of the SBS process. Feedback was provided by removal of the index-matching fluid from the fibre ends to give a nominal reflectivity of ~4% from each of the end faces.

Here we highlight sample results from an extensive experimental investigation of the nonlinear dynamics of SBS. Sustained quasi-regular oscillations at twice the fibre transit time ($2T_r$), similar to that reported earlier by Bar-Joseph et al. [6], are found to exist for only limited operating conditions and at low pump power. More generally the emission is found to comprise more than one frequency component as in Fig. 4(a) (fibre length 25 m). The power spectra of this data shows two independent frequencies and their harmonics along with a weak broad-band background which increases with pump power from the SBS threshold, indicative of the emergence of weak chaotic structure underlying the dominant quasi-periodic features. The number of such coexisting components increases with pump power, resulting in more complex time structure and random modulation of the emission. For further increase in pump power, this emission ultimately degenerates into bursts of oscillations as shown in Fig. 4(b) (fibre length of 40 m), the duration and separation of which appears random.

We further find a tendency for the bursting mode of operation to predominate for longer fibre length and also for higher feedback reflectivity. On increasing the pump level for these conditions, the duration of the individual profiles of bursting oscillation decrease while their separation increases; at the highest pump level, we further find the bursting features to revert back to sustained complex oscillations, although in the possible presence of unavoidable second order Stokes emission. In general, the dynamical features of SBS in this case are strongly dependent on the pump intensity, fibre length and cavity feedback, full details of which will be published elsewhere.

Figure 3 Characteristics of dynamical outputs of back scattered (left column) and transmitted (right column) waves in the cavityless condition. (a),(b) backscattered and transmitted signals detected respectively by D_2 and D_3 for a chopped launch power of 0.33 W where the time scale is 1 msec/division throughout and fibre length is 200 m. (c)-(h) sample time series, power spectrum and phase portraits taken fron 16 K digitized recordings of backscattered and transmitted signals without chopper: The time series (traces (c) and (d)) comprise 1000 data points (step interval 5 nsec); corresponding power spectra (traces (e) and (f)) are for 16 K data points with a vertical scale gain of 10 dB/division; traces (g) and (h) are the corresponding phase portraits comprising 2000 data points constructed with a delay of 15 and 3 step intervals respectively, where fibre length is 100 m.

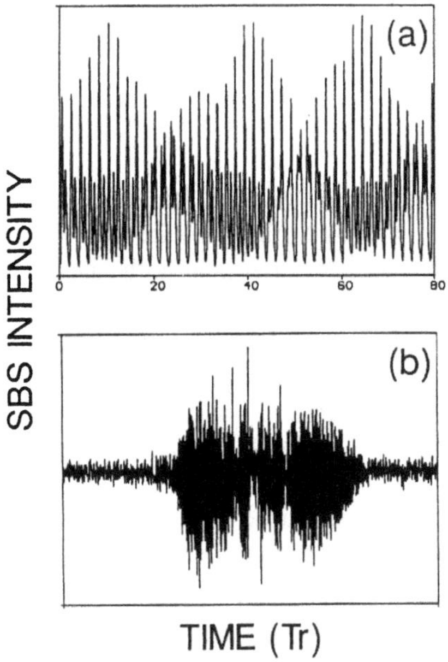

Figure 4 Two distinctive modes of oscillatory emissions of backward SBS obtained from experimental data for natural boundary conditions. (a) Quasi-periodic emissions for a fibre of 25 metres and pump power ~ 150 mW. (b) Bursting oscillations from a fibre of 40 metres and pump power ~ 200 mW. Here the time is normalised to the trip time of the medium T_r; typical duration of bursting is hundreds of T_r. Note data of trace (b) is taken at low resolution (320 ns digitisation interval) resulting in loss of high frequency information.

3. Theory

Stimulated Brillouin scattering is described as a parametric coupling between light and acoustic waves [7]. The nonlinear interaction results from electrostriction which for a sufficiently high pump intensity or long interaction length stimulates growth of the Stokes signal. With the inclusion of the nonlinear dispersion this process is described by the general coupled first order nonlinear partial differential equations [5]

$$\frac{\partial \varepsilon}{\partial \tau}+\frac{\partial \varepsilon}{\partial \xi}-iu(|\varepsilon|^2+2|\varepsilon_s|^2)\varepsilon+\frac{\beta}{2}\varepsilon=-g\varepsilon_s B \quad ,$$

$$\frac{\partial \varepsilon_s}{\partial \tau}-\frac{\partial \varepsilon_s}{\partial \xi}-iu(|\varepsilon_s|^2+2|\varepsilon|^2)\varepsilon_s+\frac{\beta}{2}\varepsilon_s=g\varepsilon B^* \quad ,$$

$$\frac{1}{\beta_A}\frac{dB}{d\tau}+B=\varepsilon\varepsilon_s^* \quad , \qquad (1)$$

where the parameters

$$\tau=\frac{ct}{nL}, \qquad \xi=z/L, \qquad u=\frac{n_2\omega L}{c}|E_0|^2, \qquad \beta=\alpha\frac{nL}{c} \text{ and } g=\frac{V_1V_3nL}{\alpha_A c}|E_0|^2$$

Here g is the small signal SBS gain (amplitude), L the interaction length, C_S is the velocity of light, n the linear refractive index, n_2 the nonlinear index coefficient related to $\chi^{(3)}$, α the absorptive coefficient of both pump and Stokes field and α_A the damping coefficient of the acoustic wave. $V_1 = \pi n^2 P_{12} c/\lambda\rho_0$, $V_3 = \gamma\pi n^5 \varepsilon_0 P_{12}/2\lambda c_s$ are the SBS coupling coefficients which are responsible for the standard SBS gain; here γ is the polarisation factor, P_{12} the longitudinal electro-optical coefficient, ρ_0 the fibre density, λ the optical wavelength and c_s the velocity of the acoustic wave. The nonlinear refraction induced by the pump and Stokes fields is described by the third term in the first and second equations of (1), which leads to a change in phase of both pump and Stokes fields in space and time. Since the velocity of the acoustic wave is much smaller than that of the optical wave, the acoustic wave as in normal treatments is highly damped, allowing the term $\partial B/\partial z$ in the material equation to be dropped.

First we demonstrate the essential role of nonlinear refraction in promoting dynamic behaviour in this process. To do so we have numerically integrated the coupled nonlinear eqn. (1) in both space and time by using the method of characteristics. Figure 5 shows the time-dependent solutions of Stokes intensity at

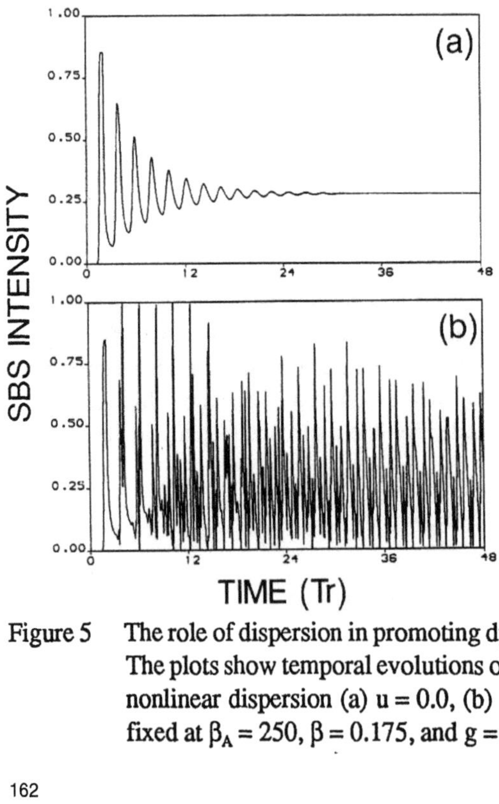

Figure 5 The role of dispersion in promoting dynamical features in cavityless case. The plots show temporal evolutions of the SBS intensity with and without nonlinear dispersion (a) u = 0.0, (b) u = 16.5. The other parameters are fixed at $\beta_A = 250$, $\beta = 0.175$, and g = 22.5.

$\xi = 0$ with and without nonlinear refraction u (traces (a) and (b)) respectively, other parameters remain unchanged. In general when u is small the long term SBS is stable and constant, approached through a transient relaxation oscillation with period $2T_r$ (trace (a)) where T_r is the single trip time of the medium. This behaviour is in accord with the early work by Johnson and Marburger [8]. On increasing u, while the amplitude of oscillation with period $2T_r$ decays, a new oscillation with period T_r grows and then saturates to form a sustained oscillation. With further increase of u, the wave-shape undergoes a series of bifurcations with the evolution of new frequencies seemingly corresponding to successive harmonics of the base frequency associated with period $2T_r$. These periodic and quasi periodic signals show, on increasing u, a gradual spectral broadening which together with a thickening in their associated phase portrait is indicative of an underlying weak chaos throughout, finally developing into fully chaotic behaviour as shown in trace (b).

In Fig. 6 we show an example of the chaotic solution for both the SBS (trace(a)) and transmitted signal intensities (trace (b)) together with their respective broad band spectra (traces (c) and (d)) and time reconstruction phase portraits (trace (e) and (f)) after the transient period. While the instabilities of the SBS signal exhibit typical modulation depths ~ 100% the transmitted signal shows instabilities superimposed on a dc background, the depth of which provides an indirect measure of the conversion efficiency of the SBS process. Here it is significant to note that the relative depths of modulation of these two signals is mainly influenced by the dispersion terms in eqn. (1) through the factor of two and according to whether ε is larger or smaller than ε_s. Comparison of the time series show qualitative evidence of temporal correlation, the signals being an inverted image of one another consistent with the parametric nature of the interaction. We note however that an exact correlation between the instabilities is not expected since the signals are spatially separated, being monitored at either end of the extended medium. Further inspection of the time series and associated broad band power spectra also show clear evidence of an underlying spectral feature associated with the round trip time $2T_r$.

Comparing these results with our experimental observations of chaotic SBS Stokes emission (see Fig. 3) under single mode conditions, we see that qualitative agreement is quite good in confirming: a) the dominance of chaotic dynamics with modulation depths ~ 100% for all pump intensities including those close to the threshold for SBS, the parametric form of these instabilities manifesting themselves in the transmitted pump signal superimposed on a stable dc background, b) broad-band spectra of these signals with spectral fine structure at T_r^{-1} and relative spectral narrowing of the transmitted pump signal and c) phase portrait reconstruction showing motion on an outward spiralling and folding trajectory, the pump signal phase portrait being an inverted form of the SBS.

As shown in our experimental data the influence of external feedback is to considerably simplify the dynamical behaviour of the SBS emission so providing an excellent basis for quantitative tests of our mathematical description. To simulate experimental conditions we consider Eqn. (1) accounting for weak feedback (~4% natural reflectivity of the fibre) namely $R_1R_2 << 1$, where R_1, R_2 are the reflectivities of

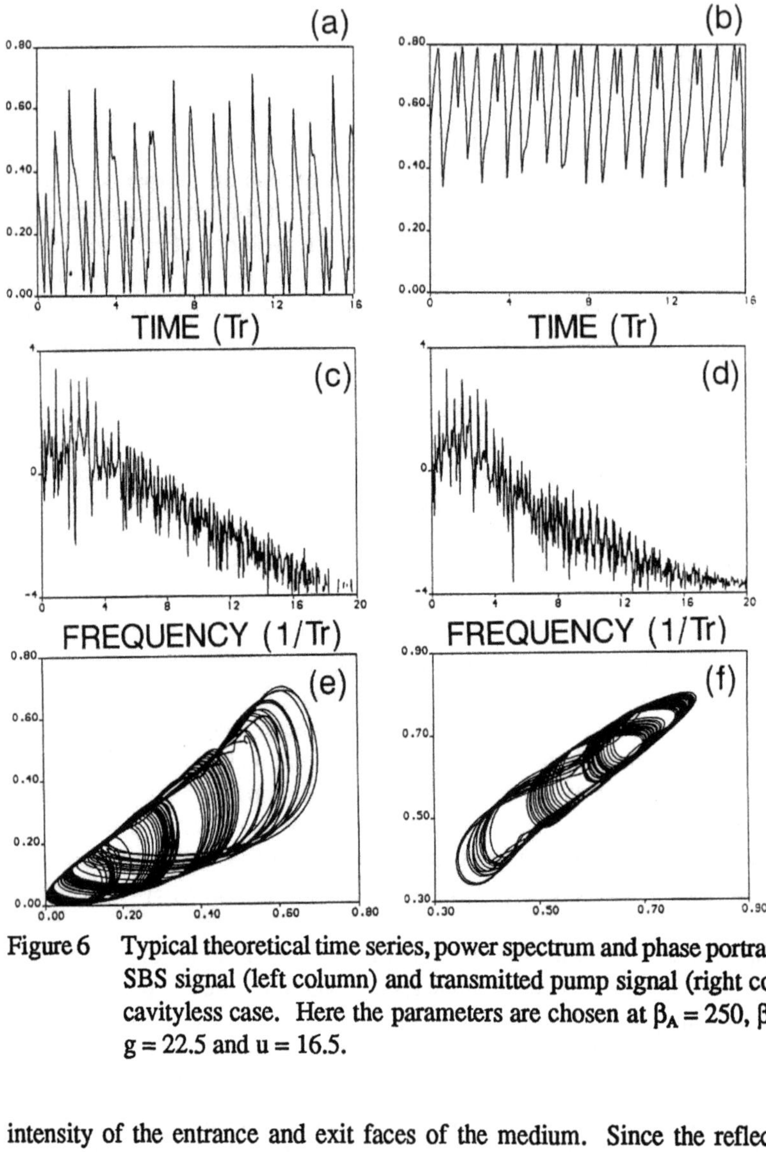

Figure 6 Typical theoretical time series, power spectrum and phase portraits for the SBS signal (left column) and transmitted pump signal (right column) in cavityless case. Here the parameters are chosen at $\beta_A = 250$, $\beta = 0.175$, $g = 22.5$ and $u = 16.5$.

intensity of the entrance and exit faces of the medium. Since the reflected field intensities from both ends are weak we neglect their interaction with the initial pump and Stokes fields already present in the interaction length in Eqn. (1). Furthermore the reflected pump, collinear with the incident pump, undergoes two reflections and is also neglected. For these conditions the boundary conditions become [5]

$$\varepsilon_{S,R}(t,0) = \sqrt{R_1}\,\varepsilon_S(t,0) \quad ,$$

$$\varepsilon_S(t,L) = \sqrt{R_2}\,\varepsilon_{S,R}(t,L) \quad , \tag{2}$$

where $\varepsilon_{S,R}$ is the forward SBS signal induced by the cavity effect, the evolution of

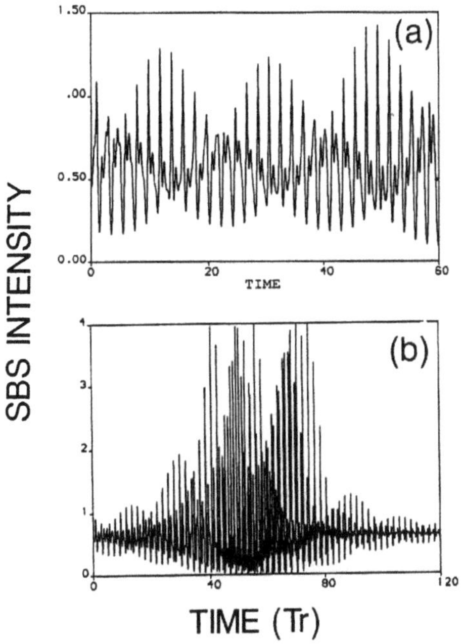

Figure 7 Two distinctive modes of oscillatory emissions of SBS with cavity obtained from theoretical solutions. (a) Quasi-periodic emissions for a fibre of 25 metres and pump power ~150 mW. (b) Bursting oscillations from a fibre of 40 metres and pump power ~ 200 mW. Here the time is normalised to the trip time of the medium T_r.

which is governed by the equation

$$\frac{\partial \varepsilon_{S,R}}{\partial \tau}+\frac{\partial \varepsilon_{S,R}}{\partial \xi}=-\frac{1}{2}\beta\varepsilon_{S,R}+iu(2\,|\,\varepsilon\,|^2+|\,\varepsilon_S\,|^2)\varepsilon_{S,R} \quad . \tag{3}$$

In contrast with cavityless operation the role of the cavity is to provide additional feedback of the SBS signal to that indigenous to the fundamental SBS interaction. It is the relative phases and amplitudes of these contributions which we conclude leads to the observed modification and simplification of the dynamics of the SBS emission when compared to that of cavityless operation.

In Fig. 7 we show two representative examples for direct comparison with the experimental data of Fig.4. Results are in good agreement with our experimental findings both regarding their temporal features and the pump scaling required to promote the change in dynamics from sustained to bursting modes of operation. Our theoretical results further confirm the eventual suppression of bursting features at the highest pump powers and the emergence of complex sustained oscillations.

4. Conclusions

In conclusion, our experimental and theoretical findings establish chaotic spatio-temporal behaviour to be an integral and generic feature of SBS. Furthermore the general form of our analysis points to the existence of similar rich dynamical behaviour in other stimulated scattering and three wave mixing processes, providing opportunity for extending investigations of nonlinear dynamics to this broad range of basic phenomena. We also note that inclusion of external optical feedback modifies the form of the chaotic dynamics and further results in a rich variety of classifiable precursor dynamical features. Finally we recognise the implications of our findings to various applications of nonlinear optics; notably optical fibre communication and technology.

5. Acknowledgements

We acknowledge the Science and Engineering Research Council for support of the work.

6. References

[1] Instabilities and Chaos in Quantum Optics,edited by F.T. Arecchi and R.G. Harrison, Spring Series in Synergetics Vol. 34 (Spring-Verlag, Berlin, 1987).

[2] Measures of Complexity and Chaos, edited by Neal B. Abraham, Alfonso M. Albano, Anthony Passamante and Paul E. Rapp, NATO ASI Series B, Phys. Vol. 208 (Plenum Press 1989).

[3] R.G. Harrison, J.S. Uppal, A. Johnston and J.V. Moloney, Phys. Rev. Lett. 65, 167 (1990).

[4] D. Cotter, Opt. Commun. 4, 10 (1983), and reference therein.

[5] A. Johnstone, Weiping Lu, J.S. Uppal and R.G. Harrison, Optics Commun. (in press).

[6] I. Bar-Joseph, A.A. Friesem, E. Lichtman and R.G. Waarts, J. Opt. Soc. Am. B 2, 1606 (1985).

[7] Y.R. Shen, The Principles of Nonlinear Optics (Wiley, New York 1984).

[8] R.V. Johnson and J.H. Marburger, Phys. Rev. A4, 1175 (1971).

Optical Chaos in a Laser Diode Pumped Nd Doped Fibre Laser

D. Derozier, S. Bielawski, and P. Glorieux

Laboratoire de Spectroscopie Hertzienne associé au CNRS,
Université des Sciences et Techniques de Lille F.A.,
Bat. P5, F-59655 Villeneuve d'Ascq Cedex, France

Abstract : Modulated pumping of a Nd^{3+} doped silica monomode fibre laser has been investigated experimentally. For small modulation amplitudes, the laser exhibits a resonance at the relaxation oscillation frequency in the region of 10 kHz. As the pump modulation is increased, period-doubling and optical chaos are observed.

Rare earth doped single mode silica fibres are promising to become the basis of an important new class of amplifiers and lasers for application in optical communication and instrumentation [1]. We have investigated the dynamic behaviour of a Nd doped fibre laser when subjected to modulated pumping. A wide range of phenomena has been observed : hysteresis, generalized bistability, period-doubling, optical chaos... . The fibre laser is noticeably different from previous lasers whose dynamics have been investigated under pump modulation. Because of the fibre dimensions, it is operated in a single transverse mode but laser oscillation usually occurs for many longitudinal modes since the cavity free spectral range (7.10^{-4} cm^{-1}) is a thousand times smaller than the homogeneous broadening (few cm^{-1}) which is eighty times smaller than the inhomogeneous one (450 cm^{-1}).

The experimental arrangement is shown in Fig.1. The pump laser is a GaAlAs laser diode (RTC CQL 44) operating at 820 nm. The pump beam is coupled into the fibre via two microscope objectives (10X, N.A.=0.25) through a mirror that is highly transmitting at the pump wavelength but highly reflecting at the laser wavelength of 1.08 μm. A 5 metre length of fibre doped with 300 ppm Nd^{3+}(core diameter =5.8 μm) absorbs most of the pump light (absorption coefficient : 0.33 m^{-1} at

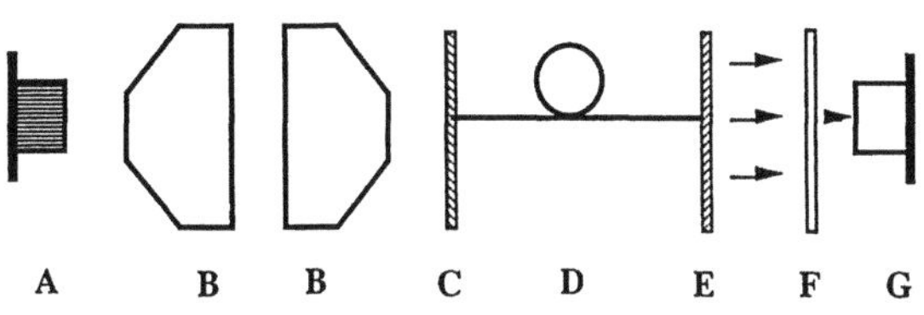

Figure 1 : Experimental arrangement for a diode pump modulated monomode fibre laser. Notation used : A - laser diode, B - microscope objectives, C - high reflector, D - Nd doped fibre, E - output coupler, F - bandpass filter, G - detector.

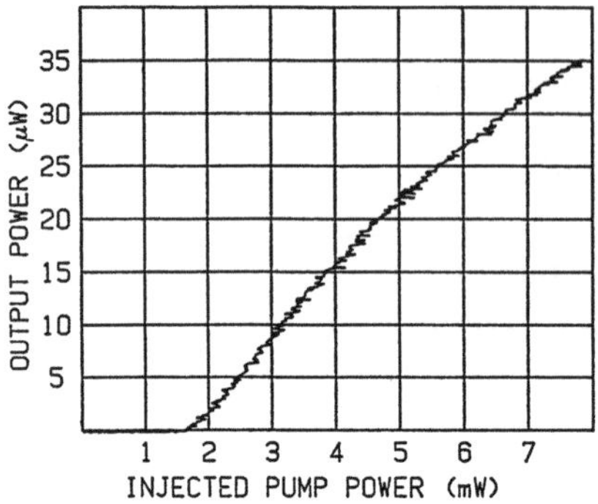

Figure 2a : Laser power versus injected pump power.

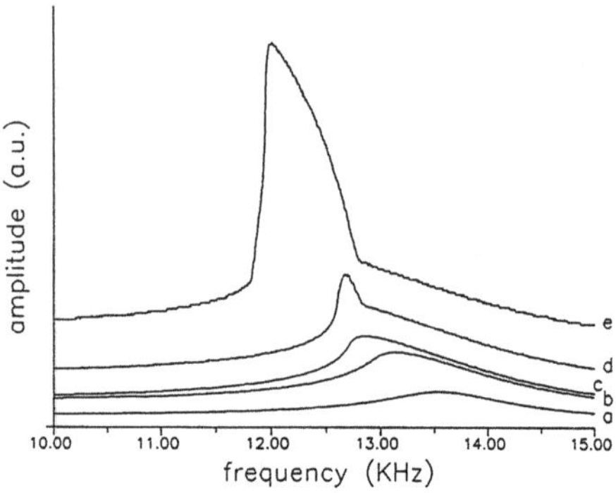

Figure 2b : Resonance curves for different modulation amplitudes (average pump power 4.6 mW). a) 0.18 mW b) 0.36 mW c) 0.45 mW d) 0.50 mW e) 0.72 mW.

the pump wavelength). The cavity is completed by a plane output mirror of 99.5 % reflectivity. The fibre ends are set against the mirror to minimize the losses. A bandpass filter could be placed in front of the silicium photodiode detector which monitors the fibre laser output.

With this arrangement, the injected pump power at threshold could be as low as 2mW (Fig. 2a). Although the high reflectivity coupling mirror reduced the output power to about 50 µW, this situation was favourable for dynamic experiments since pump parameters (ratio between the average pump power and the threshold pump power) up to 4 should be reached. The pump laser power could be sinusoidally modulated up to a frequency of about 100 kHz. For low values of modulation depth and when the laser is continuously pumped above threshold, the resonance curves shown in Fig. 2b could be obtained. The laser exhibits a low frequency resonance at about 13 kHz. The resonances broaden and shift towards low frequency as the modulation is increased. By measuring the evolution of this frequency and the width of this resonance versus the average pump power, it has been possible to evaluate the cavity losses (8 %) and the upper level relaxation time (380 µs) within a single mode two-level model. It predicts that the square of resonance frequency and the full width at half maximun are linear functions of the pump power [2], in good agreement with the experimental findings. Under high pump modulation, the laser exhibits several non-linear effects, including harmonic and subharmonic resonances and hysteresis in the response to modulation.

For large modulation amplitudes a period-doubling cascade leading to chaos has been observed as in a previous work [3] but contrary to [3], it was not necessary to modulate so strongly the pump power since in our experiment, the instantaneous pump power always remained above threshold. Fig. 3 shows the evolution of the response of the laser

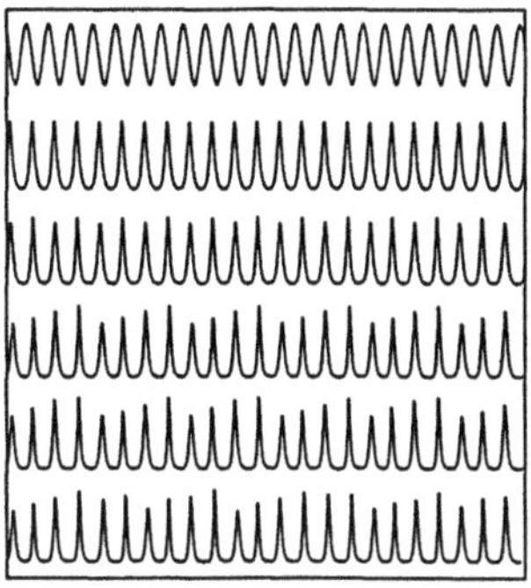

Figure 3 : Evolution of the response of the laser versus the modulation frequency. Average pump power 4.6 mW, threshold 2 mW, modulation amplitude 2.5 mW, : a) modulation; laser responses : b) 14 kHz-T, c) 13.5 kHz-2T, d) 12.7 kHz-4T, e) 12.25 kHz-8T, f) 11.8 kHz-chaotic regime.

as the modulation frequency is swept with constant amplitude of modulation. The response of the laser evolves in a T, 2T, 4T, 8T, 16T sequence and eventually becomes erratic. The occurrence of an erratic regime after a succession of such period-doublings is a good indication of deterministic chaos.

The evolution of the laser ouput may be clearly seen on a bifurcation diagram such as that used in the CO_2 laser experiments [4]. A sample and hold module synchronised on the modulation stores a signal proportional to the laser output power at some defined time of the modulation period. This sampling unit delivers a single valued output when the laser signal is T-periodic and n different values when its period is nT. Fig. 4 shows an example of a bifurcation diagram obtained by this technique when the modulation frequency was swept from high to low frequencies around the resonance with a constant amplitude. A period-doubling cascade T, 2T, 4T, 8T and 16T appears on the high frequency side between 14 and 12 kHz leading to a chaotic regime from 12 to 11.75 kHz. At that frequency, a 10T window occurs from 11.75 to 11.5 kHz and a second chaotic region is observed between 11.5 and 10.6 kHz. It is interrupted at 10.6 kHz by a 4T-periodic response. As there is no continuity between the branches of the chaotic region and those of the 4T region, we could conclude that the chaotic attractor disappeared through a boundary crisis [5].

The dynamical effects exhibited by this laser present some strong similarities with those of the single mode CO_2 laser under periodic loss or frequency modulation [4] but significant differences appear at large modulation amplitudes. Additional experiments and modelization of the fibre laser are required to check whether these differences were due to the spatial inhomogeneity of the pump, to the large number of longitudinal modes or to the fact the fibre laser could be better described by a four-level model.

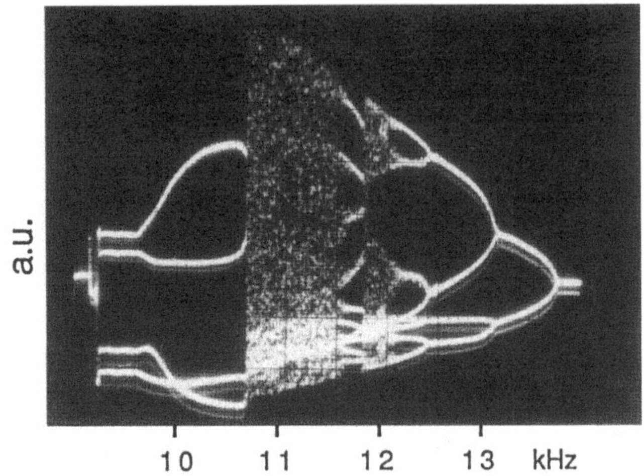

Figure 4 : Bifurcation diagram obtained using a decreasing sweep of the modulation frequency . Average pump power 4.6 mW, amplitude of modulation 2.5 mW, threshold 2 mW.

References

[1] L. REEKIE, R.J. MEARS, S.B. POOLE and D.N. PAYNE, IEEE J. Lightwave Technol. **LT-4** (1986) 956.

[2] O. SVELTO, Principles of lasers, 1st Ed., Plenum Press, New York (1976).

[3] M.W. PHILLIPS, H. GONG, A.J. FERGUSON and D.C. HANNA, Opt. Comm., **61** (1987) 215.

[4] D. DANGOISSE, P. GLORIEUX and D. HENNEQUIN, Phys. Rev. A, **36** (1987) 4775.

[5] C. GREBOGI, E. OTT and J.A. YORKE, Phys. Rev. Lett., **48** (1982) 1507. C. GREBOGI, E. OTT and J.A. YORKE, Physica (Amsterdam), **7D** (1983) 181.

Amplitude and Phase Dynamics in a Four-Level Model of a Coherently Pumped Laser

M. Arjona[1], *F. Laguarta*[1], *J. Pujol*[1], *and R. Corbalán*[2]

[1]Departament d'Optica i Optometría, Universitat Politècnica de Catalunya, E-08222 Terrassa (Barcelona), Spain
[2]Departament de Física, Universitat Autònoma de Barcelona, E-08193 Bellaterra (Barcelona), Spain

Abstract. We study the amplitude and phase dynamics in a coherently-pumped laser model that takes the vector character of the light fields into account. As typical in FIR laser experiments, we consider linearly polarized pump and generated laser beams. Phase diagrams as a function of cavity detuning and amplitude of the resonant pump field have been obtained, which show domains of stationary, oscillatory and chaotic output. We show that the laser is much more stable when the polarizations of the pump and generated laser beams are parallel than when they are orthogonal. In the latter case the laser exhibits Lorenz-like dynamics.

1. Introduction

Recent experimental studies [1] of self pulsing in an optically pumped single-mode FIR ammonia laser using heterodyne detection of the laser field have allowed us to study the dynamics not only of the amplitude (i.e. the intensity) but also of the phase. These studies have stimulated several analys of the amplitude and phase dynamics in different laser models. Thus, since the ammonia laser showed Lorenz-like behavior in a certain range of parameters, Zeghlache et al. [2] and Vilaseca et al. [3] have analyzed the predictions of the five-equations complex Lorenz-Haken model of an incoherently pumped single-mode laser operating with a homogeneously broadened two-level medium. However, since the ammonia laser is a priori a much more complex system than the one considered in the Lorenz-Haken model, Corbalán et al. [4-6] have analysed a more realistic 217-equations model of a Doppler-broadened coherently pumped three-level laser. The predictions of this high-dimensional model showed a remarkable qualitative agreement with the experimental observations of Lorenz-like behavior in the real laser. However, in spite of the success of this model, it is unable to describe the influence that another important factor, namely the polarization of the pump and generated laser fields, may have on the dynamics of the ammonia laser. In the real laser [1] the pump and generated fields, which couple to adjacent transitions in a Λ-type three-level scheme, are orthogonally polarized fields, and this aspect cannot be represented by three-nondegenerate-level laser models [4-6]. Any model that takes the vector character of the light field into account necessitates that level degeneracy related to angular momenta of the molecular states should be incorporated.

Springer Proceedings in Physics, Vol. 55
Nonlinear Dynamics and Quantum Phenomena in Optical Systems
Editors: R. Vilaseca, R. Corbalán

Owing to the high angular momenta of the states involved in the FIR ammonia laser it seems extremely hard to describe the experimental system. Instead, we recently initiated the study of the influence of light polarization on the dynamics of coherently pumped lasers using the simplest molecular transition scheme which can give us a physical insight into the light-polarization-induced effects [7]. We focus here on the amplitude and the phase dynamics predicted by the four-level laser model presented in [7].

2. Model

Figure 1 illustrates the $J_a=0 \to J_b=1 \to J_c=0$ transition scheme assumed for the homogeneously-broadened gain medium of an optically pumped ring laser. It is the simplest level scheme that enables one to consider interaction with pump $E_2(z,t)$ and generated $E_1(z,t)$ fields composed by arbitrary superpositions of right and left circulary polarized components. Both fields are considered as uniform plane monochromatic waves propagating along the optical (z) axis of the ring resonator

$$\boldsymbol{E}_j(z,t)=\sum_{\mu=\pm} e^{\mu} A^{\mu j}(t)\exp\{i[k_j z-\nu^{\mu j}t-\phi^{\mu j}(t)]\}/2+c.c \qquad (j=1,2)$$

where the unit vectors $e^{\pm} = \mp(e_x \pm i\, e_y)/\sqrt{2}$. The z axis is also chosen to be the quantization axis, so that only four levels are involved (the sublevel $(J_b=1,0)$ does not couple to the fields), which we shall simply denote as a, +, - and c. The A_2^{μ}, ν_2^{μ}, φ_2^{μ} characteristics of the pump beam are considered as constant in time control parameters; the unknown amplitudes A_1^{μ} and phases φ_1^{μ} of the generated beam are slowly varying functions of time. We assume that the pump beam drives the a-b transition only, and the generated field the b-c transition only. The field-matter coupling is characterized by four real Rabi frequencies (see Fig. 1) defined as $\beta^{\mu}=d_2A^{\mu}_2/2\hbar$ and $\alpha^{\mu}=d_1A^{\mu}_1/2\hbar$ with d_j $(j=1,2)$ being molecular transition electric dipole moments.

In the rotating wave approximation the dynamics of the system molecules plus pump and generated fields in the ring cavity is governed by a set of eighteen real first-order differential equations given by equations (2) and (3) in [7], which we solve numerically using a seventh-eighth-order Runge-Kutta routine.

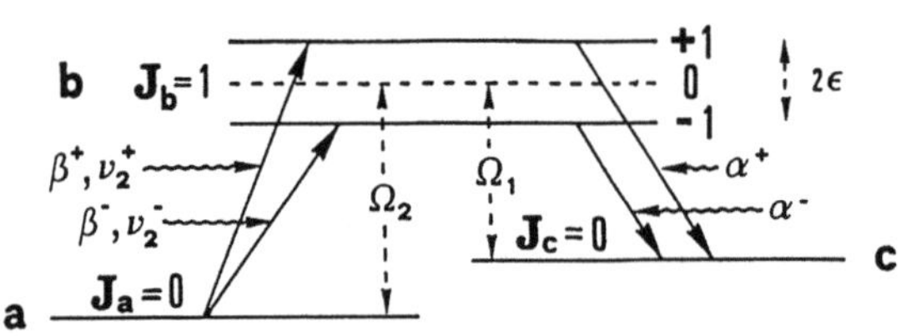

Figure 1. Schematic representation of the level system and laser fields considered in this work. $2\beta^{\pm}$ represent the pump Rabi frequencies and $2\alpha^{\pm}$ the generated ones.

3. Numerical Results and Discussions

As in the experiment [1] we will restrict ourselves here to studying the case of linearly polarized pump and generated laser fields, with either parallel or orthogonal polarizations. For these cases one has $\epsilon=0$, $v^{+}_{j}=v^{-}_{j}$, $\varphi_j^{+} = \varphi_j^{-} = \Phi$ $(j=1,2)$; $\alpha^{+}=\alpha^{-}=\alpha$, and either $\beta^{+}=\beta^{-}=\beta$ for the parallel case, or $\beta^{+}=-\beta^{-}=\beta$ for the orthogonal case. The case with an on-resonant pump and line-centre emission was presented in [7], where the analysis was limited to intensity. However, due to the absence of dispersive effects in the resonant case, and taking into account the heteroclinic character of the attractor (see Figs. 3 and 4 in ref. [7]), we can conclude that in a frame rotating at the frequency ω_c of the empty cavity (that we will adopt in the following as the reference frequency, i.e. $v^{\pm}_{1}=\omega_c$) the dynamics of the phase in the resonant case is limited to alternating jumps of π radians [3]. In this paper we extend our previous study [7] by considering the influence on the amplitude and phase dynamics of the detuning $\Delta^{c}=\omega_c-\Omega_1$ between cavity and amplifying transition line-centre frequencies. We consider the cavity detuning Δ^{c} and the pump field amplitude β as the control parameters. The same value is adopted for the relaxation rate γ_{+-} of the coherence between the two upper levels and for the relaxation rate $\gamma_{\perp}$ of the coherences associated to the b-c lasing transition. All the other parameters are fixed at the values reported in [7], which were appropriate for the experimental conditions of Weiss and Brock [8] in their study of the 81-μm emission from a $^{14}NH_3$ laser operating at a pressure of 10-Pa.

At $\Delta^{c}=0$ we found two attractors whose bifurcation diagrams are shown in Fig. 2(a) of ref. [7] for the case of orthogonal polarizations for pump and generated laser fields, and in Fig. 2(c) for parallel polarizations. One of these attractors appears at the first laser threshold as a stable fixed point associated to the steady lasing state and, if β is increased adiabatically, remains so for all the range of β values here investigated. Thus, all of the dynamics of our system is associated with the second attractor. The laser reaches this second attractor starting from the nonlasing $\alpha=0$ state and switching abruptly the pump beam.

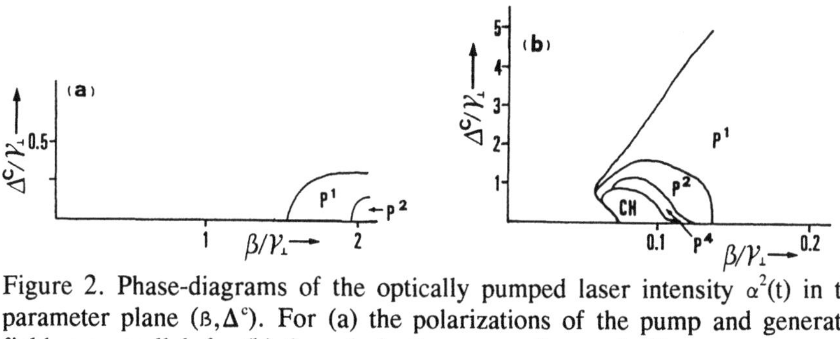

Figure 2. Phase-diagrams of the optically pumped laser intensity $\alpha^{2}(t)$ in the parameter plane (β,Δ^{c}). For (a) the polarizations of the pump and generated fields are parallel; for (b) the polarizations are orthogonal. The parameters used in the calculations are referred to in the text. CH represents chaotic emission and P^{n} represents periodic pulsing of period n.

The continuation in the (β,Δ^c) parameter plane of the boundaries between regions of different dynamic behavior associated with the second attractor is shown in Fig. 2(a) for the case of parallel polarizations of the pump and generated fields, and in Fig. 2(b) for the case of orthogonal polarizations. A comparison of Fig. 2(a) and 2(b) shows that the optically pumped laser is much more stable operating with parallel polarizations than with the orthogonal configuration. In fact, while the instability pump threshold is $\beta/\gamma_\perp=0.0743$ at $\Delta^c=0$ for chaotic emission (CH) in Fig. 2(b), it is increased to $\beta/\gamma_\perp=1.54$ at $\Delta^c=0$ for regular period-one pulsing (P^1) in Fig. 2(a). A global feature of Fig. 2 is a simplification of the dynamics when the cavity detuning is sufficiently increased. This prediction of the present model agrees with well-known results obtained for two-level lasers [2] and for coherently pumped three-level lasers [9]. However, the optically pumped laser with orthogonally polarized fields appears to be less sensitive than these other lasers to the stabilizing effect of detuning. In fact, note that the range of Δ^c values considered in Fig. 2(b) is about one order of magnitude larger than in the other cases (see also Fig. 2 in ref. [9]). Another remarkable feature in Fig. 2(b) is the unstabilization influence of small detuning. Thus, the instability pump threshold is reduced from the value previously mentioned for $\Delta^c=0$, to $\beta/\gamma_\perp=0.0562$ at $\Delta^c/\gamma_\perp=0.75$. This

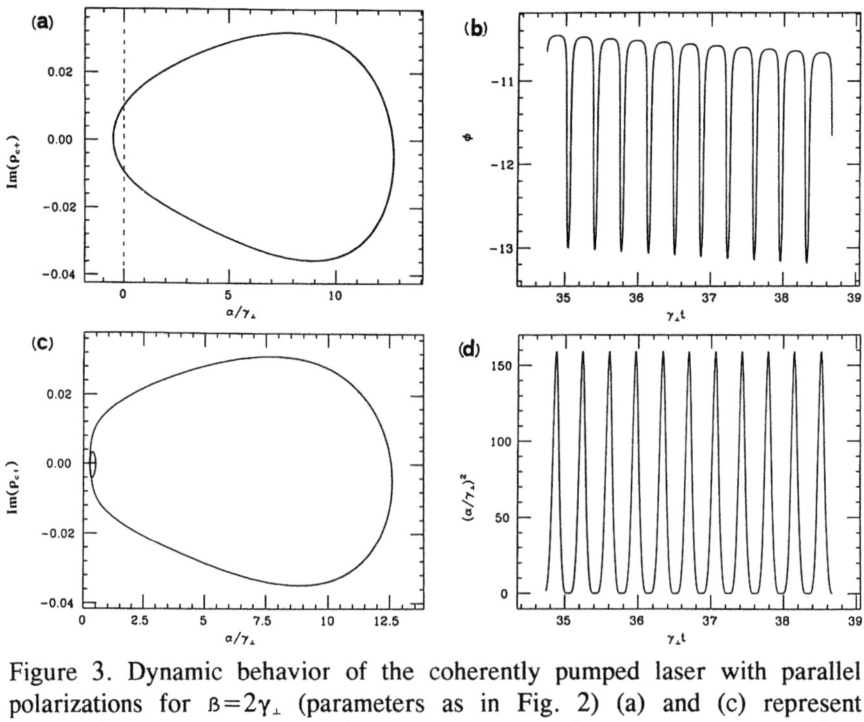

Figure 3. Dynamic behavior of the coherently pumped laser with parallel polarizations for $\beta=2\gamma_\perp$ (parameters as in Fig. 2) (a) and (c) represent attractor's projections on the $(\alpha, \mathrm{Im}(\rho_{c+}))$ plane, (b) and (d) represent time evolution of Φ and α^2, respectively. In (a) $\Delta^c=0$, in (b)-(d) $\Delta^c=0.1\gamma_\perp$.

phenomenon was also found in the three-level laser [9], but does not appear in the two-level laser [2] nor in the parallel configuration (see Fig. 2(a)).

Figure 3 displays results for the parallel configuration. Fig. 3(a) shows that the regular period-two pulsing regime (P^2) of Fig. 2(a) at ($\beta/\gamma_\perp=2$, $\Delta^c=0$) corresponds to an asymmetric single-loop solution (i.e. with unequal domains of positive and negative α values) of the model equations. Keeping the pump amplitude fixed at $\beta/\gamma_\perp=2$ and introducing a small cavity detuning, $\Delta^c=0.1\gamma_\perp$, the intensity continues exhibiting a period-two pulsing regime [Fig. 3(d)]; the dynamic evolution of the phase [Fig. 3(b)] shows a small global slope, which is a consequence of dispersive effects, and sharp variations of about 0.8π radians which occur each time the amplitude $\alpha(t)$ comes very close to zero. Note that the corresponding projection of the attractor [Fig. 3(c)] shows now two loops, with $\alpha(t)$ remaining always positive. However, according to the physical interpretation given in [3], the attractors at $\Delta^c=0$ and $\Delta^c=0.1\gamma_\perp$ have the same structure. The fact that $\alpha(t)$ does not change sign in the detuned case is compensated by the sharp phase variations.

In the case of orthogonal polarizations Fig. 2(b) shows that keeping the pump amplitude β fixed at a value well beyond the instability threshold, and increasing the detuning from $\Delta^c=0$, the system follows an inverse period-doubling sequence out of chaos. This sequence of bifurcations is also characteristic of the Lorenz-Haken model [2]. Several intensity pulsation waveforms at different steps of this sequence are shown in Figs. 4(d), 4(e), and 4(f). The respective two-dimensional (α, $\mathrm{Im}(\rho_{c+})$) projections of the attractor are given in Figs. 4(a), 4(b), and 4(c). The associated phase dynamics are displayed in Figs. 4(g), 4(h), and 4(i), respectively. The case with $\Delta^c=0$ was presented in Fig. 3 of ref. [7], where the attractor was of the heteroclinic-type [1] or Lorenz-type, and also the sequence of the bifurcations as a function of pump amplitude showed remarkable similarities with Lorenz's model predictions, although they appeared in our model at a reduced pump threshold.

Although, the shape of the attractor's projections in Fig. 4 apparently indicates that it is not another of the heteroclinic type, again the existence of phase jumps, which decrease from π radians by increasing detuning, assure [3] the heteroclinic character of the dynamics also in the detuned case. One can notice in Fig. 4 that, again, as in the Lorenz-Haken model [2], the steep transitions in the phase evolution occur between pulses and depend on the height of the preceding intensity peak, with larger jumps occurring after larger peaks.

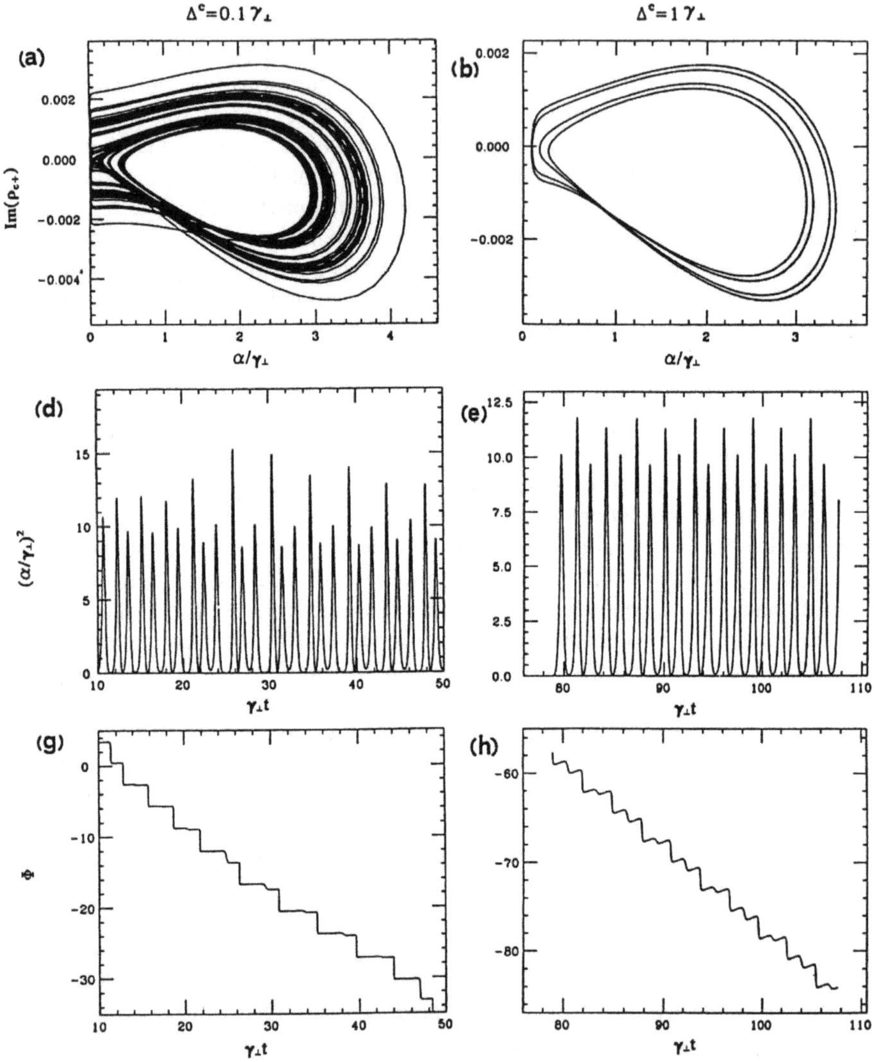

Figure 4. Dynamic behavior of the coherently pumped laser, when the polarizations of the pump and generated fields are orthogonal, for $ß=0.09\gamma_\perp$ as the cavity tuning is swept (parameters as in Fig. 2). The values of $\Delta^c/\gamma_\perp$ are 0.1 dor (a), (d) and (g) 1 for (b), (e) and (h); and 1.25 for (c), (f) and (i). (a)-(c) represent attractor's projections on the (α, Im(ρ_{c+})) plane. (d)-(f) and (g)-(i) represent time evolutions of α^2 and Φ, respectively.

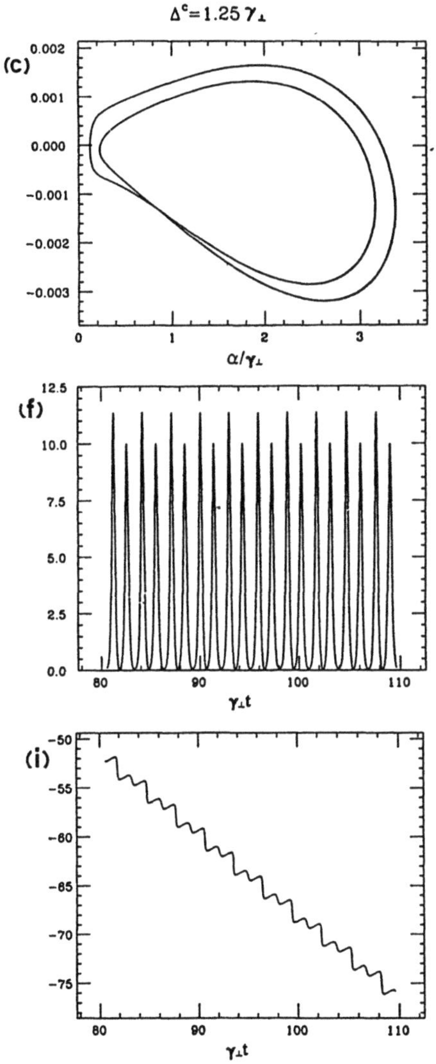
Δ^c=1.25 γ⊥
(c)
α/γ⊥
(f)
γ⊥t
(i)
γ⊥t

4. Acknowledgements

We would like to thank R. Vilaseca for helpful discussions. This work was supported by the Dirección General de Investigación Científica y Técnica (Spain). Project N° PB86-0650-C03-01.

5. References

1. C.O. Weiss, N.B. Abraham, and U. Hübbner, Phys.Rev.Lett.61 (1988) 1587
2. H. Zeghlache and P. Mandel, J.Opt.Soc.Am.B4 (1985) 18; H. Zeghlache, P. Mandel, N.B. Abraham, and C.O. Weiss, Phys.Rev.A 38 (1988) 3128
3. R. Vilaseca, G.J. de Valcárcel, and E. Roldán, Phys. Rev. A 41 (1990) 5269
4. R. Corbalán, F. Laguarta, J. Pujol, and R. Vilaseca, Opt.Commun.71 (1989) 290
5. E. Roldan, G.J. de Valcárcel, R. Vilaseca, F. Silva, J. Pujol, R. Corbalán, and F. Laguarta, Opt.Commun.73 (1989) 506
6. R. Vilaseca, E. Roldán, G.J. Valcárcel, and R. Corbalán, "Comparison between two-level and inhomogeneously-broadened FIR three-level laser dynamics" in "OSA Proceedings on Nonlinear Dynamics in Optical Systems", edited by N.B. Abraham, E.M. Garmire, and P. Mandel (Optical Society of America, Washington, D.C.) in press
7. M. Arjona, R. Corbalán, F. Laguarta, J. Pujol, and R. Vilaseca, Phys.Rev.A 41 (1990) 6559
8. C.O. Weiss and J. Brock, Phys.Rev.Lett.57 (1986) 2804
9. J. Pujol, F. Laguarta, R. Corbalán, and R. Vilaseca, J.Phys.D:Applied Physics 21 (1988) S180

Linear Stability Analysis of a Resonant Cascade Laser

E. Roldán[1], *G.J. de Valcárcel*[2], *and R. Vilaseca*[1]

[1]Departament d'Optica, Universitat de València, Dr. Moliner 50, E-46100 Burjassot (València), Spain

[2]Departament de Física Aplicada, Universitat Politècnica de València, E-46020 València, Spain

Abstract. The linear stability analysis of the stationary solution of a cascade laser, in which two fields coupled to two adjacent transitions are generated, is performed in the particular case of exact resonance and equal parameters for both fields and both transitions. Analytical expressions are obtained which show that the second laser threshold can be lower than for the single-field two-level laser.

In the last few years the dynamic behaviour of different classes of lasers has been investigated in detail. In particular, two-level lasers, optically-pumped lasers and lasers with saturable absorber, parameter modulation, feedback or injected signal show a large variety of periodic and chaotic regimes [1].

In the present work the linear stability analysis of a new class of laser, which to our knowledge has not been investigated so far, is reported. We consider a cascade laser in which two fields $\mathbf{E}_1$ and $\mathbf{E}_2$, coupled through electric dipole interaction to transitions 0-1 and 2-0, respectively (Fig.1), are generated inside the laser cavity. Each level i (i=0,1,2) is populated through an incoherent pumping mechanism according to a constant rate λ_i and is affected by relaxation processes characterized by a rate γ_i. Inhomogeneous broadening is ignored.

Considering single-mode unidirectional emission (ring cavity) and in the usual plane-wave and uniform-field limits [2], the total field inside the laser cavity can be described, within the semiclassical theory, as

$$\mathbf{E}(z,t) = \mathbf{E}_1(z,t) + \mathbf{E}_2(z,t) \tag{1}$$

where z is the propagation direction along the laser cavity, and

$$\mathbf{E}_i(z,t) = 1/2\ \mathbf{e}_i E_i(t)\ \exp[-i(\upsilon_i t+\phi_i(t)-k_i z)] + \text{c.c.} \tag{2}$$

(i=1,2), where $\mathbf{e}_i, E_i(t), k_i, \upsilon_i/2\pi$ and $\phi_i(t)$ represent the polarization vector, slowly varying real field amplitude, wavenumber, frequency and phase of the field $\mathbf{E}_i$. The detuning of $\mathbf{E}_2$ with respect to the transition 2-0 of frequency $\omega_{20}/2\pi$ is denoted as $\delta=\upsilon_2-\omega_{20}$, and that of the two-photon transition with respect to the transition 2-1 (of frequency $\omega_{21}/2\pi$) by $\varepsilon=\upsilon_1+\upsilon_2-\omega_{21}=\delta+(\upsilon_1-\omega_{01})$. (Fig.1). We neglect cross interaction between the field $\mathbf{E}_1$ and the transition ω_{20} and between $\mathbf{E}_2$ and the transition ω_{01} (non-degenerate case).

By using the density-matrix formalism and in the usual SVEA and RWA approximations [2], the laser equations can be written in the following way:

Springer Proceedings in Physics, Vol. 55
Nonlinear Dynamics and Quantum Phenomena in Optical Systems
Editors: R. Vilaseca, R. Corbalán

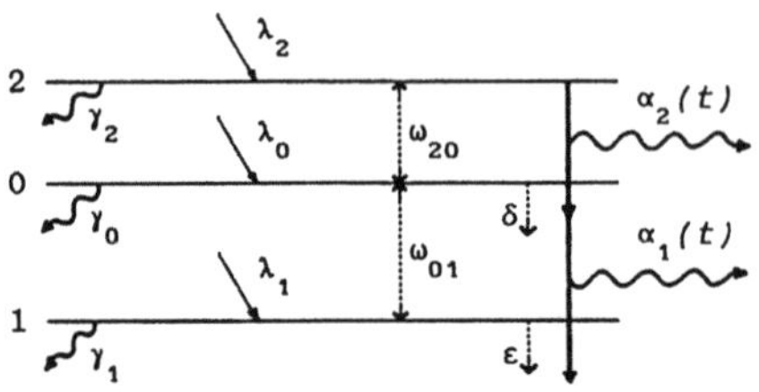

Fig.1.- Scheme of the level system and generated fields. See text for notation.

$$
\begin{aligned}
\dot{\rho}_{00} &= -\gamma_0\rho_{00} + \lambda_0 + i\alpha_2(\rho_{20}-\rho_{02}) - i\alpha_1(\rho_{01}-\rho_{10}) \\
\dot{\rho}_{11} &= -\gamma_1\rho_{11} + \lambda_1 + i\alpha_1(\rho_{01}-\rho_{10}) \\
\dot{\rho}_{22} &= -\gamma_2\rho_{22} + \lambda_2 - i\alpha_2(\rho_{20}-\rho_{02}) \\
\dot{\rho}_{20} &= -[\Gamma_{20} - i(\dot{\phi}_2+\delta)]\,\rho_{20} - i\,\alpha_2(\rho_{22}-\rho_{00}) - i\,\alpha_1\rho_{21} \\
\dot{\rho}_{01} &= -[\Gamma_{01} - i(\dot{\phi}_1+\varepsilon-\delta)]\,\rho_{01} - i\,\alpha_1(\rho_{00}-\rho_{11}) + i\,\alpha_2\rho_{21} \\
\dot{\rho}_{21} &= -[\Gamma_{21} - i(\dot{\phi}_1+\dot{\phi}_2+\varepsilon)]\,\rho_{21} + i\,\alpha_2\rho_{01} - i\,\alpha_1\rho_{20} \\
\dot{\alpha}_1 &= -\kappa_1\alpha_1 - g_1 \mathrm{Im}\rho_{01} \quad ; \qquad \dot{\phi}_1 = -g_1 \mathrm{Re}\rho_{01}/\alpha_1 \\
\dot{\alpha}_2 &= -\kappa_2\alpha_2 - g_2 \mathrm{Im}\rho_{20} \quad ; \qquad \dot{\phi}_2 = -g_2 \mathrm{Re}\rho_{20}/\alpha_2
\end{aligned}
\tag{3}
$$

where ρ_{ii} represents the population of level i and ρ_{ij} the coherence associated with the transition i-j (i,j=1,0,2); Γ_{ij} is the relaxation rate for the coherence ρ_{ij} ($\Gamma_{ij} \geq (\gamma_i+\gamma_j)/2$); 2 α_i is the Rabi frequency for the field $\mathbf{E}_i$:

$$\alpha_i(t) = (\mu_{0i}.e_i/2h)\ E_i(t) \tag{4}$$

(where μ_{0i} represents the electric dipole matrix element between levels 0 and i, which can be chosen to be real); and κ_i and g_i represent the cavity losses rate and gain parameter, respectively, for the field $\mathbf{E}_i$. Thus, the first three (real) equations of (3) describe the level population evolution, the next two (complex) equations describe the electric dipole induced on each allowed transition, the sixth (complex) equation describes the two-photon coherence and the last four (real) equations describe the field evolution.

Since the equations for ϕ_1 and ϕ_2 can be introduced into the equations for ρ_{20}, ρ_{01} and ρ_{21}, the system (3) represents a set of eleven coupled real equations. In the limit of small δ and ε (Fig.1) the system can be properly called a cascade laser, whereas in the limit of large δ ($|\delta| >> \Gamma_{20},\Gamma_{01}$) and small ε it represents a two-photon laser. In this work we deal only with the first case.

To obtain simple analytical expressions let us consider the particular case of a fully-tuned cascade laser, $\delta=\varepsilon=0$, with identical parameters for both 2-0 and 0-1 transitions, $\gamma_0=\gamma_1=\gamma_2\equiv\gamma_{||}$, $\Gamma_{01}=\Gamma_{20}\equiv\gamma_\perp$ and $\Gamma_{21}\equiv\Gamma$, and for both fields, $\kappa_1=\kappa_2\equiv\kappa$

and $g_1=g_2\equiv g$. If we define new variables:

$$D^{\pm} \equiv (\rho_{22}-\rho_{00}) \pm (\rho_{00}-\rho_{11})\ ;\ \lambda^{\pm} \equiv (\lambda_2-\lambda_0) \pm (\lambda_0-\lambda_1)$$
$$y^{\pm} \equiv \mathrm{Im}\,\rho_{20} \pm \mathrm{Im}\,\rho_{01}\ ;\ \alpha^{\pm} \equiv \alpha_1 \pm \alpha_2\ ;\ x_{21} \equiv \mathrm{Re}\,\rho_{21} \tag{5}$$

then eqs.(3) reduce to the following set of seven real equations:

$$\dot{\alpha}^+ = -\kappa\alpha^+ - gy^+\ ;\ \dot{\alpha}^- = -\kappa\alpha^- - gy^-$$
$$\dot{y}^+ = -\gamma_\perp y^+ - (\alpha^+D^+ - \alpha^-D^-)/2 - x_{21}\,\alpha^-$$
$$\dot{y}^- = -\gamma_\perp y^- + (\alpha^+D^- - \alpha^-D^+)/2 + x_{21}\,\alpha^+$$
$$\dot{x}_{21} = -\Gamma x_{21} + (\alpha^-y^+ - \alpha^+y^-)/2 \tag{6}$$
$$\dot{D}^+ = -\gamma_{||}D^+ + \lambda^+ + (\alpha^+y^+ + \alpha^-y^-)$$
$$\dot{D}^- = -\gamma_{||}D^- + \lambda^- - 3(\alpha^-y^+ + \alpha^+y^-)\ .$$

The rest of the variables, i.e. $\mathrm{Re}\rho_{20}$, $\mathrm{Re}\rho_{01}$, $\mathrm{Im}\rho_{21}$, ϕ_1 and ϕ_2, can be taken as zero. In the case of equal pumping for both transitions, i.e. $\lambda^-=0$, the nontrivial stationary solution of eqs.(6) is given by

$$\alpha^+ = [2\gamma_{||}\gamma_\perp(r-1)]^{1/2}\ ,\quad (\text{with } r = \lambda^+ g/2\kappa\gamma_{||}\gamma_\perp)$$
$$D^+ = 2\gamma_\perp\kappa/g\ ;\ y^+ = -\kappa\alpha^+/g\ ;\ \alpha^- = y^- = D^- = x_{21} = 0\ . \tag{7}$$

This solution exists for $r \geq 1$ ($r=1$ is the first laser threshold) and entails an equality between both generated fields, namely $\alpha_1=\alpha_2=\alpha^+/2$. This solution for each field is formally identical to the stationary solution for a homogeneously broadened single-field two-level laser [2,3]. Its stability conditions, however, are different, as we show next.

To perform a linear stability analysis of the solution (7) -let us denote this solution as $X^o\equiv(\alpha^+, y^+, D^+, \alpha^-, y^-, D^-, x_{21})$- we study, as usual, the time evolution of an arbitrarily small perturbation $\delta X(t)=\delta X(0)\exp(\mu t)$ added to X^o. By introducing $X^o+\delta X(t)$ into the eqs.(6) and retaining only the linear terms with respect to δX, the following linear algebraic equation is obtained:

$$\begin{bmatrix} -\kappa & -g & 0 & 0 & 0 & 0 & 0 \\ -D^+/2 & -\gamma_\perp & -\alpha^+/2 & 0 & 0 & 0 & 0 \\ y^+ & \alpha^+ & -\gamma_{||} & 0 & 0 & 0 & 0 \\ 0 & 0 & 0 & -\kappa & -g & 0 & 0 \\ 0 & 0 & 0 & -D^+/2 & -\gamma_\perp & \alpha^+/2 & \alpha^+ \\ 0 & 0 & 0 & -3y^+ & -3\alpha^+ & -\gamma_{||} & 0 \\ 0 & 0 & 0 & y^+/2 & -\alpha^+/2 & 0 & -\Gamma \end{bmatrix} \begin{bmatrix} \delta\alpha^+ \\ \delta y^+ \\ \delta D^+ \\ \delta\alpha^- \\ \delta y^- \\ \delta D^- \\ \delta x_{21} \end{bmatrix} = \mu \begin{bmatrix} \delta\alpha^+ \\ \delta y+ \\ \delta D^+ \\ \delta\alpha^- \\ \delta y^- \\ \delta D^- \\ \delta x_{21} \end{bmatrix} \tag{8}$$

Hence μ represents the eigenvalues of the 7x7 matrix of eq.(8). Since this matrix is composed of two diagonal submatrices of 3x3 and 4x4 elements, the characteristic equation is of the form A·B=0, where the polynomials A and B can be expressed as

$$A = \mu^3+c_1\mu^2+c_2\mu+c_3$$
$$B = \mu^4+d_1\mu^3+d_2\mu^2+d_3\mu+d_4 \qquad (9)$$

where the constant coefficients c_i and d_i can be easily deduced from eq.(8).

When all the real parts of μ, solutions of A·B=0, are negative, the stationary solution will be stable, whereas when at least one of these real parts is positive, then the solution will be unstable. By applying the Hurwitz criterion to A and B separately, the following necessary conditions for the appearance of unstable behaviour are obtained:

a) From A: The same conditions as in the Lorenz-Haken model [3] are found:

$$\kappa > \kappa_A \doteq \gamma_{\|}+\gamma_{\perp} \qquad (10\ a)$$

$$r \geq r_A = (\kappa/\ \gamma_{\perp})(\kappa+\gamma_{\|}+3\ \gamma_{\perp})\ /\ (\kappa-\gamma_{\|}-\ \gamma_{\perp}) \qquad (10\ b)$$

Expr. (10 a) represents the "bad cavity" condition, and r_A in (10 b) defines the "second laser threshold" or Lorenz threshold.

b) From B: Two independent conditions are found:

B1.–

$$\kappa > \kappa_{B1} = (3\ \gamma_{\|}+4\ \gamma_{\perp}+\Gamma)/2 \qquad (11\ a)$$

$$r \geq r_{B1} = 1+ (\gamma_{\|}+\Gamma)\ \frac{\gamma_{\|}\Gamma+(\kappa+\gamma_{\perp})(\kappa+\gamma_{\|}+\gamma_{\perp}+\Gamma)}{\gamma_{\|}\gamma_{\perp}(2\kappa-3\gamma_{\|}-4\gamma_{\perp}-\Gamma)}\ . \qquad (11\ b)$$

Since $\kappa_{B1} \geq 2\kappa_A$ and $r_{B1} > r_A$, these conditions are less restrictive than conditions (10) and therefore they do not increase the domain of instability of the laser.

B2.- This condition is given by

$$H_3 = a(r-1)^2 + b(r-1) + c \quad < 0 \qquad (12)$$

where a is proportional to ($\kappa_{B1}-\kappa$) and b and c are functions of $\gamma_{\|}$, $\gamma_{\perp}$, Γ and κ (c is always positive).

There are two possibilities for expression (12) to be verified as a function of a, b and c, namely :

(i) if $\kappa>\kappa_{B1}$ (a<0) -condition (11 a)- one finds instabilities if $r>r_{B2-}$ (independently of the values of b and c)

(ii) if $\kappa<\kappa_{B1}$ (a>0), b<0 and $b^2>4ac$, the condition for instabilities reads

$$r_{B2-} \leq r \leq r_{B2+}\ . \qquad (13)$$

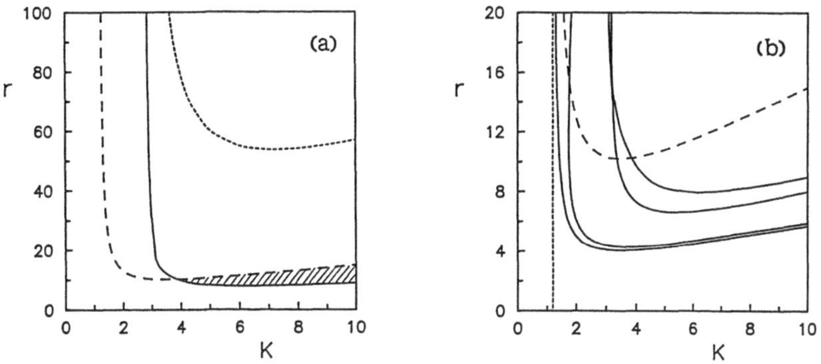

Fig.2.- (a) Dependence of the thresholds r_A (lower dashed line), r_{B1} (upper dashed line) and r_{B2} (continuous line) on the cavity losses $\kappa=\kappa/\gamma_\perp$, for $\gamma_\parallel=0.2\gamma_\perp$ and $\Gamma=\gamma_\perp$. In the domains above any one of these curves the stationary solution can be unstable. (b) Influence of $\Gamma=\Gamma/\gamma_\perp$ on the threshold r_{B2} (continuous line) for $\gamma_\parallel=0.2\gamma_\perp$. The four curves correspond, from top to bottom, to Γ=1,2,10 and 10^5, respectively. The dashed line corresponds to the Lorenz threshold r_A. The vertical line corresponds to the bad cavity limit.

If b is positive, the condition (12) cannot be fulfilled (we do not give expressions for $r_{B2\pm}$ for the sake of brevity).

Fig.2a shows the dependence of the three obtained thresholds, r_A, r_{B1} and r_{B2} on κ, for a particular set of parameters. Compared to the single-field two-level laser, two features can be pointed out. First, the bad-cavity condition still holds, and second, a domain exists (lined area) in which instabilities can appear for a pump strength below the standard second laser threshold r_A. Fig.2b shows the influence of Γ on the threshold r_{B2}. It is found that r_{B2-} diminishes when Γ increases. In the limit $\Gamma/\gamma_\perp\rightarrow\infty$ it is given by

$$r_{B2-} = (r_A+2)/3 \tag{14}$$

which for $\gamma_\parallel=0$ reaches its minimum value, $r_{B2-}=11/3$, which is 27/11 times smaller than the minimum value -9- of r_A.

We have also obtained the expression of the frequency of oscillation, $\Omega/2\pi$, at the Hopf bifurcation associated to the threshold r_{B2-}:

$$\Omega^2 = \frac{\gamma_\parallel\Gamma(\kappa+\gamma_\perp) + \gamma_\parallel\gamma_\perp(6\kappa+3\Gamma+\gamma_\parallel)(r_{B2-}-1)}{\gamma_\parallel+\gamma_\perp+\Gamma+\kappa} \tag{15}$$

which in the case $\Gamma/\gamma_\perp\rightarrow\infty$ (adiabatical elimination of the two-photon coherence) coincides with that of the Lorenz-Haken model [3].

Another significant feature in the behaviour of the cascade laser is that a solution with $x_{21}=0$ exists, which for $\lambda^-=0$ is identical, as eqs.(6) show, to the

Lorenz-Haken solution. Thus, the typical chaotic solution of the Lorenz model also applies to the cascade laser. Presumably, however, in the present case this solution is unstable against fluctuations in x_{21} for a significant domain in the control parameter space. This problem and the study of dynamical solutions above the instability thresholds will be the object of future work, as well as the extension of the model to the pure two-photon laser.

References

1. See, for instance:
   Instabilities and Chaos in Quantum Optics, ed. by F.T.Arecchi and R.G.Harrison (Springer-Verlag, Berlin, 1987).
   C.O. Weiss and R. Vilaseca, *Laser Dynamics*, (Physik-Verlag VCH, Weinheim, 1991)
2. M. Sargent III, M.O. Scully, and W.E. Lamb, *Laser Physics* (Addison-Wesley, Reading, 1974).
3. H. Haken, *Synergetics: An Introduction* (Springer-Verlag, Berlin, 1985).

Phase Anholonomy and Quasiperiodicity in Optical Systems with Intensity Pulsations

C.Z. Ning and H. Haken

Institut für Theoretische Physik und Synergetik, Universität Stuttgart, Pfaffenwaldring 57/4, W-7000 Stuttgart 80, Fed. Rep. of Germany

Abstract We show that some lasers and nonlinear optical devices have in common certain characteristics, which lead to the phase anholonomy and quasiperiodicity of the field amplitude when the intensity undergoes a periodic pulsation. The amount of the phase anholonomy accumulated in a period of the pulsation is defined generally for such systems and further calculated for a subclass of such systems near the Hopf bifurcation point. The quasiperiodicity and the simultaneous occurrence of the two oscillations are analytically shown. By constructing a simple Hamiltonian, we show that the phase anholonomy in these dissipative sytems can be interpreted as the Berry phase.

1. INTRODUCTION

The phase problem of the electric field in lasers and some other nonlinear optical devices operating in the regime where the electric field intensity shows pulsation has received considerable attention and interest recently [1, 2, 3, 4, 5, 6]. In this regime the phase of the electric field, which is an arbitrary constant for the CW operation in a reference system rotating with the CW-frequency, shows a slope with respect to time besides the part oscillating with the frequency of the intensity pulsation. This slope is a manifestation of a frequency shift from the CW-frequency. The phenomenon has been observed in lasers with inhomogeneously [1] and homogeneously [2] broadened atomic lines, in the four wave mixers, and in the optical parameter oscillators [5]. The renewed interest of the phase problem is due to a possible analogy, pointed out in [3, 4], between this phase behaviour and the Berry phase [7].

The existence of the phase anholonomy in a wide range of optically dissipative systems leads us to ask the following questions: What do all known systems showing such behaviour have in common? What general conclusions can one draw about the phase anholonomy? Can one indeed establish an analogy between a certain phase in dissipative systems and the Berry phase in conservative systems? The present work aims to answer these questions.

2. CALCULATION OF THE PHASE ANHOLONOMY

We first note that a common property of all the above-mentioned systems including the detuned 2-photon lasers [8] and the bidirectional ring lasers [6] is

Springer Proceedings in Physics, Vol. 55
Nonlinear Dynamics and Quantum Phenomena in Optical Systems
Editors: R. Vilaseca, R. Corbalán

the possibility that their equations, after appropriate transformations, can be written in the following general form:

$$\dot{x}_i = F_i(x_1, x_2, ...x_n, \phi), \tag{1}$$
$$\dot{\phi} = h(x_1, x_2, ...x_n), \tag{2}$$

where a *static* part of the frequency corresponding to $x_i = 0$ (the CW state) has been removed from the phase equation. Therefore any constant part occurring in $\dot{\phi}$ would be due to the oscillation of x_i. Note that the ϕ variable does not occur on the RHS of the ϕ-equation. Another property of these systems is the existence of the periodic solutions of amplitude variables x_i for some parameters.

An immediate consequence of the above properties is the periodicity of $h(x_1, x_2...x_n)$. That is, if we consider the case where x_i's oscillate periodically, $f(t) \equiv h(x_1(t), x_2(t), ...x_n(t))$ itself will be a periodic function of the time, i.e., $f(t+T) = f(t)$ with $T = 2\pi/\Omega$ and Ω is the frequency of the amplitude variables. This suggests we expand f into a Fourier series,

$$\dot{\phi} = \frac{a_0}{2} + \sum_{n=1}^{\infty}(a_n cos\frac{2n\pi t}{T} + b_n sin\frac{2n\pi t}{T}), \tag{3}$$

where a_n's and b_n's are given by the well-known coefficients' formulas. The constant part (the average) is given by

$$\frac{a_0}{2} = \frac{1}{T}\int_0^T f(t)dt \equiv \omega_2. \tag{4}$$

Now it is clear that the phase contains two parts: one contains various harmonics of the periodic solution of the *amplitude* variables and the other is a constant part, which gives a phase slope during time evolution. After a whole cycle, all amplitude variables and the second part of the phase variable return to their starting values, but the overall phase does not return due to the first part in (3), giving rise to a phase anholonomy. The amount of the phase anholonomy (the phase accumulation in a period of T) is given by

$$\Delta\phi = \omega_2 T. \tag{5}$$

Obviously this result is consistent with that given in Eq. (26) in [3] for a detuned laser. But here the result is valid generally for the system (1, 2).

A simplification can be made, if we confine ourselves to the case where the RHS of (1) does not depend on ϕ, a case without back coupling of the phase to the amplitude equations [9]. This is the case for free-running devices, i.e., without externally driven field, such as detuned one- and two-photon lasers, and the complex Lorenz equations. In this case, an elimination of the fast variable and a successive transformation of (1) to a normal form have been performed for an arbitrary dimensional system near a Hopf bifurcation point in [10]. The resulting equations for the order parameters in the polar coordinate system can be written as

$$\dot{r} = \alpha r + g_{3,r} r^3 + ..., \quad (6)$$
$$\dot{\psi} = \beta + g_{3,i} r^2 + ... \equiv \Omega, \quad (7)$$

where α and β are real and imaginary parts of the eigenvalue corresponding to the complex order parameter. During the eliminations and the transformations that reduce (1) to (6,7), equation (2) undergoes the following changes. An elimination of the fast variables changes it into

$$\dot{\phi} = h(x_1, x_2, x_3(x_1, x_2), ... x_n(x_1, x_2)) \equiv f_c(x_1, x_2), \quad (8)$$

where $x_1 = x_2^*$ is the order parameter. All the other variables x_i, $(i = 3, 4, ..n)$ corresponding to eigenvalues with negative real part have been eliminated. The further nonlinear transformation [10] (with $v = r \exp(i\psi)$)

$$x_1 = v + P_2(v, v^*) + P_3(v, v^*) + ... \equiv x_1(v, v^*) \quad (9)$$

casts the order parameter equations for x_1, x_2 into the form of (6,7). In (9), P_n's are given by the nth order of polynomials of v and v^* and have been explicitly worked out up to $n = 3$ in [10].

Now suppose that $f_c(x_1, x_2)$ is generally of the form

$$f_c(x_1, x_2) = c_{10}x_1 + c_{01}x_2 + c_{20}x_1^2 + c_{11}x_1x_2 + c_{02}x_2^2 + ... \quad (10)$$

In the normal form "coordinates" v, it can be written as

$$\begin{aligned} f_c \quad & (x_1(v, v^*), x_2(v, v^*)) = c_{10}v + c_{01}v^* + (c_{10}b_{20} + c_{01}b_{02}^* + c_{20})v^2 \\ + \quad & (c_{10}b_{11} + c_{01}b_{11}^* + c_{11})vv^* + (c_{10}b_{02} + c_{01}b_{02}^* + c_{02})v^{*2} + ... \\ \equiv \quad & \overline{f}_c(v, v^*). \end{aligned} \quad (11)$$

According to the general formula (4),the frequency shift due to the oscillation with frequency Ω is given by

$$\begin{aligned} \omega_2 &= \frac{\Omega}{2\pi}\int_0^{\frac{2\pi}{\Omega}} \overline{f}_c(v, v^*)dt = \frac{\Omega}{2\pi}\int_0^{\frac{2\pi}{\Omega}} \overline{f}_c(r_0 e^{i\Omega t}, r_0 e^{-i\Omega t})dt \\ &= \frac{\Omega}{2\pi}(c_{10}b_{11} + c_{01}b_{11}^* + c_{11})r_0^2\frac{2\pi}{\Omega} + ... \equiv K r_0^2 + ... \end{aligned} \quad (12)$$

where $K = c_{10}b_{11} + c_{01}b_{11}^* + c_{11}$. Suppose now $g_{3,r} < 0$ for $\alpha = 0$, then for a positive small α and for a long enough time, r will reach r_0 defined by $\alpha r_0 + g_{3,r} r_0^3 = 0$. At this stage, r in (12) can be substituted by r_0. The whole asymptotic motion of the systems near the bifurcation point is described by the eqn. (7) and

$$\overline{\dot{\phi}} = \omega_2. \quad (13)$$

The quasiperiodical behaviour and the simultaneous occurrence of the two oscillations become analytically understandable. A similar result has been obtained by us in [3] for the detuned lasers. Here we see, however, that the result is quite general and can be established on a better mathematical ground.

The amount of the phase accumulated in a period of T is, according to (5) and (12), given by

$$\Delta\phi \approx \frac{K}{\Omega} 2\pi r_0^2 \equiv 2\frac{K}{\Omega} A, \tag{14}$$

where $A = \pi r_0^2$ is the area surrounded by the limit cycle in the manifold tangential, at the origin, to the subspace of the order parameters. In the higher order approximation, the area-dependence of $\Delta\phi$ will be generally replaced by a polynomial of even powers of r_0. This is evident from (11) where an average over one period leaves only terms like $v^j v^{*j}$. This fact indicates that even in the higher order the amount of the phase anholonomy is still determined by the size of the periodic oscillations.

3. COMPARISON WITH THE BERRY PHASE

Now it is in order to show that the phase anholonomy that occurred above can indeed be interpreted as a manifestation of the Berry phase. We note first that the electrical field in the above systems (or even other systems) is written in the form $E \sim \epsilon(t)\exp(i\phi)$ where the phase due to the CW frequency has been removed from ϕ and therefore the CW solution corresponds to $\epsilon = 1$ and $\phi = const.$ In the regime of pulsations $\epsilon = 1 + \rho(t)$ with $\rho(t) = \rho(t+T)$ and T is the period of the pulse. ϕ becomes simultaneously a time dependent quantity with $\phi(T+t) = \omega_2 T + \phi(t)$, which can be derived from (3). To construct a Hamiltonian, we note that $\exp(i\phi)/\sqrt{(2\pi)} \equiv \psi_1$ is an eigenfunction of the Hamiltonian $H(t) = -\epsilon(t)\frac{\partial^2}{\partial\phi^2}$ with eigenvalue $\epsilon(t)$. Thus the temporal behaviour of the ϕ can be equally well described by this eigenfunction of the Hamiltonian. For the CW state, the Hamiltonian is static and the eigenfunction corresponds to the CW field. In the periodical pulse regime with period T, $H(t+T) = H(t)$ is a periodic operator. After a whole cycle the system (the Hamiltonian) returns to its original form, the phase of this eigenfunction, however, generally need not return, which results in a phase anholonomy. According to the formula of Berry, the Berry phase is given by

$$\gamma_1 = -i\int_0^T dt(\psi_1, \frac{d}{dt}\psi_1) = -i\int_0^T dt[\int_0^{2\pi} d\phi\psi_1^*\dot{\psi}_1] = \int_0^T dt\dot{\phi} = \omega_2 T. \tag{15}$$

The last step is due to the substitution of (3) and (4). Therefore, the phase accumulation during a period of the pulsation can be indeed interpreted as a Berry phase resulting from a cyclic Hamiltonian. Note that the adiabatic condition is not needed here because there is no transition probability between any pair of eigenstates induced by the change of the Hamiltonian as can be seen from the expressions for the eigenstates.

To conclude, we have shown that all known lasers and nonlinear optical systems showing phase anholonomy fall into a unified class of dissipative systems. A useful procedure for calculating such phase anholonomy is put forward and demonstrated. A general mathematical connection with, and the interpretation of this phase as Berry's phase is realized. This is a first identification of the Berry phase in dissipative systems and at the same time leads to another way of realizing the Berry phase by electric field. All these in turn will help to gain further understanding of the Berry phase or the phase anholonomy in different systems.

This work was supported by the Deutsche Forschungsgemeinschaft.

References

[1] M.F.H.Tarroja, N.B.Abraham, D.K.Bandy and L.M.Narducci, Phys. Rev. **34**A, 3148(1986)

[2] H.Zeglache, P.Mandel, N.B.Abraham, and C.O.Weiss, Phys. Rev. **38**A, 3128(1988)

[3] C.Z.Ning and H.Haken, Z.Phys. B **81**,457(1990)

[4] R.Vilaseca, G.J.de Valcarcel and E.Roldan, Phy.Rev., A **41**, 5269(1990)

[5] P.Mandel, P.Galatola, L.A.Lugiato and K.Wang, (1990, preprint)

[6] L.M.Hoffer, G.L.Lippi and N.B.Abraham and P.Mandel, Opt. Commun., **66**,219(1988), G.D'Alessandro, A.Politi and F.T.Arecchi, Phy.Rev.,A **40**,5464(1989)

[7] M.V.Berry, Proc.Roy.Soc.London,A **392**,45(1984), J.Phys. A, Math.Gen., **18**,15(1985), Nature, **326**, 277(1987), Sci.Am., Dec.,1988

[8] C.Z.Ning, Z.Phys.,B**71**, 247(1988)

[9] P.Mandel and H.Zeghlache, Opt.Commun.,**47**, 146(1983), H.Zeghlache and P.Mandel, J.Opt.Soc.Am. B **2**, 18(1985)

[10] C.Z.Ning and H.Haken, Phys.Rev.A,**41**,3826(1990)

Local and Global Bifurcations in Opto-thermal Bistability with Localized Absorption

J.I. Rosell[1], *F. Pi*[2], *F. Boixader*[3], *J. Farjas*[3], *J. Massaneda*[3], *and G. Orriols*[3]

[1]Departament de Física i Enginyeria Nuclear, Universitat Politècnica de Catalunya, E-08017 Barcelona, Spain
[2]Departament d'Òptica i Optometria, Universitat Politècnica de Catalunya, E-08222 Terrassa (Barcelona), Spain
[3]Departament de Física, Universitat Autònoma de Barcelona, E-08193 Bellaterra (Barcelona), Spain

Abstract. We consider the dynamics of opto-thermal bistable cavities with one absorbing mirror and a transparent multilayer spacer of opposite thermo-optic materials. We report a variety of local and global instabilities and, in particular, the occurrence of Shil'nikov chaos in a trilayer device.

1. Introduction

Thermo-Optical Bistability with Localized Absorption (BOITAL) deals with etalons in which the light absorption occurs in one of the reflective coatings and the thermo-optic spacer is transparent. The separation of functions within the bistable system introduces the possibility of complex structures with more than one absorbing film and/or with various layers of different thermo-optic materials [1]. These kinds of structures possess composite feedback mechanisms and may be used to study dynamic instabilities in passive optical systems [2].

We are investigating the dynamics of BOITAL cavities with an absorbing mirror and a multilayer spacer of alternatively opposite thermo-optic materials [3,4]. Such dynamical systems exhibit homoclinic bifurcations and their effective dimension is determined by the number of layers in the spacer. For bilayer systems, a variety of homoclinic bifurcations without complicated orbit structures have been observed both theoretically and experimentally [4]. For systems of higher dimension, the homoclinicity may appear accompanied by a rich dynamics associated with the saddle-focus nature of the invariant set at which the homoclinic orbit makes tangency. The work of several authors suggests the important role of homoclinic orbits in determining the global properties in many problems in which complicated motion has been observed [5].

The aim of this paper is to show the occurrence of Shil'nikov chaos [6] in the numerically simulated dynamics of a BOITAL trilayer system.

Springer Proceedings in Physics, Vol. 55
Nonlinear Dynamics and Quantum Phenomena in Optical Systems
Editors: R. Vilaseca, R. Corbalán

2. Local bifurcations

The opto-thermal bistable system consists of an N-layer structure of alternatively opposite thermo-optic materials between two reflective films and is illuminated by a plane wave of intensity I_i. The input mirror is partially absorbing and, including the interference effects, the absorption as a function of the roundtrip phase shift ψ is described by $A(\psi)$. If the mirror films are assumed very thin, the dynamics may be then described by a set of homogeneous heat equations, one per each layer, subjected to the corresponding continuity and boundary conditions, as detailed in ref. 4. The phase shift is affected by the thermo-optic effect and, in the steady state, it may be shown that it satisfies the following equation

$$\psi = \psi^o + A(\psi)\,\beta\, I_i \qquad (1)$$

where the parameter $\beta = \beta_1 + \beta_2 + .. + \beta_N$ includes all the dependences on the material properties and sizes and it characterizes the thermo-refractive nonlinearity of the global spacer. The β_j describes the contribution of the jth layer by taking into account its relative position within the system. In the steady state, β appears only as a scale factor for the intensities and it is convenient to introduce the normalization $\psi_i = \beta I_i$.

Fig. 1a shows the branching diagram obtained from Eq. 1 for a given function $A(\psi)$ and, as typical in dispersive optical bistabiliy, it exhibits successive S-shaped curves. The slope of the curve at a given (ψ_i, ψ) point is given by $A(\psi)/[1-p]$, where p is equal to $\psi_i[\partial A/\partial\psi]$ and its value has been schematically indicated in the diagram of fig. 1a.

The linear stability analysis of the stationary solution yields the following main conclusions [7]:

a) The local stability of a stationary solution (ψ_i, ψ) is fully characterized by the corresponding value of p.

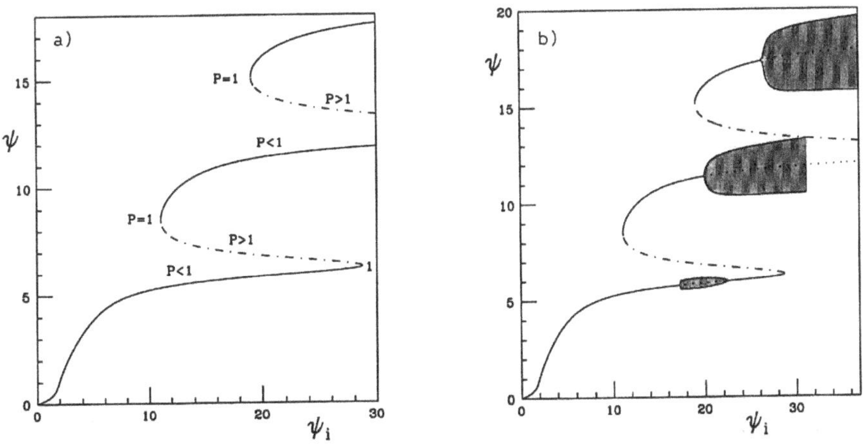

Fig. 1 a) Stationary branching diagram for an N-layer system.
b) Oscillating and stationary branches for a given bilayer.

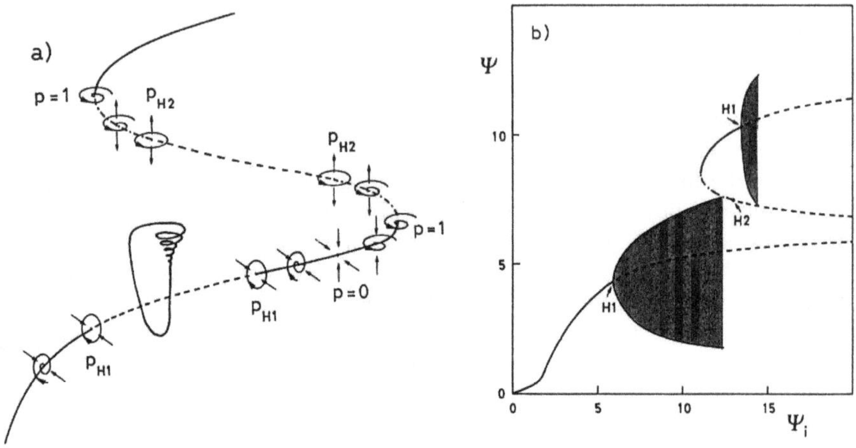

Fig. 2 a) Schematic evolution of significant eigenvalues for a trilayer system.
b) Branching diagram for a given trilayer system.

b) The maximum number of eigenvalues approaching the imaginary axis is determined by N. So, the effective dimension of the problem is equal to N and a reasonable phase space is the one defined by the partial phase shifts associated with the various layers, i.e., $(\psi_1, \psi_2, .., \psi_N)$.
c) For N=1, a real eigenvalue crosses the imaginary axis at p=1. It denotes the saddle-node bifurcation occurring at the turning points of the stationary diagram and the node (saddle) nature of the positive (negative) slope branches.
d) For N=2, besides the saddle-node bifurcation at p=1 there is a Hopf bifurcation at a p value lower than 1. As it is shown in fig. 1b for a particular bilayer system, each node branch presents an unstable region bounded by two Hopf bifurcations and where the system oscillates.
e) For N=3, a second Hopf bifurcation originating oscillations of higher frequency may occur at a p value higher than 1. The typical situation is shown in fig. 2a, where the evolution of the three significant eigenvalues is schematized over the stationary branching diagram. Since the oscillating frequencies are associated with bifurcations of different fixed points, the phase portraits will be as indicated in fig. 2a.

3. Global instabilities

The fig. 1b also points out a bifurcation of global nature occurring in the second node-branch when the oscillation amplitude becomes so large that it reaches a neighboring saddle point, i.e., the limit cycle makes homoclinic tangency to the saddle and disappears. A variety of homoclinic bifurcations without complicated orbit structure have been numerically and experimentally observed in bilayer systems [4].

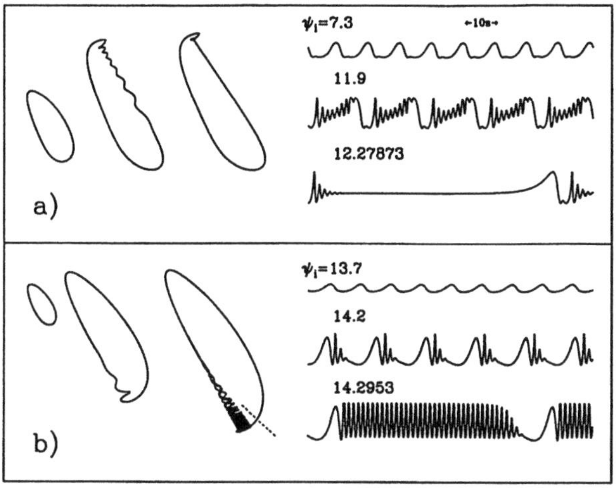

Fig. 3 (ψ_2,ψ_3) phase portraits and time evolution of reflected intensity for different input intensities in the trilayer case of fig. 2b.

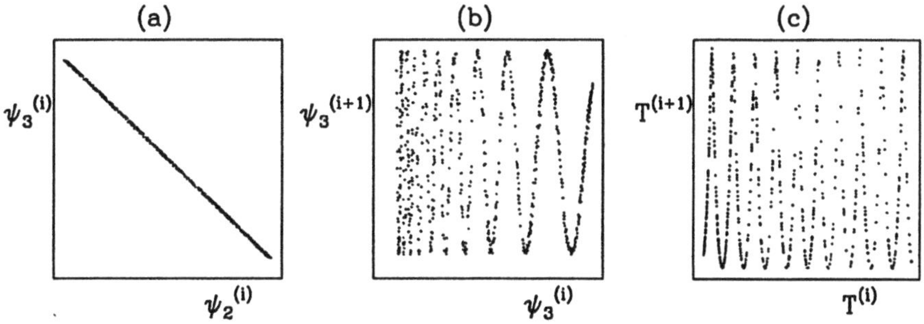

Fig. 4 Poincaré section and first return maps for the last case of fig. 3.

In the trilayer case, the homoclinicity may appear accompanied by a rich dynamics if the saddle-focus is near to the Hopf bifurcation point and, in this way, the Shil'nikov's conditions for chaos are fullfilled. Figure 2b shows the branching diagram numerically obtained for a given trilayer system. In this case, the first homoclinic bifurcation takes place where the saddle-focus is far from the Hopf point and the Shil'nikov's condition is not satisfied. On the contrary, the homoclinicity for the second oscillating branch takes place when the saddle point has become a saddle cycle. These two situations are respectively described in figs. 3a and 3b, which show the phase portraits and corresponding time evolutions of the reflected intensity for different input intensities. Notice that, in the case of fig. 3a, quasi-homoclinicity is achieved with the orbit remaining periodic. In the case of fig. 3b, the phase portraits point out the limit cycle in the saddle invariant set and the occurrence of aperiodic motion near homoclinicity.

The fine structure of the phase portraits near homoclinicity has been analyzed in the Poincaré plane indicated in fig. 3b. The Poincaré sections are smooth curves like the one shown in fig. 4a and the correponding one-dimensional first-return maps for the phase coordinate and for the time are represented in figs. 4b and 4c, respectively. The approach to homoclinicity is evidenced by an indefinite increase of the number of humps in the one-dimensional maps. This kind of map points out the occurrence of homoclinic chaos [5].

Acknowledgment

This work has been supported by the Dirección General de Investigación Científica y Tecnológica, under project no. PB87-0783.

References

1. G. Orriols, C. Schmidt-Iglesias, F. Pi, Opt. Commun. 63 (1987) 66-72.
2. F. Pi, G. Orriols, in Interaction of Radiation with Matter. A Volume in honour of Adriano Gozzini, G. Alzetta, E. Arimondo, F. Bassani and L. Radicati, eds. (Scuola Normale Superiore, Pisa, 1986) 239-255.
3. J.I. Rosell, F. Pi, C. Schmidt-Iglesias, G. Orriols, in Dynamics of Non-Linear Optical Systems, L. Pesquera and F.J. Bermejo, eds. (World Sc., 1989) 212-218.
4. J.I. Rosell, F. Pi, F. Boixader, R. Herrero, J. Farjas, G. Orriols, in Technical Digest on Nonlinear Dynamics in Optical Systems, 1990 (O.S.A., 1990) 119-121; Opt. Commun., to be published.
5. A. Arneodo, P.H. Coullet, E.A. Spiegel, C. Tresser, Physica 14D (1985) 327-347; P. Gaspard, R. Kapral, G. Nicolis, J. Stat. Phys. 35 (1984) 697-727.
6. L.P. Shil'nikov, Sov. Math. Dakl. 6 (1965) 163-166.
7. G. Orriols, J. I. Rosell, F. Pi, J. Farjas, to be published.

Induced Superradiance and Dynamical Chaos in a Thin Film of Two-Level Atoms

Yu.A. Logvin, A.M. Samson, and S.I. Turovets

Institute of Physics, BSSR Academy of Sciences,
SU-220602 Minsk, USSR

Abstract. The interaction between ultrashort light pulses and an inverted thin film of two-level atoms is investigated. It is shown that the induced superradiance effect occurs due to the initial excited-state atoms. The behaviour of the bistable thin film forced by a field with a modulated amplitude is considered, taking into account the relaxation times of the matter. The regimes of chaotic response appearing in the Feigenbaum scenario have been found numerically.

1. Introduction

A thin film (TF) of two-level atoms is a model demonstrating a wide range of nonlinear optical phenomena and, due to this, attracting the attention of many researchers [1-4]. In this paper we present within the scope of the TF model a theoretical study of the following problems.

First we consider coherent interaction between short light pulses and the TF, which is initially in the inverted state. The geometry of the model permits the reduction of the solution of the Maxwell-Bloch equations to the study of some analytically solvable dynamical system [1]. By this means, TFs may be considered as exactly integrable systems in optics, which are essential for theoretical investigation of the interaction mechanisms between light and matter.

Another problem is the investigation of non-stationary processes associated with switching in the TF bistable element [4]. Exactly these processes determine the speed of a possible optical computer. In addition, the investigation of different regimes of a nonlinear system is interesting from the point of view of the theory of dynamical systems.

Springer Proceedings in Physics, Vol. 55
Nonlinear Dynamics and Quantum Phenomena in Optical Systems
Editors: R. Vilaseca, R. Corbalán

2. Model equations

It is assumed that a plane electromagnetic wave with a slowly varying amplitude $E_i(t)$ and frequency ω is incident on the TF placed in vacuum between planes x=0 and x=L. The film thickness L is assumed to be much smaller than the incident radiation wavelength. The consideration of the system of Maxwell equations with suitable boundary conditions leads to the following expressions for the amplitudes of the reflected E_r and transmitted E_t waves [1]:

$$E_r(t) = \hbar V(t)/pT_R, \quad E_t(t) = \hbar V(t)/pT_R + E_i(t), \tag{1}$$

where V(t) is the off-diagonal density matrix element of the two-level atom, p is the transition dipole moment, $T_R=\hbar c/2\pi NL\omega p^2$, N is the atomic concentration in the TF. In relations (1) exact resonance is assumed between the incident wave and the atomic transition. The reflected field E_r is identical with the wave emitted by atoms in the backward direction while the transmitted light is the superposition of the incident and secondary fields.

The system of material equations for polarization (P(t)= pNV(t)) and population inversion W(t) is as follows [1]:

$$\dot{V} = W(pE_i/\hbar + V/T_R) - V/T_2 , \tag{2}$$

$$\dot{W} = -V(pE_i/\hbar + V/T_R) - (W + 1)/T_1 , \tag{3}$$

where T_1 and T_2 are the times of longitudinal and transverse relaxation, respectively.

By introducing the Bloch angle Θ the system of equations (2)-(3) takes, in the limit $T_{1,2}= \infty$, the form of a damped pendulum equation, which is basic in our treatment of induced superradiance,

$$\dot{\Theta} - \sin\Theta = pT_RE_i(\tau)/\hbar , \quad \tau = t/T_R . \tag{4}$$

Eq. (4) has been solved by different authors [1,2] considering the noninverted TF. Here an emphasis is placed on the effects which appear due to the initial atomic excited state.

3. Induced superradiance

Considering the incidence of a rectangularly shaped pulse with the dimensionless duration τ_0 and amplitude F < 1 and integrating equation (4) for initial condition $\Theta = 0$ we have

$$\Theta(\tau) = 2 \operatorname{arctg} F \; [(1 - F^2)^{1/2} \operatorname{ctg}(\tau(1 - F^2)^{1/2}/2) - 1]^{-1}. \quad (5)$$

In this case the expression for the transmitted field takes the form

$$\frac{pT_R E_t(\tau)}{\hbar} = \frac{F(1 - F^2)}{\operatorname{ch}(\tau\sqrt{1 - F^2}) - \sqrt{1 - F^2}\cdot \operatorname{sh}(\tau\sqrt{1 - F^2}) - F^2} . \quad (6)$$

Initial atomic excited states give rise to the sech-like shape of the dependences $E_t(\tau)$, thus manifesting the induced superradiance. At some moment $\tau=\tau^*$ the system passes the lower stable equilibrium position $\Theta=\pi$, giving $E_t=E_i$ and $E_r=0$. The value τ^*, as well as τ_M, being the delay time of the induced superradiance pulse, may be easily found from expression (6), $\tau_M=\tau^*/2 = \operatorname{arcsh}((1 - F^2)^{1/2}/F)/(1 - F^2)$. At $F\longrightarrow 1$ we obtain $\tau_M \longrightarrow 1$ or in dimensional form $\tau_M = T_R$, differing by a factor of lnN from the quantum-electrodynamic delay time $T_D = T_R \ln N$ [5].

In systems whose sizes are much less than the wavelength, an observation of cooperative effects is hindered by short relaxation times due to strong atom interaction at close distances. If the superradiance is induced by an external signal, then the emission of the superradiation pulse may be considerably accelerated and thus conditions for its experimental observation are facilitated.

4. Modulated input

The standard dependence between E_t and E_i describing the behaviour of bistable optical devices is obtained from (1) - (3) on equating the derivatives $\dot{V}$ and $\dot{W}$ to zero, $y = x + 2Cx/(1 + x^2)$, where $x = pT_2E_t/\hbar$, $y = pT_2E_i/\hbar$ and $C = T_2/2T_R$. The S-shaped characteristic arises at $C > 4$ [4]. Here we theoretically study the dynamics of the bistable TF when a light wave with a modulated amplitude is incident on it, $E_i = E_{io} + E_{i1}\cos\Omega t$, where $\Omega \ll \omega$. System (2) - (3) can be rewritten in a form permitting a more vivid analysis of its behaviour,

$$\begin{aligned}\ddot{V} + 2\dot{V} + (2C + y^2 + 1)V + 4CyV^2 + 4C^2V^3 &= \\ -y + (\dot{V} + V)\frac{d}{d\tau}\ln(2CV + y), \quad \tau &= t/T_2. \quad (7)\end{aligned}$$

Eq. (7) is an equation of a damped nonlinear oscillator excited both parametrically and by an external action because y is

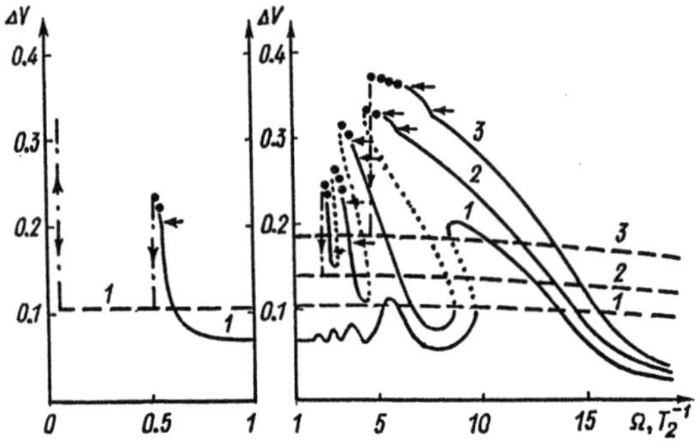

Fig.1. The AFC of oscillator (7) (C=20, $pT_2E_{io}/\hbar$=16,76) at the input modulation $pT_2E_{i1}/\hbar$ = 4.5(1); 6.0(2); 8.0(3). The solid curves correspond to oscillations on the lower branch and the dashed ones on the upper branch of the bistable characteristic. The arrows point to the sites of period doubling bifurcations, the circles point to chaotic parts. The dash-dot lines with arrows indicate jumps from one branch to another.

time-dependent. In the theory of dynamical systems the driven nonlinear oscillator is one of the simplest models in which dynamical chaos has been observed [6]. It gives reason to suppose that in a given system there exist complicated (including chaotic) regimes, which has been confirmed by numerical experiment.

Fig.1 illustrates the amplitude-frequency characteristics (AFC) of oscillator (7) . It is seen that with increasing E_{i1} the resonance structure for oscillations near one of the steady states becomes more pronounced while the amplitude of oscillations on the other branch weakly depends on the modulation frequency. More than that, the AFC on the "dispersion" branch at large modulation amplitudes has its own bistable part due to the "λ-shaped form" of the main resonance. It can be seen also that chaotization of oscillations by the Feigenbaum scenario takes place at larger modulation values. Thus, in general, there appear multistable states in the transmission and reflection of the signal's alternating component. Such a "dichotomy" of response to the modulated signal is likely to be possible in other models of optical bistability as well and can be used for testing and diagnosing bistable elements.

References

1. M.G. Benedict and E.D. Trifonov, Phys. Rev. A 38, 2854 (1988).
2. V.I. Rupasov and V.I. Yudson, Sov. J. Quantum Electron. 12, 1415 (1982).
3. A.M. Samson, Yu.A. Logvin and S.I. Turovets, Optics Comm. 78, 208 (1990).
4. S.M. Zakharov and E.A. Manykin, Poverhnost (Moscow) 2, 137 (1988).
5. R. Bonifacio and L. Lugiato, Phys. Rev. A 11, 1507 (1975).
6. A.M. Samson, Yu.A. Logvin and S.I. Turovets, Izv. vyssh. uchebn. zaved. Radiofizika 33, 49 (1990).

Analysis of a Modulated LIS Model

H. Zeghlache, V. Zehnle, and P. Glorieux

Laboratoire de Spectroscopie Hertzienne, associé au CNRS,
Université des Sciences et Techniques de Lille, Bat. P5,
F-59655 Villeneuve d'Ascq Cedex, France

Abstract: We have analyzed a model of a laser with an injected signal (LIS). In order to perform analytical calculations as far as possible, we have used a simplified model which exhibits a Hopf bifurcation. Standard mathematical methods are used to investigate the vicinity of this critical region and a complete description of the LIS behavior is given in the parameter space. In the presence of modulation of the external field amplitude, frequency lockings between the external modulation frequency and the internal frequency of the LIS are analytically studied.

I. The LIS model

In a two-level model of a LIS, the Maxwell-Bloch equations usually involve the complex output field, the atomic polarisation, and the population inversion [1]. When adiabatic elimination of the atomic variables is valid, which is the case for dye lasers for example, the Maxwell-Bloch equations can be reduced to a single equation for the complex field [2,3]. This can be written as

$$\frac{dX}{dt} = -i\,\Omega\, X + \left(\frac{R}{1+XX^*} - 1\right) X + \mathcal{A}, \tag{1}$$

where X is the rescaled complex output field, $\mathcal{A}$ the amplitude of the external field, R the pump parameter, and Ω is the detuning of the external field frequency from the cavity frequency. This simple LIS model is two-dimensional and governed by three parameters.

II. The steady states and linear stability analysis

In cartesian coordinates X=u+iv, the steady states (u_s, v_s) are the zeros of the time derivatives of u and v, and given by: $u_s = -(I_s/\mathcal{A})\left(\frac{R}{1+I_s} - 1\right)$ and $v_s = -\Omega I_s/\mathcal{A}$. The steady intensity I_s obeys the implicit equation

$$\mathcal{A}^2 = \left(\frac{R}{1+I_s} - 1\right)^2 I_s + \Omega^2 I_s\,. \tag{2}$$

Springer Proceedings in Physics, Vol. 55
Nonlinear Dynamics and Quantum Phenomena in Optical Systems
Editors: R. Vilaseca, R. Corbalán

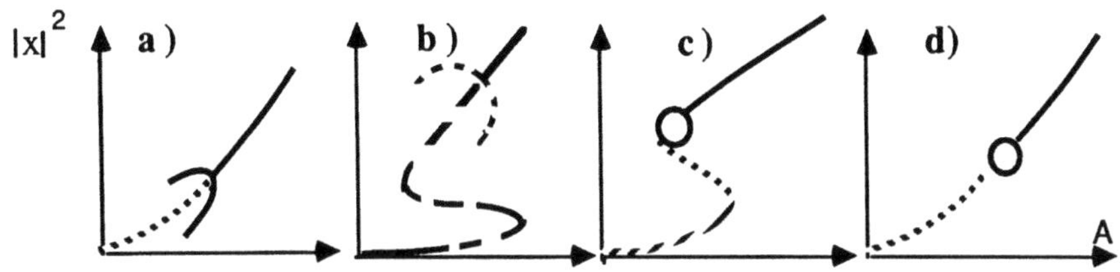

Fig.1: The differents LIS bifurcation diagrams. Full (dashed) line: stable (unstable) steady state. In bistable cases (b,c), eqs(3) hold and the upper branch destabilises via a Hopf bifurcation (b) when eq.(5) is fulfilled.

This cubic equation describe an S-shaped curve for I_s versus A when the following inequalities for the remaining parameters (R,Ω) hold

$$1+\Omega^2 < \frac{(R+2)^3}{27\,R} \text{ and } R>1. \tag{3}$$

Otherwise eq.(3) has only one solution and the LIS has a monostable behavior with respect to the external field amplitude A.

In order to test the stability of the steady state (u_s,v_s), a perturbation of the form: $u = u_s + \delta u\, e^{\lambda t}$ and $v = v_s + \delta v\, e^{\lambda t}$, where $\delta u \ll u_s$ and $\delta v \ll v_s$, is injected in eq.(1). At first order in $(\delta u, \delta v)$, the eigenvalues of the characteristic equation are

$$\lambda_{1,2} = \frac{R}{(1+I_s)^2} - 1 \pm \sqrt{\frac{R^2 I_s^2}{(1+I_s)^4} - \Omega^2}\,. \tag{4}$$

As a parameter is varied, a change of sign of $Re(\lambda)$ corresponds to a Hopf bifurcation and under suitable conditions small amplitude oscillations of the intensity may settle. A real negative λ becoming positive leads to a destabilisation of the steady state towards large amplitude pulsations.
A Hopf bifurcation exists if:

$$\Omega > (\sqrt{R} - 1). \tag{5}$$

It occurs for an A value given by $A^2 = I_h(\Omega^2 + I_h^2)$, where $I_h = (\sqrt{R} - 1)$ is the corresponding value of the intensity. The frequency at the Hopf point is then given by $\Omega_h^2 = \Omega^2 - I_h^2 = (v_h^2 - u_h^2)(I_h^2/u_h^2)$ which assumes $v_h > u_h$. The Hopf point is also characterized by: $u_h = -I_h^2/A$ and $v_h = -\Omega I_h/A$.
Combining eqs. (2) and (3), different bifurcation diagrams appear in the (R, Ω^2) space. Fig. 1 shows all possible behaviors described by the linear stability analysis versus A. As expected, for large A values, the steady state is stable since the system just follows the strong driving field.

III. Construction of the small amplitude periodic solutions

We now describe the solutions in the vicinity of the critical Hopf point, and again use A as a bifurcation parameter. A standard perturbative method is used [4]. It consists of developing eq.(1) in the neighbourhood of the critical point a_h as $A = a_h + \varepsilon^2 a_2$ with $\varepsilon << 1$. A slow time scale defined by $T = \varepsilon^2 t$ is introduced such that $d_t = \partial_t + \varepsilon^2 \partial_T$.
The complex variable X is then developed as

$$X(t,T) = X_h + \varepsilon X_1(t,T) + \varepsilon^2 X_2(t,T) + O(\varepsilon^3) \tag{6}$$

where $X_h = (u_h, v_h)$, and it is reinjected in eq.(1). At first order in ε, the components of X_1 are given by

$$\begin{Bmatrix} x_1(t,T) \\ y_1(t,T) \end{Bmatrix} = \begin{Bmatrix} A(T) \\ B(T) \end{Bmatrix} e^{i\Omega_h t} + \text{c.c}\,, \tag{7}$$

where $B(T) = \frac{u_h}{v_h}\left(-1 + i\,\frac{\Omega_h}{v_h^2 - u_h^2}\right) A(T)$.

The solvability condition falls at third order with the generic equation

$$\partial_T A = (\alpha + i\beta)\, A + (\gamma + i\delta)\, A|A|^2 \tag{8}$$

where in our case

$$\alpha + i\beta = \frac{4\, u_h\, a_2}{v_h^2 - u_h^2}\left(\frac{1}{1+I_h} + i\,\frac{I_h\,(1-I_h)}{2\,\Omega_h(1+I_h)}\right)$$

$$\gamma = \frac{4\, I_h^2}{(v_h^2 - u_h^2)^2}\,\frac{2v_h^2 - 1}{(1+I_h)^2}. \tag{9}$$

$$\delta = -\frac{4 I_h^2}{3\Omega_h (v_h^2 - u_h^2)^2}\left\{6v_h^2 - u_h^2 - \frac{I_h}{1+I_h}(21 v_h^2 - u_h^2) + \frac{I_h^2}{(1+I_h^2)^2}(17 v_h^2 + 3u_h^2)\right\}.$$

In polar coordinates $A = \rho e^{i\theta}$, the steady states are: $\rho_s = 0$ and $\rho_s^2 = -\alpha/\gamma$. The finite small amplitude periodic solution occurs with a frequency

$$\omega = \Omega_h + \varepsilon^2\left(\beta - \delta\,\frac{\alpha}{\gamma}\right). \tag{10}$$

A linear stability analysis of these solutions leads to a stability exchange between the zero and finite amplitude solution for $\alpha = 0$, and the zero solution is stable when $\alpha < 0$. The small amplitude oscillations exist and are stable (the Hopf bifurcation is supercritical) when $\alpha > 0$ and $\gamma < 0$. One can find that it corresponds to $a_2 < 0$, and

$$\Omega^2 > \frac{(\sqrt{R} - 1)^2}{2\sqrt{R} - 3}. \tag{11}$$

In fig.2, the existence and stability domains of the small amplitude oscillations are represented in the parameters space (R, Ω^2).

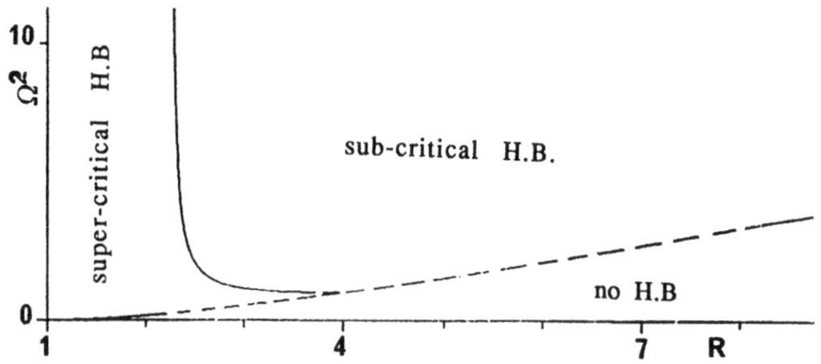

Fig.2: H.B. stands for Hopf Bifurcation. Solid (dashed) line represents eq.(11) (resp. eq.(5)).

IV. The LIS with a modulated amplitude

We analyse eq.(1) for an amplitude modulation, $\mathcal{A} \longrightarrow \mathcal{A}(1 + m\cos(\Omega' t))$, Ω' and m are the modulation frequency and amplitude respectively.
The autonomous system studied in the preceding sections presents only fixed points or limit cycles. Adding modulation introduces an extra degree of freedom and the system may now exhibit a much wider class of dynamical behaviors.
If the autonomous LIS displays self oscillations of frequency ω, and if modulation is added, the system may lock to the external frequency or behave quasi-periodically. A numerical study has been performed when $\mathcal{A}$, R, Ω are kept in the vicinity of the Hopf bifurcation . As Ω' is varied, the system shows locked regions, or Arnold tongues, in the (Ω', m) plane: the forced LIS is oscillating at some period $T = nT'$ (n=1,2,...) multiple of the external period $T'=2\pi/\Omega'$. We now characterize analytically [4] the main resonance ($\Omega' \approx \omega$) in the limit of small modulation ($m \ll 1$). As in section III, the parameters $\mathcal{A}$, R, Ω are kept near their super-critical Hopf values and a perturbation development is performed:

$\mathcal{A} = a_h + \varepsilon^2 a_2 + a_3 \varepsilon^3 \cos(\Omega' t)$, with $\Omega' = \Omega_h + \varepsilon^2 \omega_2$.

Two time scales are introduced $\tau = \Omega' t$ and $T = \varepsilon^2 t$. The development in eq.(6) remains. At first order in ε, we still find eq.(7) with t replaced by τ and $\exp(i\Omega_h t)$ by $e^{i\tau}$.
The solvability condition now reads:

$$\partial_T A = \Gamma + (\alpha + i\beta - i\omega_2)A + (\gamma + i\delta) A|A|^2 \cdot \qquad (12)$$

Γ is directly related to the modulation amplitude by: $\Gamma = \frac{m}{4}\left(1 - i\,\frac{v_h{}^2 - u_h{}^2}{\Omega_h}\right)$.
When stable, the steady states of eq.(13) lead to oscillating solutions synchronous with Ω'. The amplitudes $|A_s| = \sqrt{r_s}$ obey the implicit equation:

$$|\Gamma|^2 = r_s \{ (\alpha + \gamma r_s)^2 + (\beta' + \delta r_s)^2 \}. \qquad (13)$$

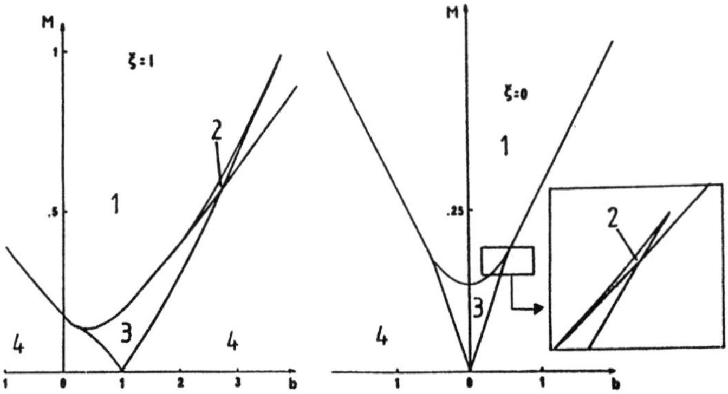

Fig.3 : Tongue 1/1 obtained analytically for ξ=1 and ξ=0.

with $\beta' = \beta - \omega_2$. The linear stability analysis applied to these steady states leads to the following characteristic equation:

$$\lambda^2 - 2\lambda\,(\alpha + 2\,\gamma\, r_s) + \alpha^2 + \beta'^2 + 4\,(\,\alpha\gamma + \beta'\delta)\, r_s + 3\,(\gamma^2 + \delta^2)\, r_s^2 = 0. \qquad (14)$$

The analysis of eqs.(13) and (14) is simplified if we introduce a rescaled modulation amplitude $M = \sqrt{|\Gamma|^2\,|\gamma|\,/\alpha^3}$ and replace ω_2 by a shifted and rescaled frequency $b = (\beta - \omega_2)/\alpha$. It can be shown that the stable regions for A_s in the (M, b) plane are functions of $\xi = \delta/\gamma$ only.
As an illustration, we consider two situations, $\xi = 0$ and $\xi = 1$, typical of the super-critical Hopf region. The corresponding tongues are given in figs.3. In both cases, and as M-->0, the tongues begin at abscissa $b=\xi$ and the value $\omega_2 = \beta - \alpha\,(\delta/\gamma\,)$ is recovered. In the region denoted by 1, there is one stable stationary value A_s . In both regions 2 and 3, three solutions coexist but only one is stable in region 3, while bistability between two stable solutions occurs in region 2. Finally, in region 4 there are no stable solutions at all and the modulated system behaves quasi-periodically. As ξ departs from zero, the tongue becomes asymmetrical but the qualitative description remains. The same features are found when a similar analysis is performed with a modulated pump parameter R. More details will be presented in a future publication.

Conclusions

We have analyzed the simplest model describing a LIS, and defined completely the behavior of the system in the parameters space. We also present some analytical calculations performed for the construction of the small amplitude self-pulsations. In the case of a modulated LIS, we have treated the case of the driving field amplitude modulation, analyzed the resonance 1/1, and described more precisely the related Arnold tongue.

References

[1] D.K. Bandy, L.A. Lugiato and L.M. Narducci, " *Instabilities and Chaos in Quantum Optics*". (F.T. Arecchi and R.G. Harrison eds.), Springer Verlag (Berlin, 1987).

[2] T. Yamada, R. Graham, Phys. Lett. 82A, 321(1981).

[3] H.J. Scholtz, T. Yamada, H. Brand and R. Graham, Phys. Rev. Lett. 45 , 1322(1980).

[4] G. Ioos and D.D. Joseph, "*Elementary Stability and Bifurcation Theory*", Springer Verlag (Berlin, 1980).

[5] A. Ito, Prog. Theo. Phys. 61, 45 (1979).

Role of the Earth's Magnetic Field and Atomic Coupling in Polarization Instabilities in Quasi-Isotropic Lasers

J.C. Cotteverte, F. Bretenaker, and A. Le Floch

Laboratoire d'Electronique Quantique-Physique des Lasers,
U.A. CNRS 1202, Université de Rennes I,
F-35042 Rennes Cedex, France

Abstract. We show here theoretically and experimentally that the Earth's magnetic field and the atomic coupling play a determining role in the polarization instabilities that can occur in quasi-isotropic lasers. On the one hand, it is proved that a weak atomic coupling laser can be very sensitive to low magnetic fields and exhibits a sinusoidal Adler-type dynamics with respect to the azimuth of its axis. On the other hand, a strong atomic coupling forbids such instabilities but leads to two possible vectorial bistability mechanisms for the circularly polarized eigenstates of the laser.

1. Introduction

The study of polarization effects in Zeeman lasers has been an early subject of interest in laser physics [1]. More recently, there has been renewed interest in the dynamics of laser eigenstates, especially to build all optical logic gates with low switching energies [2]. Moreover, polarization instabilities and chaos have been reported when an AC and a DC external magnetic field are applied to the active medium [3]. A new instability threshold has also been predicted in the case of a non-weak atomic coupling laser [4]. Finally, very recently [5], polarization instabilities have been observed in the case of weak atomic coupling high-gain quasi-isotropic lasers without any applied external magnetic field. The authors of Ref. 5 attributed the observed instabilities to the role of atomic dynamics and ruled out the influence of the Earth's magnetic field. One may consequently wonder what is the exact role of the Earth's magnetic field in such instabilities and what is the influence of the atomic coupling. Consequently, the aim of this paper is to investigate the role of small magnetic fields in the dynamics of quasi-isotropic lasers' eigenstates, as well in the case of weak atomic coupling as in the case of non-weak atomic coupling. In the first part of this paper, we deliberately consider a weak atomic coupling laser submitted to the Earth's magnetic field and compare the predicted polarization dynamics with our experiments. In the second part of this work, we show theoretically and experimentally that this dynamics is drastically modified in the case of the circularly polarized eigenstates of a non-weak atomic coupling laser.

2. Influence of the Earth's Magnetic Field on the Polarization Dynamics of a Weak Atomic Coupling Quasi-Isotropic Laser

Let us consider the quasi-isotropic laser containing an x-y-type loss anisotropy shown in Fig. 1. If we consider the role of the Earth's magnetic field, only its axial component B_a is of interest because the transverse one does not induce any Faraday

Springer Proceedings in Physics, Vol. 55
Nonlinear Dynamics and Quantum Phenomena in Optical Systems
Editors: R. Vilaseca, R. Corbalán

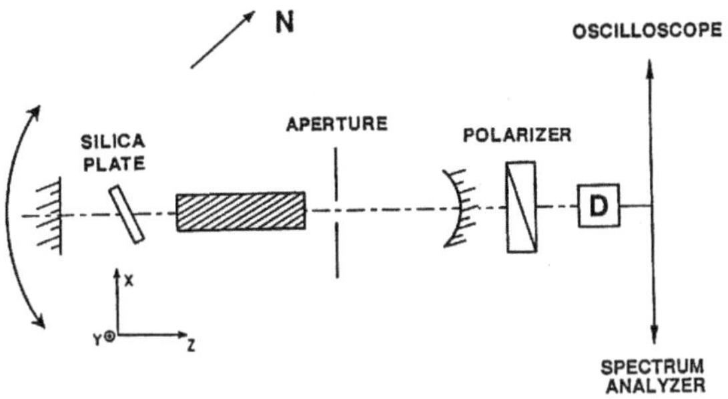

Figure 1: Experimental setup used to study the influence of the Earth's magnetic field.

rotation. If we assume that a linear polarization making an angle θ with the x axis exists in the laser cavity, the evolution equation for θ is obtained considering one round-trip in the cavity :

$$\frac{d\theta}{dt} = \frac{c}{2L}\left(\frac{t_x}{t_y} - 1\right)\sin 2\theta + \frac{c}{L}\phi \quad , \tag{1}$$

where c is the velocity of light, L is the length of the cavity and t_x and t_y are the transmission coefficients of the plate. ϕ represents the Faraday rotation of the polarization for one pass through the active medium :

$$\phi = Vl\, B_a \quad , \tag{2}$$

where V is the Verdet constant of the active medium of length l and B_a is the axial component of the Earth's magnetic field given by $B_a = B \cos \psi$ where B is the horizontal component of the Earth's magnetic field and ψ is the azimuth of the laser propagation axis. Equation (1) is a typical Adler-type [6] first order differential equation, leading to two possible behaviors for the polarization angle θ. If $|B_a| < |B_c|$, where $B_c = (1 - t_x / t_y) / 2Vl$ is a critical value for the magnetic field, the system is "locked" and eq. (1) exhibits a stable stationary solution θ_0. On the contrary, for $|B_a| > |B_c|$, the Faraday rotation ϕ overcomes the drawback effect of the loss-anisotropy and the polarization rotates periodically at frequency :

$$\Omega = \frac{c}{2L}(1 - t_x/t_y)\left(\frac{B^2}{B_c^2}\cos^2\psi - 1\right)^{1/2} \quad . \tag{3}$$

Our predictions about the role of the Earth's magnetic field can consequently be summarized in the following way. On the one hand, for low values of the axial component of the Earth's magnetic field (i.e. around the East-West orientation of the laser), the system lies in the locking region and remains stable. On the contrary, if B_a becomes important enough for the Faraday rotation to overcome the loss anisotropy, polarization instabilities occur. These predictions are checked with a 39-

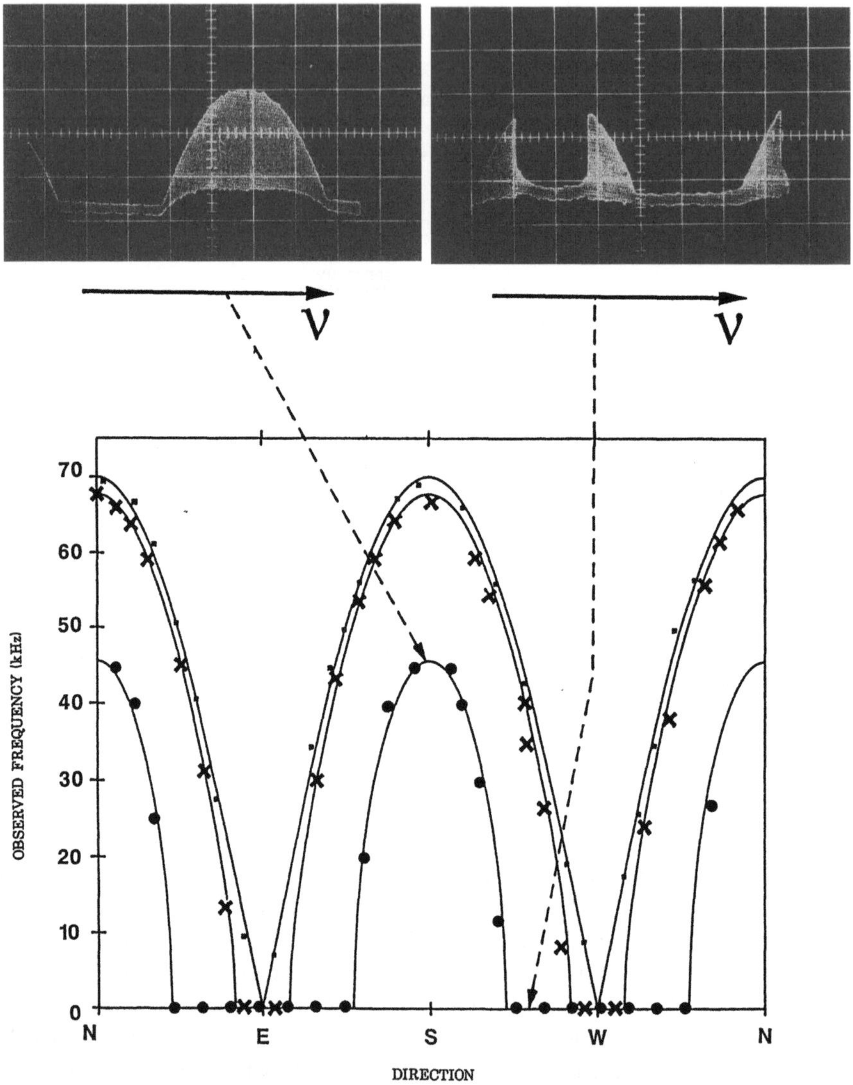

Figure 2: Experimental evolution of the frequency of the instabilities vs azimuth of the laser axis for three values of the loss anisotropy. The corresponding theoretical curves are in full line. The photographs display the output power vs frequency profiles detected through the polarizer for two different orientations of the laser axis (ν axis: 59 MHz per division).

cm-long 3.39-μm ($J = 1 \rightarrow J = 2$) ^3He-^{20}Ne laser with a 19-cm-long discharge tube. This monomode and weak atomic coupling laser is set on a bench that can rotate in the horizontal plane. The frequency detected at line center through a linear polarizer is then plotted vs the azimuth of the laser axis for three values of the intracavity loss-anisotropy (see the points in Fig. 2). These points perfectly fit the theoretical sinusoidal Adler-type curves computed from eq. (3). As expected, the polarization exhibits a locking region around the East-West direction that increases with the intracavity loss-anisotropy. Now, when the cavity frequency is tuned, one can observe through the linear polarizer output power profiles such as those shown in the photographs of Fig. 2. Such profiles show that the sensitivity of our laser to the Earth's magnetic field varies with the lasing frequency. This effect is due to the saturation of the Faraday rotation. This Faraday rotation is indeed proportional to the difference $n^+ - n^-$ between the σ^+ and σ^- refractive indices. Each of these refractive indices can be obtained from a rate equation model :

$$n(\omega) - 1 = \frac{g_0}{k}\left\{\frac{\omega}{\sqrt{\pi}\,ku} - \frac{1}{2}\left(1 - \frac{A}{B}\right)\frac{\omega}{\sqrt{(A+B)^2 - 4\omega^2}}\right\} , \qquad (4)$$

where $A = (\omega^2 + \gamma^2)^{1/2}$, $B = (\omega^2 + \gamma^2(1 + 2S))^{1/2}$; g_o is the unsaturated gain coefficient, k the wave number, ku the Doppler linewidth, γ the homogeneous linewidth, S the saturation parameter and ω the detuning of the considered σ component. The evolution of the rotation rate Ω can be related to the relative excitation η through eqs. (3) and (4) and :

$$\eta^{-1} = \exp\left[-(\omega/ku)^2\right](1 + A/B)\,\gamma\left[(A+B)^2 - 4\omega^2\right]^{1/2} . \qquad (5)$$

The evolution of Ω with respect to the detuning is shown in Fig. 3 for three values of η. Here again the experimental measurements are in agreement with the theoretical calculations. Keeping the relative excitation constant, the sensitivity of our laser to the Earth's magnetic field can be increased by increasing the losses of the cavity, as seen from eq. (4). In the experiment, we introduce glass plates in the cavity to increase the losses and observe the evolution of the rate of rotation of the polarization for a constant relative excitation η. The results are shown in Fig. 4 and show again a good agreement with the predictions [7].

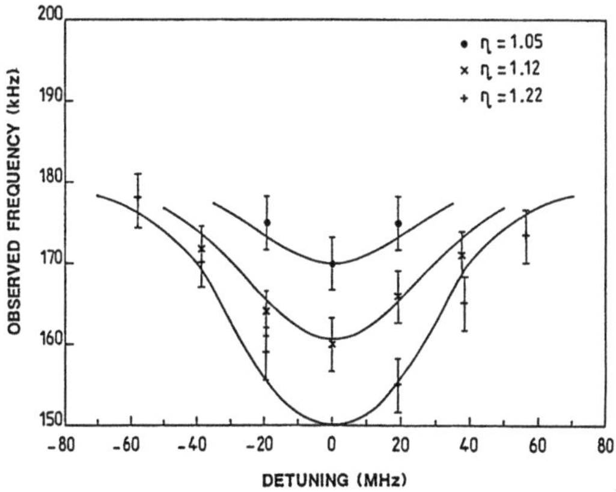

Figure 3: Evolution of the observed frequency vs detuning for three different relative excitations. The theoretical curves are in full line.

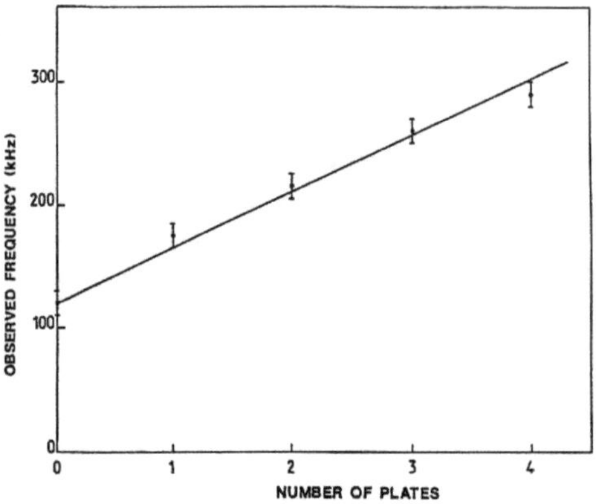

Figure 4: Evolution of the observed frequency vs the number of glass plates introduced inside the cavity for a constant relative excitation. The theoretical curve is in full line.

3. The Dynamics of the Circularly Polarized Eigenstates of a Non-Weak Atomic Coupling Quasi-Isotropic Laser

Up to now, we have shown the predominant role of the Earth's magnetic field in the polarization instabilities of a weak atomic coupling quasi-isotropic laser. We are now going to focus our attention on the case of the non-weak atomic coupling and consider the cavity shown in Fig. 5. This cavity contains a Faraday rotator providing a non-reciprocal dephasing $\Delta\phi_{+-}$ between an incident σ^+ and an incident σ^- waves. The eigenstates of such a cavity are a σ^+ and a σ^- eigenstates with eigenfrequencies $\nu_\pm = c/2L\left(k \pm \Delta\phi_\pm/\pi\right)$. The evolution of the amplitudes E_+ and E_- of these eigenstates can be described by self-consistent equations in a form derived from Lamb's theory [8] :

$$\dot{E}_+ = E_+\left(\alpha_+ - \beta_+E_+^2 - \theta_{+-}E_-^2\right) \quad , \tag{6a}$$

$$\dot{E}_- = E_-\left(\alpha_- - \beta_-E_-^2 - \theta_{+-}E_+^2\right) \quad , \tag{6b}$$

where α_+ and α_- are the net gain coefficients, β_+ and β_- the self-saturations and θ_{+-} the cross saturation. The coupling constant $C = (\theta_{+-})^2/(\beta_+\beta_-)$ of our two eigenstates contains the atomic coupling that in the case of a $J = 1 \rightarrow J = 1$ transition becomes slightly greater than 1, like for a $J = 1 \rightarrow J = 0$ transition. Consequently, for low

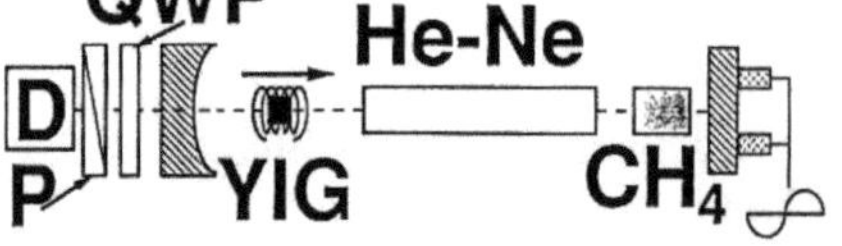

Figure 5: Experimental setup used to study the dynamics of circular eigenstates.

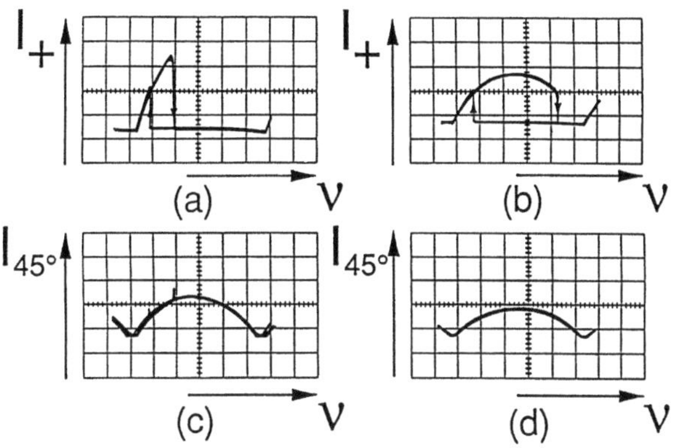

Figure 6: The two basic flipping mechanisms. (a) and (c) are the σ^+ output powers vs cavity tuning profiles. (b) and (d) are the output powers vs cavity tuning profiles with the linear polarizer aligned along the axes of the quarter-wave plate. (a) and (b) correspond to the rotation mechanism ($\Delta\phi_{+-} \approx 0.25°$) and (c) and (d) to the inhibition mechanism ($\Delta\phi_{+-} \approx 0.8°$).

$\Delta\phi_{+-}$, our two eigenstates will exhibit vectorial bistability. The steady-state solutions of eq. (6) can be associated with a Landau's potential V that leads to the prediction of two possible bistability mechanisms [9], depending on the value of $\Delta\phi_{+-}$: a rotation mechanism where the polarization "rotates" continuously from σ^+ to σ^- for low $\Delta\phi_{+-}$ and an inhibition mechanism where, during the flip, one eigenstate disappears as the other one appears at its own frequency for higher values of $\Delta\phi_{+-}$. These predictions have been checked experimentally with a 66-cm-long cavity containing a 7:1 ^{3}He-^{20}Ne mixture at P = 1.1 Torr. As shown in Fig. 5, this cavity also contains a CH_4 cell that, once filled with 10 Torr of methane, restrains the $J = 1 \rightarrow J = 2$ transition at $\lambda = 3.3922$ μm from operating and leads to laser oscillation with the $J = 1 \rightarrow J = 1$ transition at $\lambda = 3.3912$ μm. The use of the YIG crystal to provide the Faraday rotation $\Delta\phi_{+-}$ allows us to avoid the decrease of C that would be induced by a magnetic field applied to the discharge itself. Some typical results are shown in Fig. 6. This figure displays the output power detected through a circular polarizer vs cavity frequency. In Figs. 6(a) and 6(c) the circular polarizer selects only the σ^+ component and in 6(b) and 6(d) the linear polarizer is aligned along the axes of the quarter-wave plate. Figs. 6(a) and 6(b) correspond to $\Delta\phi_{+-} = 0.25°$ and the peaks of Fig. 6(b) show that the polarization "rotates" continuously from one eigenstate to the other during the flip. On the contrary, for $\Delta\phi_{+-} = 0.8°$ (see Figs. 6(c) and 6(d)), these peaks disappear, showing that the flipping mechanism is now the inhibition mechanism. One can study the evolution of the hysteresis cycle with $\Delta\phi_{+-}$ in both mechanisms (see Fig. 7). One can see that the width of the cycle first increases with $\Delta\phi_{+-}$ in the rotation mechanism (Figs. 7(a-c)) to reach a limit width and that in the inhibition mechanism this evolution is reversed (Figs. 7(d, e)). As shown in Fig. 7, these results are in very good agreement with theory. The experiment also indicates that the evolution of the width of the hysteresis cycle with the excitation is also reversed in the two mechanisms. When one increases the excitation, the width of the hysteresis cycle is reduced in the

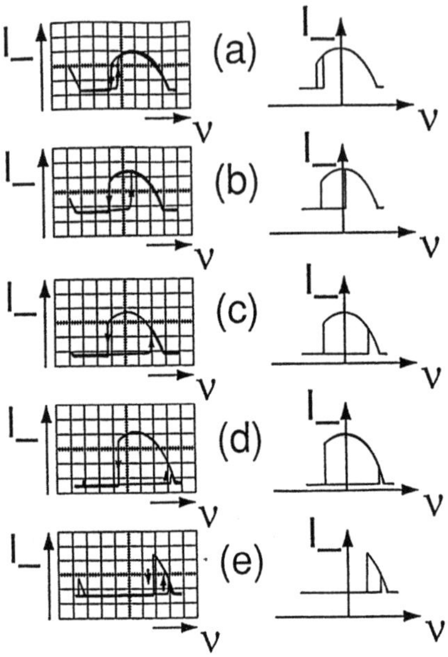

Figure 7: Experimental (left column) and theoretical (right column) evolutions of the hysteresis loop for five values of $\Delta\phi_{+-}$ (0.27°, 0. 42°, 0. 65°, 1.02° and 1.25° respectively). (a), (b) and (c) correspond to the rotation mechanism and (d) and (e) to the inhibition mechanism.

rotation mechanism and increased in the inhibition mechanism, in agreement with the predictions [9]. We consequently observe that the behavior is completely different from the case $J = 1 \rightarrow J = 2$ and that no instabilities occur.

4. Conclusion

We have shown theoretically and experimentally that the Earth's magnetic field and the atomic coupling play a crucial role in polarization instabilities that can occur in quasi-isotropic lasers. On the one hand, the Earth's magnetic field can induce polarization instabilities in a weak atomic coupling laser, especially if the finesse of the cavity is bad. This result can be interpreted without any extra involvment of the atomic dynamics, contrary to what was stated in Ref. 5. On the other hand, it has been proved that a non-weak atomic coupling quasi-isotropic laser exhibits the two basic mechanisms of vectorial bistability between its two circularly polarized eigenstates. Such a system was not observed to exhibit any particular instabilities, in spite of the predictions of Ref. 4.

References

[1] W. Van Haeringen, Phys. Rev. **158**, 256 (1967); W. J. Tomlinson and R. L. Fork, Phys. Rev. **164**, 466 (1967); A. Le Floch and R. Le Naour, Phys. Rev. A **4**, 290 (1971); A. Le Floch and G. Stephan, Phys. Rev. A **6**, 845 (1972).

[2] Y. C. Chen and J. M. Liu, Appl. Phys. Lett. **46**, 16 (1985); A. Le Floch, G. Ropars, J. M. Lenormand, and R. Le Naour, Phys. Rev. Lett. **52**, 918 (1984); G. Ropars, A. Le Floch, and R. Le Naour, Europhys. Lett. **3**, 695 (1987).

[3] P. Glorieux and A. Le Floch, Opt. Commun. **79**, 229 (1990); A. Le Floch and P. Glorieux in *Proceedings of EQEC, Hannover, Sept. 1988* and in *Laser Spectroscopy IX* (*Proceedings of the ninth international conference on laser spectroscopy*), edited by M.S. Feld, J. E. Thomas, and A. Mooradian (Academic Press, San Diego, CA, 1990).

[4] G. P. Puccioni, M. V. Tratnik, J. E. Sipe, and G. L. Oppo, Opt. Lett. **12**, 242 (1987).

[5] G. P. Puccioni, G. L. Lippi, N. B. Abraham, and F. T. Arecchi, Opt. Commun. **72**, 361 (1989).

[6] R. Adler, Proc. IRE **34**, 351 (1946).

[7] J. C. Cotteverte, F. Bretenaker, and A. Le Floch, Opt. Commun. **79**, 321 (1990).

[8] M. Sargent III, M. O. Scully, and W. E. Lamb, Jr., *Laser Physics* (Addison-Wesley, Reading, MA, 1974).

[9] J. C. Cotteverte, F. Bretenaker, and A. Le Floch, "The dynamics of circularly polarized eigenstates in lasers with non-weak atomic coupling", to be published.

Ultrashort Pulse Dynamics in Nonlinear Couplers

M. Romagnoli, S. Trillo, S. Wabnitz, and L. Zoccolotti
Fondazione Ugo Bordoni, Via B. Castiglione 59,
I-00142 Roma, Italy

Abstract. Nonlinear coupled waveguides may permit the switching of ultrashort optical pulses. We present and compare here numerical studies of the propagation of coupled solitons for different physical models of the nonlinearity. In particular, we consider first an ideal nonresonant (Kerr) nonlinearity, then we deal with the coherent interaction with two-level absorbers (self-induced transparency), and finally we consider the case of an active semiconductor material.

1. Introduction

In a directional coupler, linear coupling between the modes of the two individual waveguides is due to the overlap of their evanescent fields. The presence of a nonlinearity in the response of the material may be exploited for performing all-optical switching functions [1]. In addition, for a continuous wave (cw) the nonlinear dynamics of the modal coupling may exhibit interesting spatial instability phenomena [2]. In ultrafast optical processing applications one is interested in short optical pulses and efficient switching may only be obtained by exploiting the particle-like nature of solitons [3]. Depending on the specific mechanism for the nonlinear light-matter interaction, one may encounter different types of solitons. We compare here three important and somewhat extreme cases, which correspond to the different possible relationships between the pulsewidth and the characteristic nonlinearity response time in the nonlinear directional coupler (NLDC).

2. Nonresonant fiber couplers

Consider first the case when the pulsewidth is much longer than all relaxation times involved in the nonlinear response of the material. This occurs for example through the virtual excitation of electronic levels in a transparent dielectric like a glass fiber. The pulse propagation in a fiber coupler is governed by the nonlinear Schrödinger (NLS) equations [4]

$$i\frac{\partial u_1}{\partial z} + \frac{1}{2}\frac{\partial^2 u_1}{\partial t^2} + (|u_1|^2 + |u_2|^2)u_1 = \epsilon_1 u_2 + \epsilon_2 |u_2|^2 u_1, \qquad (1)$$

Springer Proceedings in Physics, Vol. 55
Nonlinear Dynamics and Quantum Phenomena in Optical Systems
Editors: R. Vilaseca, R. Corbalán

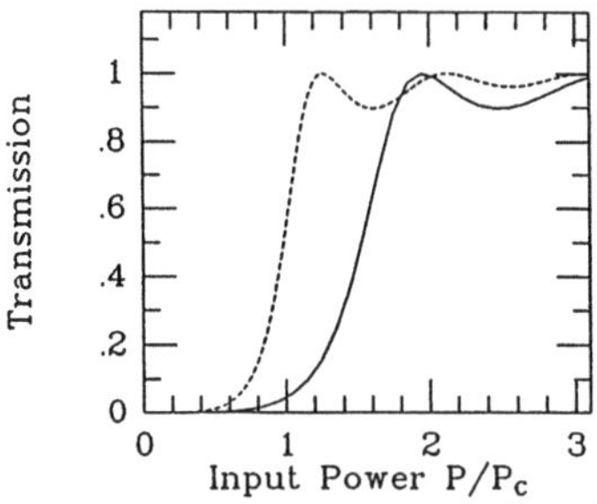

Figure 1: Transmission for a cw wave (dashed) and a soliton (here the pulse width was adjusted according to the amplitude, see eq.(2)) in a Kerr type NLDC.

$$i\frac{\partial u_2}{\partial z}+\frac{1}{2}\frac{\partial^2 u_2}{\partial t^2}+(|u_1|^2+|u_2|^2)u_2=\epsilon_1 u_1+\epsilon_2|u_1|^2 u_2.$$

Note that normalized units have been used above. Moreover, here the linear modal structure is maintained also in the nonlinear regime. The right hand side has been written as a perturbation to the coupled NLS equations of the left-hand side, which possess the vector soliton solution [5]

$$\begin{aligned} u_1 &= \eta\sqrt{p_1}sech(\eta t)e^{i\phi+i\psi/2}, \\ u_2 &= \eta\sqrt{p_2}sech(\eta t)e^{i\phi-i\psi/2}. \end{aligned} \tag{2}$$

By assuming that ϵ_1 and ϵ_2 are small, one may look for the slow evolution (i.e., on a length scale of ϵ^{-1}) of the soliton parameters [6], namely $y=p_1-p_2$ and ψ. Straightforward perturbation analysis yields the Hamiltonian

$$H(y,\psi)=\sqrt{(1-\eta^2)}\,cos(\psi)+\frac{2}{3}\eta^2 y^2 \tag{3}$$

governing the soliton dynamics (here we have set $\epsilon_2=4\epsilon_1$ so that the cw switching power of the coupler is equal to one. The Hamiltonian (3) is the same as for cw waves [7], except for an increase of 3/2 in the effective switching power. This point has been confirmed by numerical simulations: fig.1 shows the energy transmission from the incidence channel for cw (dashed curve) and soliton (solid curve) excitation.

3. Passive two-level atom coupler

It is interesting to observe that even in the extreme case whenever the pulse width is much shorter than all the relaxation times of the material re-

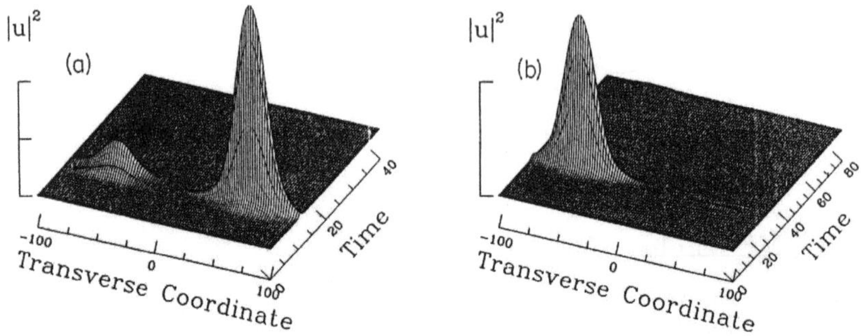

Figure 2: Output pulse switching from cross (a) to bar (b) state in two-level NLDC.

sponse, the process of soliton coupling still permits particle-like switching behavior. We assume here a coupler with single-mode slab waveguides, doped with two-level absorbing centers. The relevant Maxwell-Bloch equations read

$$\begin{aligned} 2\beta\frac{\partial E}{\partial z} &= \left(i\delta n(x)E + \frac{i}{2}\frac{\partial^2 E}{\partial x^2} + \alpha(x)P \right) \\ \frac{\partial P}{\partial t} &= -NE \\ \frac{\partial N}{\partial t} &= \frac{1}{2}\left(PE^* + EP^*\right). \end{aligned} \tag{4}$$

Here β is the linear propagation constant of the guided mode of an individual waveguide, $t = t' - z/V$ is a retarded characteristic time, x is a transverse coordinate, V is the modal group velocity of the mode, δn is the linear refractive index profile, α is the absorption coefficient, and finally E, P, and N are the electric field, the polarization and the population inversion, respectively. Figure (2) compares the intensities of output pulses that would emerge from a one linear coupling length long coupler, whenever the input pulse width and area are varied. In order to ensure digital switching characteristics, we assumed a tapered NLDC [8], that is with variable spacing between the waveguides.

4. Active semiconductor coupler

In this section we shall be concerned with ultrashort pulse switching by an active semiconductor NLDC, a case which is intermediate between those discussed above. In fact, we consider here pulses that are much shorter than the carrier relaxation time, but still longer than the polarization dephasing timescale so that the field equation is only coupled to a rate equation for the carrier density in each amplifier

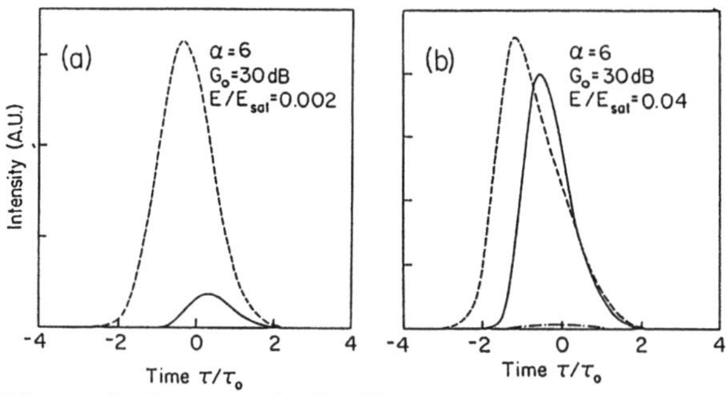

Figure 3: Same as in fig.(2), with active semiconductor NLDC.

$$
\begin{aligned}
& i\frac{\partial u_j}{\partial z} + \frac{1}{2}(i\alpha - 1)g_j u_j = \epsilon u_{3-j}, \\
& \frac{\partial g_j}{\partial t} = -\frac{g_j - g_{j0}}{\tau_c} - \frac{g_j}{E_{sat}}|u_j|^2, j = 1, 2.
\end{aligned}
\tag{5}
$$

Here α is the linewidth enhancement factor, g and g_0 are the saturated gain and its small signal value, respectively, and E_{sat} is the saturation energy of the amplifier. Figure (3) shows examples of a numerical solution of eqs.(5): by increasing the input pulse energy the interplay between differential gain saturation and its accompanying refractive index mismatch may again lead to nearly undistorted, although incomplete, switching of the input pulse.

References

[1] S.M. Jensen, IEEE J. Quantum Electron. **QE-18**, 1580 (1982).

[2] B. Daino, G.Gregori, and S. Wabnitz, J. Appl. Phys. **58**, 4512 (1985).

[3] N.J. Doran, and D. Wood, J. Opt. Soc. Am. **B 4**, 1843 (1987).

[4] S. Trillo, S. Wabnitz, E.M. Wright, and G.I. Stegeman, Opt. Lett. **13**, 672 (1988).

[5] S.V. Manakov, Sov. Phys. JETP **38**, 248 (1974).

[6] D.J. Muraki, and W.L. Kath, Phys. Lett. A **139**, 379 (1989).

[7] E. Caglioti, S. Trillo, and S. Wabnitz, Opt. Lett. **12**, 1044 (1987).

[8] A. Guzman, M. Romagnoli, and S. Wabnitz, Appl. Phys. Lett. **56**, 614 (1990).

[9] S. Trillo, S. Wabnitz, J.M. Soto-Crespo, and E.M. Wright, to be published on the J. Opt. Soc. Am. B

Deterministic Phase Diffusion, Dimensions and Entropies of Chaotic Lasing in the He-Ne Ring Laser Gyro

I.E. Zuikov[1], *P.G. Krivitsky*[1], *A.M. Samson*[2], and *S.I. Turovets*[2]

[1]Byelorussian Polytechnical Institute, SU-220027 Minsk, USSR
[2]Institute of Physics, BSSR Academy of Sciences, SU-220602 Minsk, USSR

Abstract. Theoretical prediction, experimental observations and detailed studies of instabilities and chaos in the laser gyroscope with an alternating bias or with external sinusoidal perturbations are presented.

1. Introduction

The gas ring laser as a sensitive element of measuring instruments can find itself in the conditions of narrow-band noise, e.g., vibrations or electromagnetic fields. Such periodic perturbations of the parameters lead to the effect of induced locking of the beat frequency of counterpropagating waves and to the appearance of insensitivity "shelves" on the output characteristic of the laser [1,2]. The present paper shows that alongside the locking effect in such lasers dynamical chaos must be observed.

2. Theoretical model

The dynamics of a single-mode ring gas laser obeys the well-known system of Lamb amplitude-phase equations [3]. For the sake of simplicity, we assume that external perturbations only lead to periodic oscillations of the frequency bias (the regime of bias modulation is sometimes used for the output characteristic linearization). Then, in the approximation of a weak coupling through backscattering for the phase difference of counterpropagating waves we can obtain [1]

$$\frac{d^2\varphi}{d\tau^2} + \frac{1}{\sqrt{\beta_c}} \cdot \frac{d\varphi}{d\tau} + \sin\varphi = \nu_0 + \nu_1 \cdot \sin\omega_M\tau/\sqrt{\Gamma\nu_L} , \tag{1}$$

where the dimensionless quantities $\tau=(\Gamma\nu_L)^{1/2}t$, $\beta_c=\nu_L/\Gamma$, $\nu_0=$

Springer Proceedings in Physics, Vol. 55
Nonlinear Dynamics and Quantum Phenomena in Optical Systems
Editors: R. Vilaseca, R. Corbalán

Ω_0/ν_L, $\nu_1=\Omega_1/\nu_L$ and ν_L is the "dead band" half-width, Γ is the coefficient of phase relaxation, $\Omega_0+\Omega_1\cdot\sin\omega_M t$ is the resonator frequency difference for counterpropagating waves.

In this writing, the dynamics of the phase difference is analogous to the dynamics of the rf-biased Josephson junction with the hysteresis parameter β_c for which in a number of works [4 and references therein] the possibility of the appearance of dynamic chaos has been shown. Summarizing the available data, we can distinguish in the parameter space zones with the highest probability of detecting chaos: $\beta_c>0.25$; $\omega_M\lesssim(\Gamma\nu_L)^{1/2}$; $\nu_0\lesssim1$; $\nu_1\sim1-10$. These limitations on the parameters are, to a certain extent, an inversion of the conditions for applicability of analytical methods of solving equation (1).

It should be noted that the requirement $\beta_c=\nu_L/\Gamma>1/4$, contradicts, generally speaking, the condition of weak coupling at which equation (1) was obtained. Nevertheless, a number of ring gas lasers have an output characteristic with a considerable hysteresis and a smooth transition from the regime of beats to locking and can phenomenologically be described from the point of view of model (1).

Our numerical simulations, based on the Lamb semiclassical equation as well as on phenomenological model (1), indicate that at least two types of chaos are possible in the system: "period-doubling" and "intermittency" . In the first case, the phase difference oscillations are irregular but amplitude-limited, while in the second case the phase angle difference experiences diffusive motion.

3. Experimental results

From theoretical analysis it can be concluded that to register chaos in experiment, ring lasers with a pronounced hysteresis of the output characteristic should be chosen.

Experiments were performed on a monoblock four-mirror He-Ne laser with a wavelength of 0.63 μm and a dual isotope mixture of Ne. Near the generation line center the half-width of the dead band ν_L was 700-800 Hz at the hysteresis magnitude 40-50%, which corresponds to $\beta_c\approx 6$. The operating point was chosen in the nonlinear part of the output characteristic ($\nu_0\lesssim\nu_L$).

At scanning of one of the control parameters (perturbation amplitude, constant bias, detuning from the center of the generation line) the transition from the time-periodic beat signal to chaos was observed. In this case, the phase portrait of the system took on a "smeared" form of a strange attractor, and the noise level in the signal spectrum rose jumplike by 30 dB. It is found that the boundaries of chaos in the space of control parameters are in good agreement with the results of numerical modeling.

We have also directly proved by experiment the dynamic nature of the observed irregular pulsations by means of determining the correlation dimension D_2 and entropy K_2 on the basis of the Grassberger-Procaccia algorithm.

To optimize the digitizing frequency of the analog-to-digital converter (ADC) at a finite length of experimental recording, a series of numerical calculations in the framework of theoretical model (1) has been performed. For the parameters corresponding to the experiment the correlation dimension D_2 was saturated at a sampling frequency 8–16 times higher than the modulation frequency and was equal, depending on the type of attractor, to 2.4±0.1 (intermittency) and 2.1±0.1 (Feigenbaum attractors).

In the experiment a 12-bit ADC connected to a personal computer was used. At the modulation frequency 380 Hz, which constituted 1/4 of locking band, the clock frequency was 3125 Hz. The total length of the basic set of experimental data amounted to 8000 samples for chaos with a minimal threshold. The principal results can be summarized as follows (see also Fig. 1).

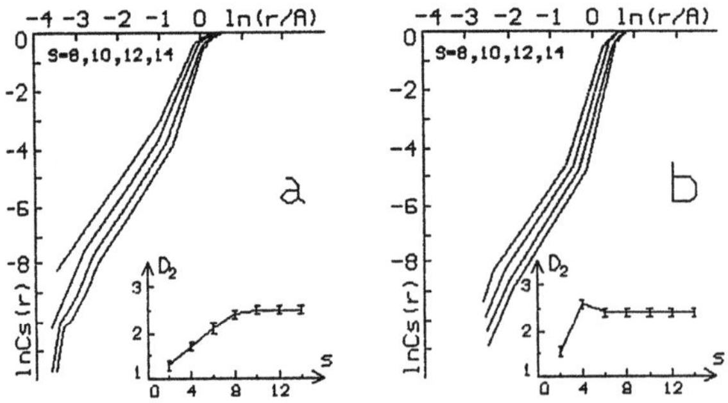

Fig.1. The correlation integral $C_s(r)$ vs the correlation radius r at a logarithmic scale: a - experiment, b - model (1) (β_c=6; ν_0=0; ν_1=1.2; ω_M/ν_L=0.3). Inserts show the corresponding dependencies of the correlation dimensions D_2 on the embedding space dimensions s. A is the signal swing.

Irregular oscillations of the beat signal are dynamic in nature (K_2 is positive and finite), motion occurs on the low-dimensional strange attractor with D_2=2.6±0.1. The dynamics of such lasers can be described by a system of no more than three autonomous first-order differential equation (D_2<3). Thus, the transition from the Lamb equations to the simpler model (1) is reasonable.

Within error, the experimental value of D_2 coincides with the value obtained in the phenomenological model which permits classifying the observable chaotic regimes as intermittency or chaos with diffusion in the phase space.

4. Conclusion

It is shown that in the ring gas laser with the hysteresis of the output characteristic at periodic perturbation of the constant frequency bias, chaotic processes are possible, and they should be taken into account to provide the optimum conditions for operation of such lasers as components of measuring instruments. We have found also that a simple modification of the first order equation for the phase angle difference leads to explanation for all of the results obtained. It is worth noting that this equation is isomorphic to a classic dynamical system which describes many physical phenomena. Alongside with the famous Lorenz-Haken case this model, we hope, will be another suitable candidate for fundamental studies of nonlinear dynamics in lasers.

The authors thank Dr. A.A. Kutsak for helpful discussions.

References

1. G.S. Kruglik. Zh. Prikl. Spectrosk. 7, 569 (1967).
2. G.S. Kruglik, E.G. Pestov, V.R. Pocrovscy, and A.A. Kutsak, Zh. Prikl. Spectrosk. 12, 432 (1970); 13, 913 (1970).
3. Wave and Fluctuation Processes in Lasers. Edited by Yu.L. Klimontovich (Nauka, Moscow, 1974).
4. F.J.Romeiras and E. Ott, Phys. Rev. 35A, 4404 (1987).

Part V

Transverse Effects

Transverse Pattern Formation in Lasers

Chr. Tamm[1], *L.A. Lugiato*[2], *M. Brambilla*[2], *A.B. Coates*[1], *C.O. Weiss*[1], *and R. McDuff*[3]

[1]Physikalisch-Technische Bundesanstalt, Postfach 3345, W-3300 Braunschweig, Fed. Rep. of Germany

[2]Dipartimento di Fisica, Università di Milano, Via Celoria 16, I-20133 Milano, Italy

[3]Department of Physics, The University of Queensland, St. Lucia, Australia

Abstract. Lasers have proven useful in studying the behaviour of non-linear systems. In the temporal domain, this behaviour may be chaotic, e.g. dynamics of a single-mode laser which obeys the Lorenz model. In spatially-extended (multimode) systems the spatial behaviour is as important as the temporal behaviour. In contrast to the single-mode case, which is described by a system of ordinary differential equations, spatially-extended systems are normally described by partial differential equations. We describe recent results with transversely extended lasers which relate to spatio-temporal chaos, pattern formation, and defects in optical fields.

1. Introduction

In the past, single-mode lasers with a stable frequency have been the predominant goal of experimental laser research. However, it has become clear in recent years that lasers can also behave in a non-static way, exhibiting complex dynamical behaviour. This behaviour may be essentially temporal, but frequently is both spatial and temporal.

Since Haken's discovery [1] in 1975 that the laser equations can be transformed into the well-known Lorenz equations, it has become clear that laser dynamics and fluid dynamics are related. This relationship has recently been strengthened by extension to multimode, spatially-varying systems. In addition we find surprising analogies between laser dynamics and the dynamics of superconductors, superfluids, solids, and even cosmic strings.

In 1984 we found chaotic dynamics in an autonomous multimode HeNe laser, which displayed the signatures of all three prominent routes to chaos: (i) period-doubling (figure 1a), (ii) quasi-periodic transition (figure 1b), and (iii) intermittent transition (figure 1c). While chaotic laser emission was thus confirmed experimentally, the quantitative theoretical understanding of this particular multimode system is still lacking.

In an attempt to simplify the experimental situation, we reconsidered the Lorenz model [2], and have been able to show that the same dynamics can describe optically-excited gas lasers [3]. As an example of the good agreement we show (a) a two-dimensional bifurcation diagram for the complex Lorenz equations, which corresponds to the laser allowing for detuning (figure 2), and the pulsing of an NH_3 laser measured at different detunings (figure 3). The measurements were

Nonlinear Dynamics and Quantum Phenomena in Optical Systems
Editors: R. Vilaseca, R. Corbalán

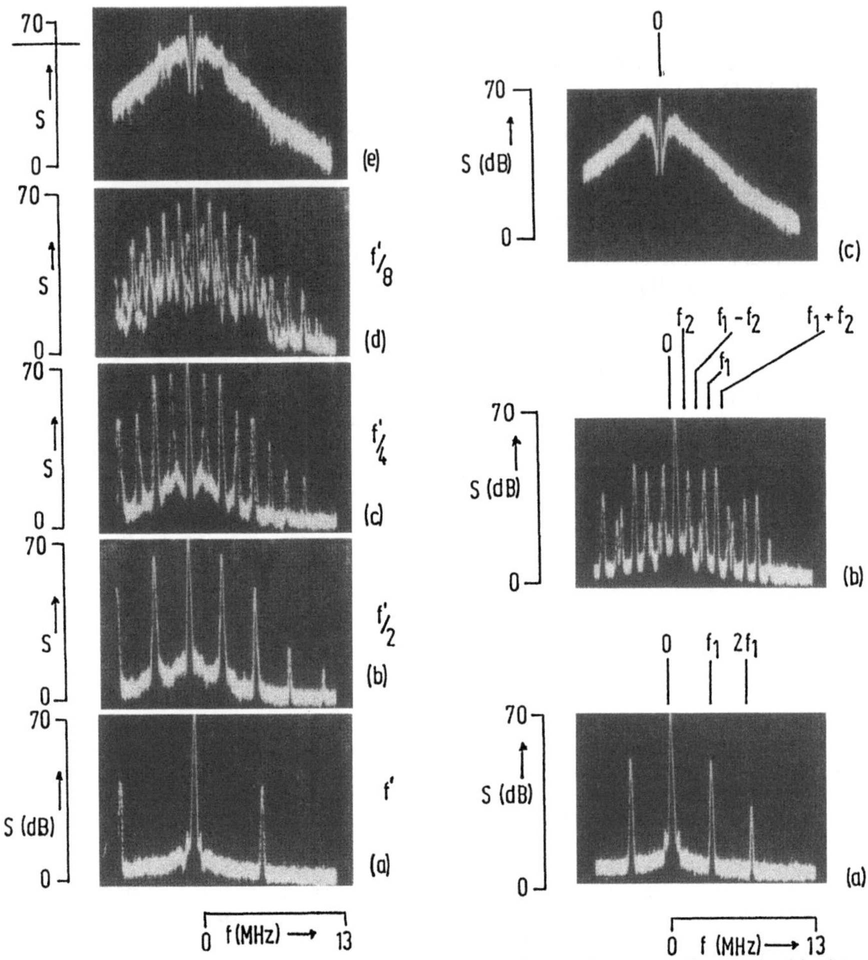

(A) via period-doubling, (B) via quasi-periodicity,

Figure 1: Laser transitions to chaos:
(spectra of laser intensity)

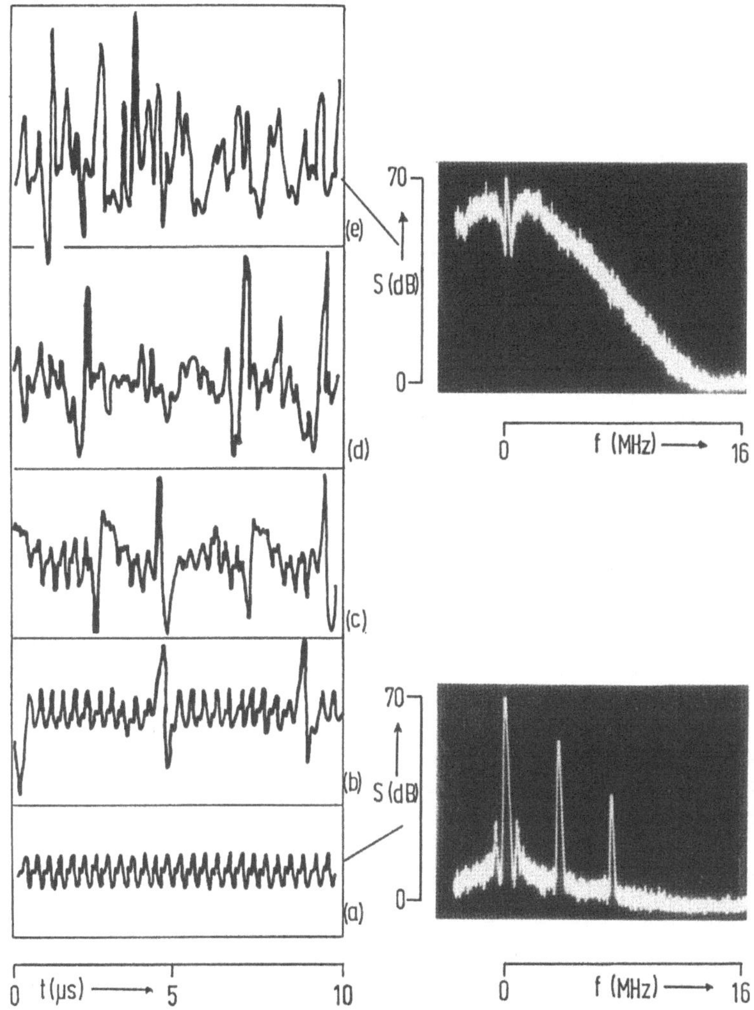

(C) via intermittency

Figure 1 (continued)

taken at the points indicated in figure 2 and show precisely what is predicted by the Lorenz equations in figure 2.

Why we find this good agreement remains unclear. The Lorenz equations are a plane-wave, homogeneous medium model, while in contrast a real laser emission field displays variation in the directions transverse to the propagation axis.

Lasers with transverse degrees of freedom may show steady-state emission (not to be confused with the emission of "modes"). To prove the existence of spatio-temporally chaotic emission from lasers, a simple experiment was carried out: a multimode Na_2 laser was observed with two detectors, each at a different position within the cross-section of the laser

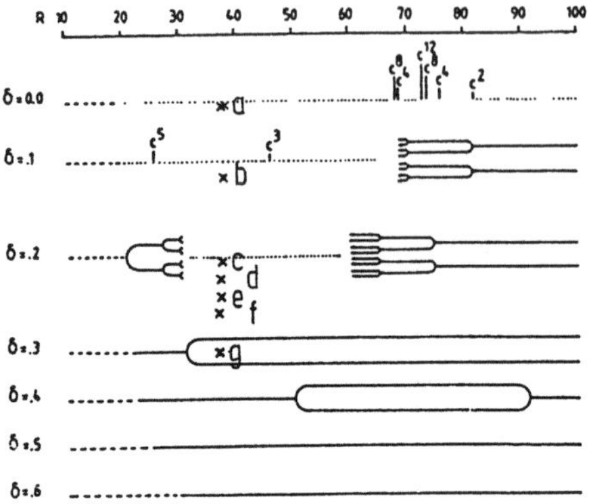

Figure 2: Bifurcation diagram for the complex Lorenz equations. R is the pump strength, S is the detuning [4].

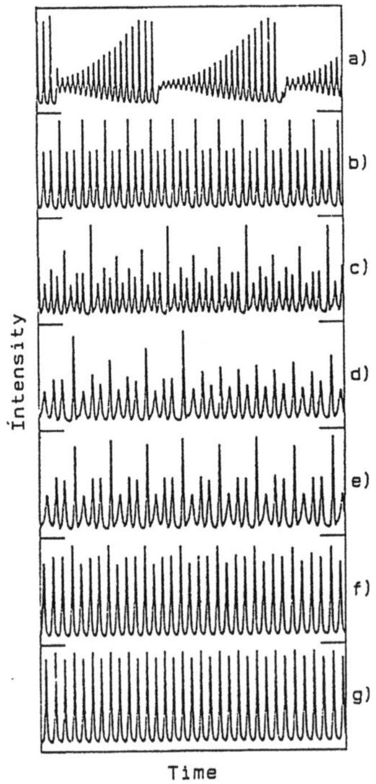

Figure 3: Intensity pulsing of an NH_3 laser at detunings marked in figure 2.

field. It was possible to find conditions under which each of the two detectors recorded chaotic emission, different for the two different detectors (figure 4). Figure 5 shows the Fourier transforms of both signals in figure 4; both display broadband spectra characteristic of chaotic dynamics.

That the two detectors display different chaotic signals appears to indicate spatio-temporal chaos, since it is clearly more complicated than temporal chaotic variation of the entire laser field. Spatio-temporal chaos implies that the two signals are different projections of one spatio-temporal attractor.

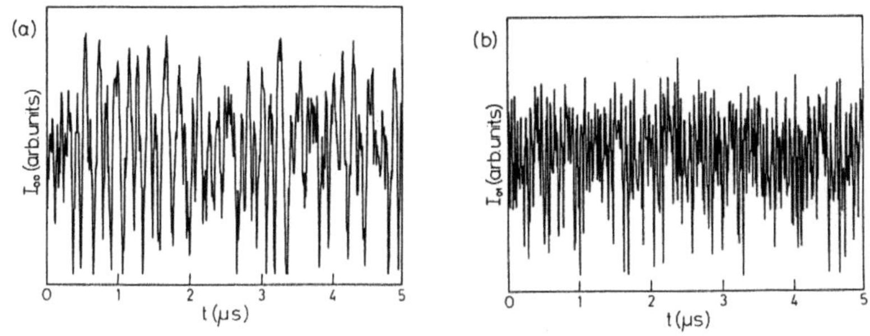

Figure 4: Output intensity (arbitrary units, 8 bit) vs time for two different positions of the detector. (a) Center of TEM_{00} mode, (b) center of upper lobe of TEM_{01} mode.

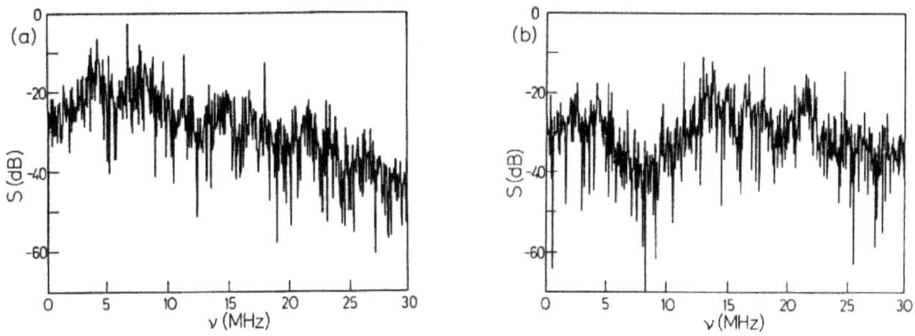

Figure 5: Power spectra calculated from the intensity data given in fig. 4.

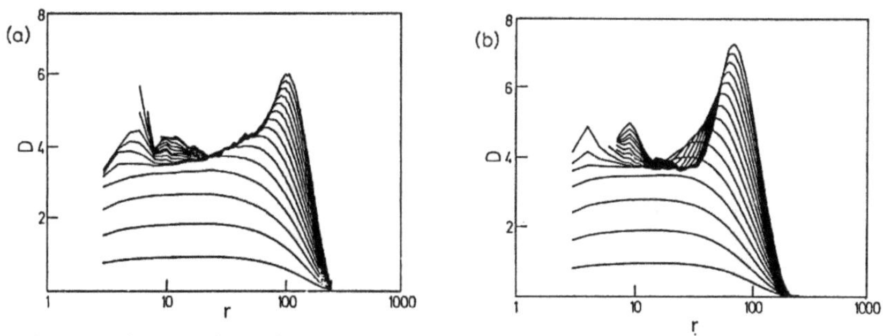

Figure 6: Derivatives of the correlation integrals C(r), representing the fractal dimension, vs r (from the data in fig. 4).

Since this is a rather general relationship, it is difficult to prove experimentally. The correlation dimensions of the two detector signals (see figure 6) have the same dimension of ≈ 4; this suggests that both could indeed be projections of the same attractor. For more evidence of a relationship between the two signals, an X-Y trace of the detector signals is shown in figure 7. This plot is not uniform, but shows some structure, which would not be expected if the two signals were independent.

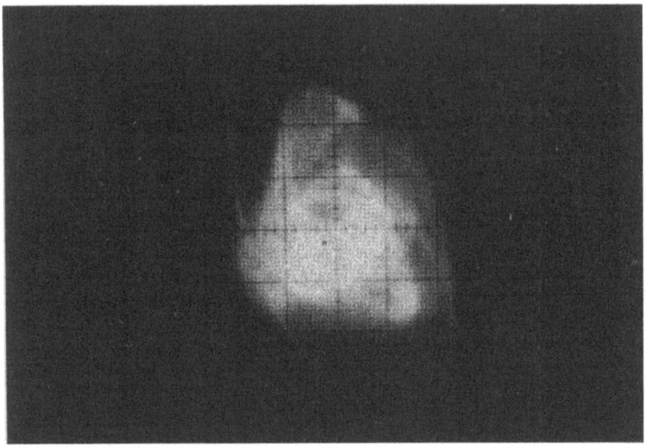

Figure 7: x-y representation of the two time signals of fig. 4 as displayed by an oscilloscope.

Although the evidence is consistent with the idea of a common attractor, a clear confirmation has not yet been possible. The mutual information transfer between the signals may provide further evidence.

Even though the dynamics of lasers with large numbers of modes may be an interesting field to study, it is too wide a field to start with directly. In [5] the spatial properties of the laser emission in the absence of temporal dynamics were considered. As a test we used the simplest case, namely the TEM_{01} and the TEM_{10} modes of a circular Fabry-Perot resonator containing an active medium. Usual resonators contain astigmatic elements like Brewster windows which destroy the frequency degeneracy of modes of equal index sum. By compensating the astigmatism, the frequencies of the TEM_{01} and TEM_{10} modes can be made to approach each other until, as predicted in [6], the two frequencies synchronise in phase to form a stationary pattern. Figure 8 shows the spectra of the intensity fluctuations of a HeNe laser which is forced to emit on first order transverse modes when its astigmatism is progressively reduced. At the largest astigmatism, (figure 8a) we see essentially a beat frequency between the two modes. The weak harmonic of the beat frequency is the first indication of the medium`s non-linearity. When the astigmatism is reduced (figure 8b) and thus the frequencies of the modes approach each other, the beat frequency is accordingly smaller. More harmonics are observed because the beat frequency comes into the range of the response time of the medium. Further reduction of astigmatism then produces a period-doubling cascade to chaos of which the first period doubling (figure 8c) and the chaotic spectrum (figure 8d) are shown. Further astigmatism reduction results in phase-locking of the two modes (figure 8e).

The locking of the TEM_{01} and TEM_{10} modes (figures 9a and 9b) results in the $TEM_{01}*$ mode or "doughnut mode" (figure 9c). The bright halves of the TEM_{01} mode are 180° out of phase. It follows, from the observation that the "doughnut

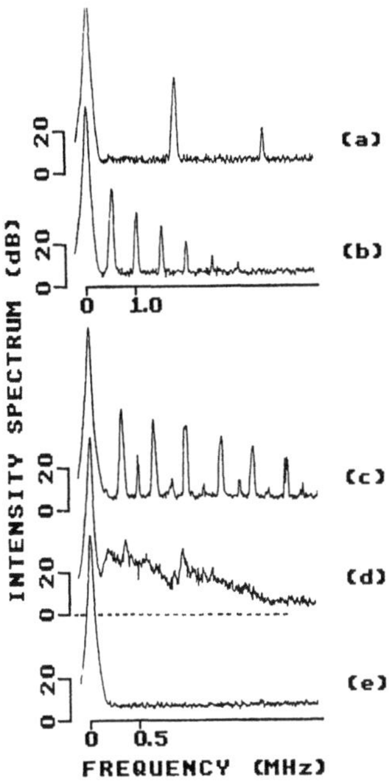

Figure 8: HeNe laser spectra for first order transverse modes. The astigmatism in (a) is progressively reduced until it disappears in (e).

mode" has no angular intensity variation, that locking of the two modes occurs with a 90° relative phase difference between the modes.

This results in a helical wave (figure 10a). This helical wave is a form of wave defect (an analogue of the screw dislocation in crystals). It is worth mentioning that other types of wave defects, analogous to defects in solids, also exist. Figure 10b shows an example of an optical edge dislocation analogue to the edge dislocations occuring in crystals [7]. It follows that the helical wave solution (doughnut mode) must be bistable in a laser, with both left-handed and right-handed helix forms which have the same frequency and intensity.

We have performed a simple experimental confirmation of the helix bistability of a laser (figure 11). The system consists of a laser emitting in the TEM_{01}* form. Part of its radiation is split off in order to observe the intensity ring pattern and to verify the phase locking of the TEM_{01} and TEM_{10} modes (by the absence of time dependence in the signal from a fast detector). Another part of the beam is split off to observe the handedness of the laser emission using an astigmatic mode converter [8]. The converter transforms the two helix waves into TEM_{01} patterns of orthogonal orientation. One detector is then used to distinguish between the left-handed and right-handed helix wave emissions. To switch between the two helix emissions, a prism angled at 45° out of the plane of the diagram rotates the helix waves by 90°. The

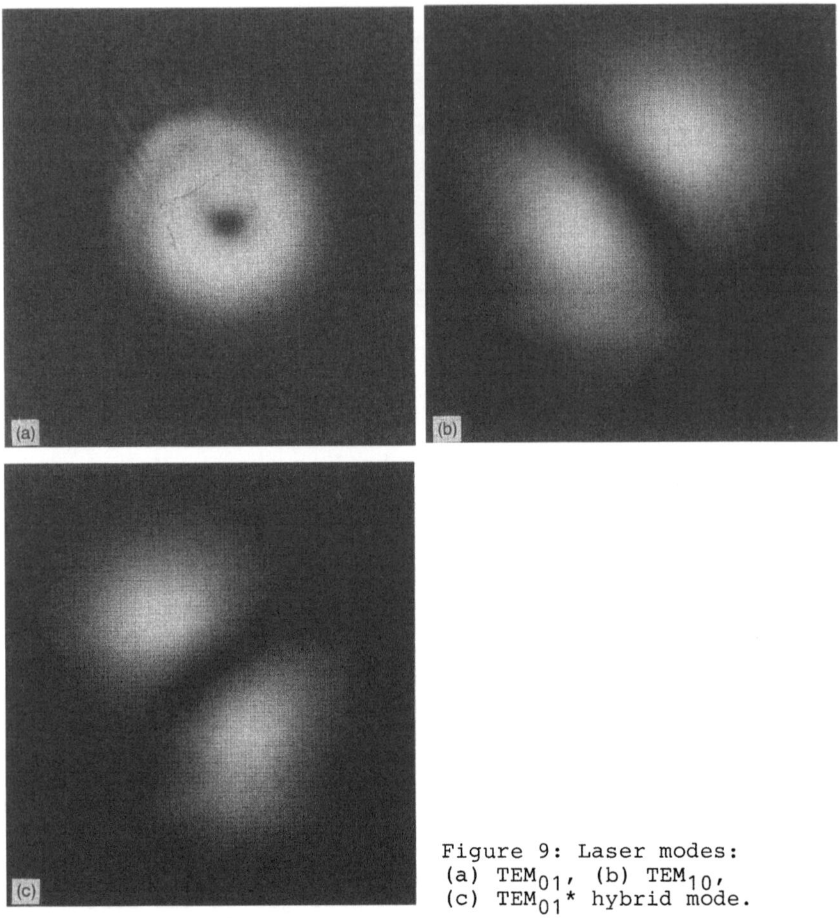

Figure 9: Laser modes:
(a) TEM_{01}, (b) TEM_{10},
(c) $TEM_{01}*$ hybrid mode.

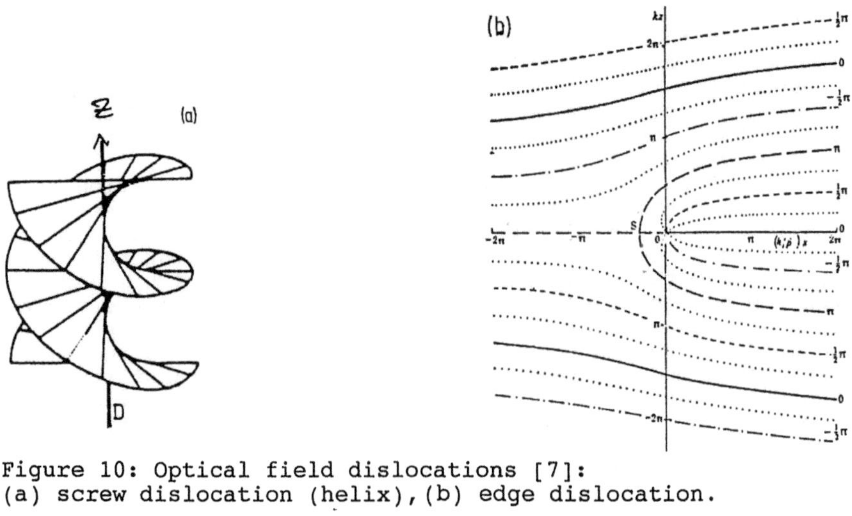

Figure 10: Optical field dislocations [7]:
(a) screw dislocation (helix), (b) edge dislocation.

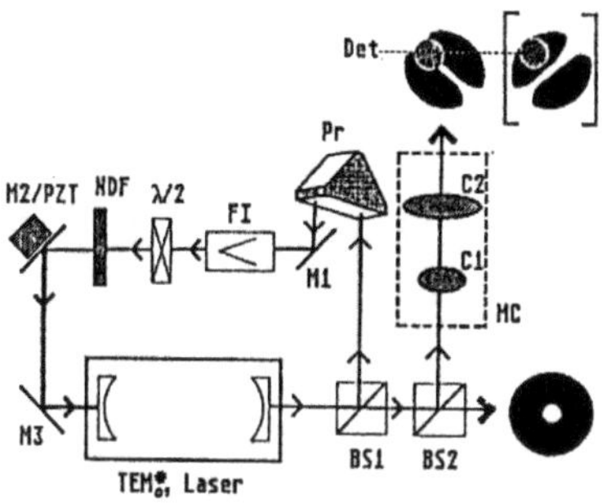

Figure 11: Schematic for laser bistability experiment. Det: detectors, Pr: prism, FI: Faraday isolator, λ/2: half-wave plate, NDF: neutral density filter, M1, M2, M3: mirrors, BS1, BS2: beam splitters, C1, C2: cylindrical lenses, MC: mode converter.

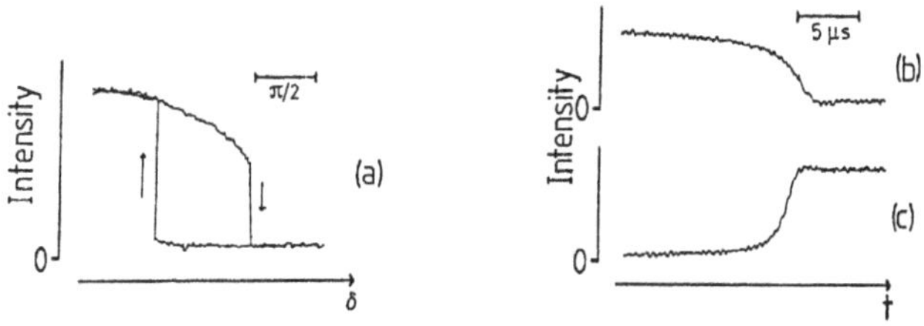

Figure 12: Mode converter output:
(a) bistability demonstration,
(b) expanded view of switching region.

rotated wave is then fed into the laser (after correction of the polarisation and attenuation). A mirror on a PZT mount is used to change the optical feed-back path length. The reflecting prism results in a path length difference between the left-handed and right-handed helix waves of half a wavelength, or 180°. The laser then emits the one helix wave which is able to interfere constructively with itself when reinjected into the laser. Switching between the emission of the helix waves occurs as the length of the feedback path is changed.

Figure 12a shows the detector intensity behind the mode converter as a function of the length of the feedback path. The bistability between the two helix wave emissions is obvious. Figures 12b and 12c show the switching between the two helix wave emissions. In this particular laser (a low gain HeNe laser) the switching time is slow, governed by the field build up time within the laser resonator.

Fields with more than one helix line or phase singularity are possible. Figure 13 shows the equiphase lines of a superposition of a TEM_{01} Laguerre mode with a helical mode. Figure 14 shows the superposition of the TEM_{12} and TEM_{21} Hermite modes, and figure 15 shows the superposition of the TEM_{45} and TEM_{54} Hermite modes. It is evident from these pictures that the equiphase lines around these helix points or phase singularity points bear a striking resemblance to electric field lines of charged particles.

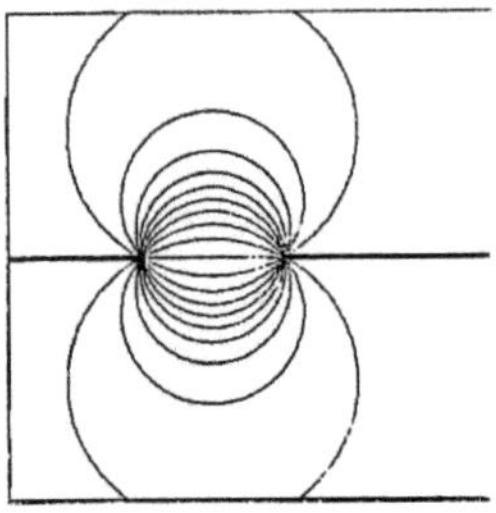

Figure 13

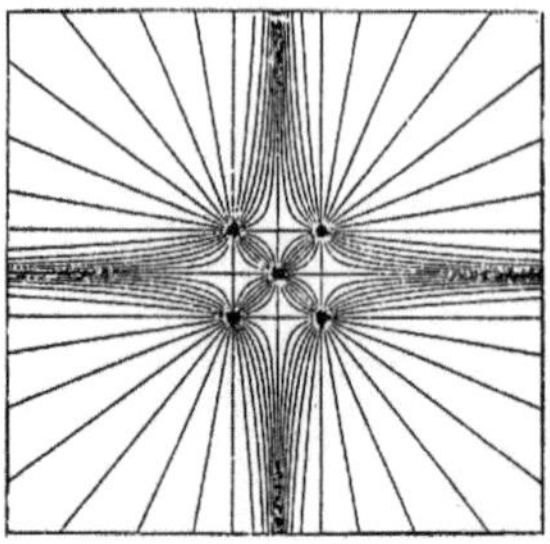

Figure 14

Figure 15

Figure 13: Equiphase lines of the superposition of a TEM_{01} Laguerre mode and a $TEM_{01}*$ hybrid mode.

Figure 14: Equiphase lines of the superposition of a TEM_{12} mode and a TEM_{21} mode.

Figure 15: Equiphase lines of the superposition of a TEM_{45} mode and a TEM_{54} mode.

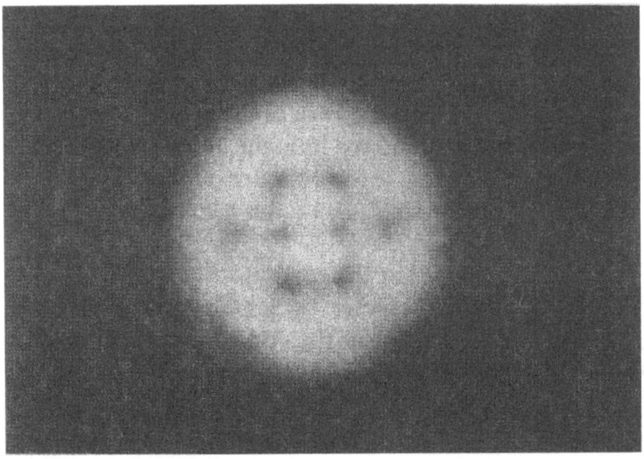

Figure 16: Photograph of a laser pattern with 8 phase singularities in a crystalline formation [9].

Figure 15 resembles the arrangement of charged particles in an ionic crystal. Even the surface effects appear correctly. Evidently figure 15 consists of right-handed and left-handed helix points. If one calculates the phase variation around the helix points, one finds that any integer multiple of 2π is possible. This integer is the "topological charge" of the singularity. In figure 15 only charges of +1 or -1 appear. Proof that such "singularity crystals" occur as laser emission fields is shown in figure 16, observed from a Na_2 laser [9]. The crystal presumably has 4 positive and 4 negative singularities. Evidently all such crystals exhibit at least bistability (simply change the signs of all the charges).

We note an advantage of this bistable system over other optically bistable systems. The helical waves differ only in their phase distribution, not in their intensity distribution; thus switching between them may not require very much addition or dissipation of energy, the source of the thermodynamic limitations which restrict other devices.

The general question arises of which patterns lasers will emit in general. Usually these will not be "modes", but combinations of frequency degenerate (or nearly degenerate) modes. The non-linearity of the active laser medium "decides" which linear combinations of the modes of a degenerate mode family the laser will emit, i.e. the non-linearity selects a set of complex coefficients for each member of the particular mode family. Recently a variational principle has been developed [10] which, by minimising the free energy of the laser, allows the determination of all possible stable laser emission patterns. In the course of these investigations it was found that some of the stable patterns co-exist, in which case the laser becomes a multistable device in the sense of "spatial multistability". The latter phenomenon might be useful for simple logic or decision functions, or for more complicated tasks such as pattern recognition or implementation of associative memories. We have recently tested the predicted stable laser fields, or patterns, experimentally. Figure 17 shows a schematic of the laser used for the observations. It is a ring laser tuned by an internal prism to a transition of Na_2 at a wavelength of 525 nm. The Na_2 vapour is generated in a heat pipe at a temperature of ≈ 500°C. The Na_2 transition is optically excited by 488 nm single-mode radiation from an Ar^+ laser [11]. This set-up makes it possible to realise most of the assumed conditions in the calculations [10]: (i) travelling wave emission, (ii) restriction of gain to a single family of modes, (iii) unidirectional emission, (iv) homogeneous broadening (since only a particular velocity group is excited by the pump laser), (v) high gain (this transition is the one with the highest gain of all known continuous laser transitions in the visible), (vi) variability of width of the pump volume (indispensable when working with spatial patterns). During operation, the laser resonator length is kept constant, tuned to a single-mode family by maximising the total output power of that family by a feedback control circuit. The astigmatism of the Brewster windows of the Na_2 cell is compensated for by the finite reflection angle of the curved

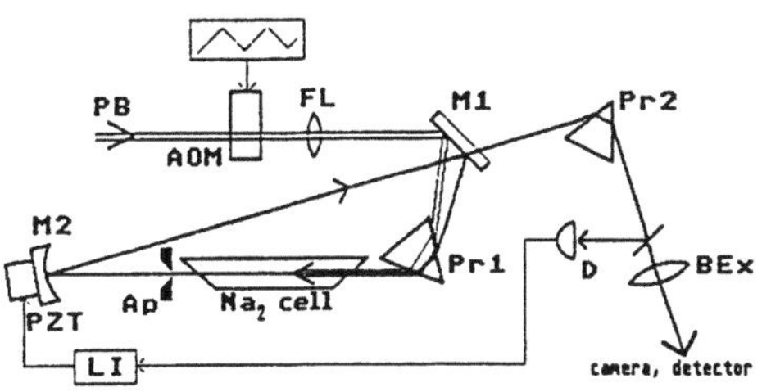

Figure 17: Schematic for transverse laser pattern experiment. PB: pump beam, FL: lens, M1, M2: mirrors, Pr1, Pr2: prisms, Ap: stop, D: detector, LI: lock-in amplifier [11].

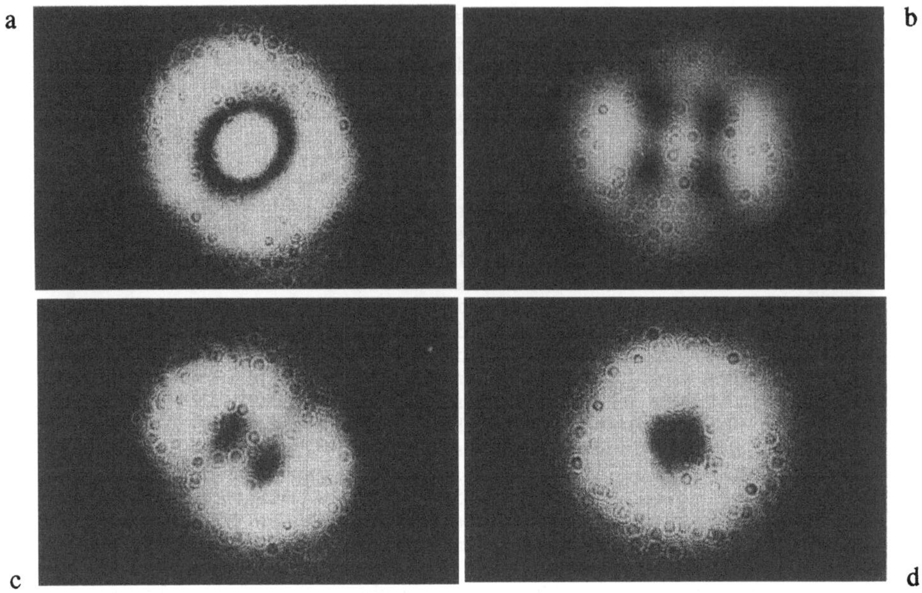

Figure 18: Mode family q=2p+l=2: stable patterns. [10].

resonator mirror, as well as by utilising a slight deviation from the angle of minimum deviation for the light path through the prism (so that the resonator is as rotationally symmetric as possible). An aperture restricts mode families of higher order from adjacent free spectral ranges.

Figure 18 shows the stable emission patterns of the 2p + l = 2 Laguerre mode family. These are, in increasing order of pump power, (a) the TEM_{01} Laguerre mode, (b) a pattern with two positive and two negative phase singularities, (c) a pattern which has two equally charged singularities, and (d) a pattern which results from the merging of the singularities in pattern (c) to form a single doubly-charged singularity.

While under the conditions for pattern (c), it is instructive to observe that it is impossible, by adjusting the resonator alignment alone, to merge its two singularities. Instead they circle about each other, behaving in the manner of equally charged particles (or current-carrying wires).

The stable emission patterns we find are exactly those that are theoretically predicted [12], including the spatial multistability phenomena. Figure 19 shows the emission patterns observed for the 2p + l = 3 Laguerre mode family. The essential difference to the patterns of the former 2p + l = 2 family is that a central phase singularity is added. In addition, figure 19b also shows the appearance of four additional outer singularities, and hence clearly displays the ionic crystal structure discussed. Figure 19d shows an example of a triply-charged singularity. It should be mentioned that figure 19 gives only a subset of the patterns in the 2p + l = 3 family. The others proved difficult to generate experimentally, due to their small domains

a b

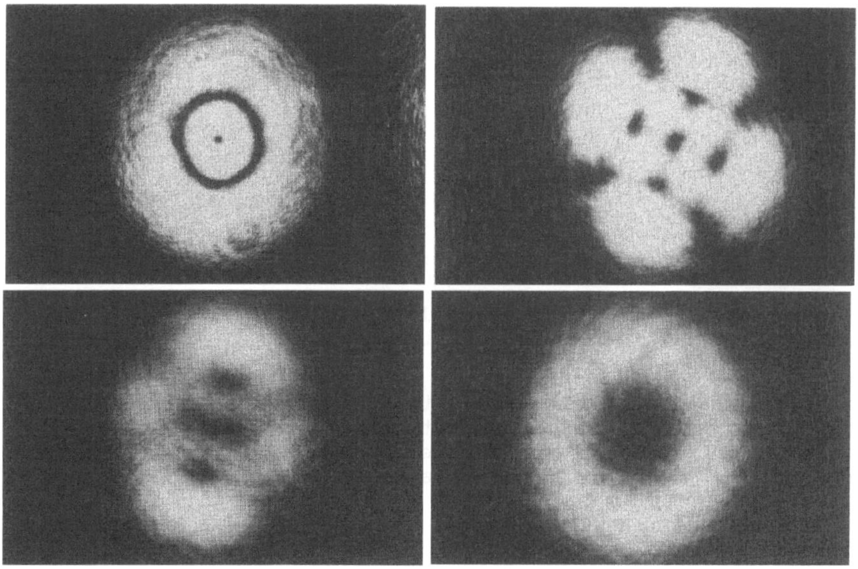

c d

Figure 19: Mode family q=2p+l=3: stable patterns [10].

of existence in the parameter space. It is evident that the complexity and number of the stable patterns in a family increases much faster than linearly with increasing family index q (= 2p + l).

One may hope that at very high mode numbers a "thermodynamic limit" is reached, where the behaviour of the optical field approaches that of an optical field in an unbounded region of the same active material (i.e. so that the effects of boundaries can be ignored). Such a result has recently been found for the case of a laser model in which over 50 000 modes could be excited [13]. A direct numerical integration of the laser partial differential equations showed, under the assumed conditions, that vortices in the optical field arise spontaneously from the vacuum fluctuations (these vortices differ globally from our phase singularities, but are the same in the neighbourhood of the defect). Vortices of opposite sign can annihilate, and the vortex pattern eventually settles into a steady-state arrangement of a more-or-less dislocated "vortex crystal". The existence of the vortices is easily understood by analogy to cosmic string models [14]. The laser modelled here has such a large cross-section that at different points in the cross-section the laser emission can build up essentially independently. Consequently the phase at different points develops independently, determined by vacuum fluctuations. The phase from then on is governed by the development of the dynamics of the system as the field equations are integrated. It is the singular points in the initial phase field which go on to become the cores of the vortices (these points are zeroes of the intensity). It is immediately evident that this vortex creation, with subsequent movement and partial annihilation of the

Figure 20: Steady-state phase lines in a laser model with a large Fresnel number [13].

vortices, can lead ultimately to very large numbers of different states, which supports the expectation of large multistabilities in high Fresnel number lasers, which could be useful for information storage and pattern recognition with lasers.

It has been found [15] that certain cellular automata, operating in a cyclic space under the influence of a nearest-neighbour interaction (resembling the action of diffraction), exhibit similar types of spontaneous vortex formation, movement, annihilation, and finally condensation (into dislocated crystal structures), as does the aforementioned numerical laser model. There is hope that a variation of these automata might generate a usable model for the laser equations which requires only a reduced amount of computational effort, and might more quickly provide experimenters with guidelines for their experiments.

2. Conclusion.

We have shown that lasers can generate a variety of transverse output patterns, and that these patterns seem to be strongly dominated by the number, orders, and positions of the phase singularities which they contain. We have also found that these patterns have properties, such as multistable families of solutions, which make them a potential basis for optical information storage and processing.

Acknowledgement: Work supported by ESPRIT (BR Action 3260).

References

1. H. Haken, Phys. Lett. **53A** (1975) 77.
2. E. Lorenz, J. Atmos. Sci. **20** (1963) 130.
3. C.O. Weiss and J. Brock, Phys. Rev. Lett. **57** (1986) 2804.
4. H. Zeghlache and P. Mandel, J. Opt. Soc. Am. **B2** (1985) 18.
5. Chr. Tamm, Phys. Rev. **A38** (1988) 5960.
6. L.A. Lugiato, C. Oldano, and L.M. Narducci, J. Opt. Soc. Am. **B5** (1988) 879.

7. M.V. Berry in "Physics of Defects", Balian et al. (Eds.), North-Holland (1981).
8. Chr. Tamm and C.O. Weiss, J. Opt. Soc. Am. **B7** (1990) 1034.
9. W. Klische and C.O. Weiss, Phys. Rev. **A39** (1989) 919.
10. M. Brambilla, F. Battipede, L.A. Lugiato, V. Penna, F. Prati, C. Tamm, and C.O. Weiss, Phys. Rev. **A** (in print)
11. Chr. Tamm and C.O. Weiss, Optics Communications **78** (1990) 253.
12. L.A. Lugiato, F. Prati, L.M. Narducci, and G.-L. Oppo, Optics Communications **69** (1989) 403.
13. P. Coullet, L. Gil, and F. Rocca, Optics Communications **73** (1989) 387.
14. W.H. Press and D.N. Spergel, Physics Today **42**, 3 (1989) 29.
15. A.K. Dewney, Scientific American **261**, 2 (1989) 88.

Principles of Transverse Pattern Formation in Lasers

M. Brambilla[1], *F. Battipede*[1], *L.A. Lugiato*[1], *F. Prati*[1], *and V. Penna*[2]

[1]Dipartimento di Fisica, Università di Milano,
Via Celoria 16, I-20133 Milano, Italy
[2]Dipartimento di Fisica, Politecnico di Torino,
Corso Duca degli Abruzzi 24, I-10129 Torino, Italy

1. Introduction

In another contribution to this book[1], we analyzed the stationary transverse patterns that emerge from the competition among the transverse cavity modes belonging to a frequency-degenerate family.
In this paper we address some questions which concern mainly the general principles that govern the phenomena of transverse spatio-temporal pattern formation and dynamics in lasers.
In Sec. 2 we formulate a variational principle, which governs the selection of the stationary patterns; we define a generalized free energy functional V of the electric field, such that the stable stationary configurations correspond to the local minima of V. The intensities and the relative phases of the modes in the steady states are fixed by this minimality requirement; the transitions from one pattern to another, which are observed by varying the control parameters, are also governed by the generalized free energy.
In Secs. 3 and 4 we establish a link between laser physics and hydrodynamics. In Section 3 we illustrate in detail the analogy between the phase singularities that we find in the transverse configuration of the electric field and the well-known vortex structures in fluids; on the basis of the concept of the order parameter, this analogy is extended to other fields such as superconductors and superfluids. Finally, in Sec. 4, we reshape the dynamical equations of the laser in the form of 'laser hydrodynamical equations' for a compressible fluid, similar to the law of mass conservation and to the Bernoulli equation.
A more detailed treatment of these problems can be found in Refs. 2 and 3.

2. The case of a frequency-degenerate family of modes. Variational principle for pattern selection

Throughout this work we will use the same notations as in Ref. 1 and therefore we will not repeat their meaning here. However, let us recall the equations which govern the dynamics of the frequency-degenerate family of modes.
The slowly-varying envelope of the electric field is written in the form

$$F(\rho,\varphi,\tau)=\sum_{pl}{}' f_{pl}(\tau)A_{pl}(\rho,\varphi) \quad , \tag{1}$$

where $A_{pl}(\rho,\varphi)$ are the Gauss–Laguerre modes described in Ref. 1. The prime indicates that the sum is restricted to the modes of the degenerate family with

Springer Proceedings in Physics, Vol. 55
Nonlinear Dynamics and Quantum Phenomena in Optical Systems
Editors: R. Vilaseca, R. Corbalán

$2p+|l|=q$. In the good cavity limit $\kappa \ll \gamma_\perp\gamma_\parallel$ the modal amplitudes f_{pl} obey the time evolution equations ($\tau=\kappa t$)

$$\frac{df_{pl}}{d\tau} = -f_{pl} + 2C\sum_{p'l'}{}' \int_0^{2\pi} d\varphi \int_0^{\infty} d\rho\, \rho \frac{A_{pl}(\rho,\varphi)A_{p'l'}(\rho,\varphi)}{1+|F(\rho,\varphi,\tau)|^2}\, \chi(\rho) \quad . \tag{2}$$

Let us now introduce the functional

$$V = \int_0^{2\pi} d\varphi \int_0^{\infty} d\rho\, \rho\Big[|F(\rho,\varphi)|^2 - 2C\chi(\rho)ln\big(1+|F(\rho,\varphi)|^2\big)\Big] \quad , \tag{3}$$

where it is understood that the field F is given by Eq. (1) . Thus, V is in fact a function of the modal amplitudes f_{pl} and of their complex conjugates f^*_{pl} with $2p+|l|=q$.
By using Eq. (1) and the orthonormality of the modes A_{pl}, one verifies easily that equations (2) can be written in the form

$$\frac{df_{pl}}{d\tau} = -\frac{\partial V}{\partial f^*_{pl}} \quad . \tag{4}$$

Hence the stationary solutions correspond to the stationary points of the functional V, which therefore plays the role of a generalized free energy in this system which lies far from thermal equilibrium[4]. A set of modal amplitudes $\{f_{pl}\}$ is a stable stationary solution if and only if it corresponds to a (local) minimum of V. Another consequence of Eqs. (4) is that one can exclude the possibility of instabilities which lead to the onset of spontaneous oscillations, because V is a monotonically decreasing function of time. As a matter of fact, we have

$$\frac{dV}{d\tau} = \sum_{pl}{}' \left[\frac{\partial V}{\partial f_{pl}}\frac{df_{pl}}{d\tau} + \frac{\partial V}{\partial f^*_{pl}}\frac{df^*_{pl}}{d\tau}\right] \tag{5}$$

so that, by using Eqs. (4) and its complex conjugate, we obtain

$$\frac{dV}{d\tau} = -2\sum_{pl}{}' \left|\frac{\partial V}{\partial f_{pl}}\right|^2 \leq 0 \quad . \tag{6}$$

The absence of oscillations arises from the fact that all the modes have the same frequency, so that frequency competition is completely eliminated.

3. Point-like vortex structure of the phase singularities

The numerical solution of Eqs. (2) describing the time evolution of the laser has shown the onset of nontrivial spatial structures of the electric field characterized by the presence of singularities in the electric phase [1,2]. In this section we thoroughly examine the nature of these singular configurations, showing that they can be analytically described as vortices in fluids.

Let us first review the analytic description of a two-dimensional, compressible, rotational fluid[5]. Let

$$\mathbf{v} = v_x \mathbf{e}_x + v_y \mathbf{e}_y \quad , \quad \mathbf{w} = \nabla \times \mathbf{v} = w \mathbf{e}_z \tag{7}$$

be the velocity field and the vorticity field, respectively. By introducing a scalar potential $\eta = \eta(x, y)$ and a vector potential $\mathbf{A} = A(x, y)\mathbf{e}_z$ we can write the velocity field as the sum of an irrotational and a divergenceless term

$$\mathbf{v} = \nabla \eta + \nabla \times \mathbf{A} \tag{8}$$

and the rotational condition $\nabla \times \mathbf{v} \neq 0$ becomes

$$-\nabla^2 A = w \quad . \tag{9}$$

Let us consider the case of N point-like vortices placed at the points (x_j, y_j) and with topological charges m_j, $j = 1...N$. In this case the vorticity w is the sum of N Dirac delta functions centered in (x_j, y_j), weighted by the topological charges m_j, and the rotational condition takes the form of a Poisson equation for the potential $A(x, y)$

$$-\nabla^2 A = q \sum_{j=1}^{N} m_j \, (x - x_j) \, (y - y_j) \quad , \tag{10}$$

where q is a real positive factor with the dimensions of vorticity.
This fact can explain the similarity between the equiphase lines of our phase singularity crystals and the field lines of the electric field generated by a corresponding set of electric charges.
We know from potential theory that the potential A can be thought of as the real part of an analytic function

$$f(z) = -\frac{2\pi}{q} A(z) + i\beta(z) \quad . \tag{11}$$

Since $f(z)$ is analytic, A and β are connected by the Cauchy conditions, which in our case are equivalent to the vector equation

$$\nabla \times \mathbf{A} = \frac{q}{2\pi} \nabla \beta \quad . \tag{12}$$

By inserting Eq. (12) into Eq. (8) we observe that the velocity field is given by

$$\mathbf{v} = \nabla \eta + \nabla \times \mathbf{A} = \nabla \eta + \frac{q}{2\pi} \nabla \beta = \frac{q}{2\pi} \nabla \Phi \quad , \tag{13}$$

where

$$\Phi = \frac{2\pi}{q} \eta + \beta \quad . \tag{14}$$

Eq. (13) is well-known in the condensed matter physics of systems with topological defects, where Φ is the phase of a complex order parameter field

$$\Psi = |\Psi| e^{i\Phi} \tag{15}$$

describing the physical state of the system at each point of the medium. In particular for superfluids and superconductors[6] Eq. (13) gives the local average velocity of 4-He superfluid atoms and the local velocity (up to the magnetic potential vector) of the frictionless charges in the eddy supercurrents respectively, while $|\Psi|^2$ is the mass density. All these systems are endowed with defects that are point-like vortices and correspond to the singularities of β.
Let us come back to the laser. In this case the 'velocity potential' Φ given by Eq. (14) is just the phase of the electric field. Therefore it is quite natural to regard the electric field as the complex order parameter field of an 'optical fluid' whose topological structures are entirely described by the β term.
With this identification it is possible to get all the relevant hydrodynamical quantities as $\mathbf{v}, \mathbf{w}$ etc. for our optical fluid. In the final section we stress further this parallelism by reducing the time evolution equation for F to a pair of hydrodynamical equations.

4. Laser hydrodynamics

In this section we make the assumption that the electric field is a linear combination of the whole set of transverse modes which correspond to a given value of the longitudinal index. In the good cavity limit the equation that governs the dynamical evolution of the electrtic field is[7]

$$\frac{\partial F}{\partial \tau} = -\left[1 - ia\left(\frac{1}{4}\nabla_\perp^2 - \rho^2 + 1\right)\right] F + 2C\chi(\rho)\frac{(1-i\Delta)F}{1+\Delta^2+|F|^2} \quad , \tag{16}$$

where a is the detuning between two adjacent families of transverse modes and Δ is the atomic detuning. In order to reformulate Eq. (16) in the form of 'hydrodynamical equations' we put

$$F = |F| e^{i\Phi} \tag{17}$$

and we introduce the quantities

$$\sigma = |F|^2 \quad , \quad \mathbf{v} = \frac{a}{2}\nabla\,\Phi \tag{18}$$

where σ plays the role of the density of the fluid in motion and $\mathbf{v}$ that of the velocity field, consistently with the analysis of the previous section.
With some algebraic manipulations, Eq. (16) and its complex conjugate can be cast in the form

$$\frac{\partial \sigma}{\partial \tau} + \nabla \cdot (\sigma \mathbf{v}) = 2\sigma\left[\frac{2C\chi(\rho)}{1+\Delta^2+\sigma} - 1\right] \tag{19}$$

$$\frac{\partial \Phi}{\partial \tau} = -\frac{a}{4}|\nabla\,\Phi|^2 + a\left[\frac{\nabla^2\sqrt{\sigma}}{4\sqrt{\sigma}} + 1 - \rho^2\right] - \frac{2C\chi(\rho)\Delta}{1+\Delta^2+\sigma} \quad . \tag{20}$$

We can compare Eqs. (19) and (20) with the fundamental equations of hydrodynamics, the equation of continuity for the mass density σ

$$\frac{\partial\sigma}{\partial\tau} + \nabla\cdot(\sigma\mathbf{v}) = 0 \tag{21}$$

and the Bernoulli equation for the velocity potential

$$\frac{\partial\Phi}{\partial\tau} = -\frac{q}{4\pi}|\nabla\,\Phi|^2 + \mathcal{P}\frac{2\pi}{q} \tag{22}$$

which involves a pressure term $\mathcal{P}$. From the comparison between these two pairs of equations we can conclude that in the optical case the total 'mass' is not conserved because of the presence of the two dissipative terms that appear in the r.h.s. of Eq. (19), while the analogy with Bernoulli equation is complete if we put $q = \pi a$ and define a 'pressure'

$$\mathcal{P} = \frac{a^2}{2}\left[\frac{\nabla^2\sqrt{\sigma}}{4\sqrt{\sigma}} + 1 - \rho^2\right] - \frac{a}{2}\frac{2C\chi(\rho)\Delta}{1+\Delta^2+\sigma} \quad . \tag{23}$$

We hope that the analysis of this section will be useful for the discussion of the relation between optical and hydrodynamical turbulence.

Acknowledgments

This work was carried out in the framework of the ESPRIT Basic Research Action on 'Transverse Optical Patterns' (TOPP).

References

1) F. Battipede, M. Brambilla, L. A. Lugiato, F. Prati and V. Penna, "Phase Singularity Crystals in Transverse Laser Patterns", this volume
2) M. Brambilla, F. Battipede, L. A. Lugiato, V. Penna, F. Prati, C. Tamm and C. O. Weiss, Phys. Rev. **A**, to appear
3) M. Brambilla, L. A. Lugiato, V. Penna, F. Prati, C. Tamm and C. O. Weiss, Phys. Rev. **A**, to appear
4) H. Haken , *"Synergetics: an Introduction"*, Springer-Verlag Berlin 1977
5) P. D. Mc Cormack and L. Crane, *"Physical Fluid Dynamics"* Academic Press, New York 1973
6) D. R. Tilley and J. Tilley, *"Superfluidity and Superconductivity"* Adam Hilger, Bristol 1986
7) L.A. Lugiato, G.L. Oppo, J. R.Tredicce and L.M. Narducci in Feature issue on *"Transverse Effects in Nonlinear Optical Systems"* N. B. Abraham and W. J. Firth eds. , Journ. Opt. Soc. Am. B June 1990

Phase Singularity Crystals in Transverse Laser Patterns

F. Battipede[1], *M. Brambilla*[1], *L.A. Lugiato*[1], *F. Prati*[1], *and V. Penna*[2]

[1]Dipartimento di Fisica, Università di Milano, Via Celoria 16, I-20133 Milano, Italy

[2]Dipartimento di Fisica, Politecnico di Torino, Corso Duca degli Abruzzi 24, I-10129 Torino, Italy

Introduction. We analyze the interaction and the competition of a set of transverse cavity modes, which belong to a frequency-degenerate family. The laser exhibits coexistence among different stationary spatial patterns . These structures display the presence of *phase singularities* analogous to the vortex structures known in such fields as, for example, hydrodynamics, superconductivity, superfluidity. These singularities are arranged in the form of regular *arrays* , and the equiphase lines of the field exhibit a notable similarity to the field lines of the electrostatic field generated by a corresponding set of point charges.
The theoretical results are in agreement with the experimental observations obtained from a Na_2 laser.

The studies of the transverse profile of the electromagnetic field propagating in a nonlinear optical system have provided interesting results over the last few years concerning nonlinear spatio-temporal dynamics, pattern formation and optical information processing.
We have investigated the structure of the transverse profile of the field in the case of a homogeneously broadened 2-level active medium in a ring cavity with spherical mirrors. In this case the model presents an intrinsic cylindrical symmetry, thus a natural choice of eigenmode basis for the e.m. field in the cavity is provided by the Gauss-Laguerre basis, whose analytical form in the paraxial approximation is

$$A_{pl}(\rho,\varphi) = \sqrt{\frac{2}{\pi}}(2\rho^2)^{|l|/2}\left[\frac{p!}{(p+|l|)!}\right]^{1/2} L_p^{|l|}(2\rho^2)e^{-\rho^2}\, e^{il\varphi} \quad . \tag{1}$$

Here ρ is the radial variable scaled to the beam waist, φ is the angular variable, p and l are respectively the radial and angular indices of the modes. for $l = 0$ the modes are cylindrically symmetrical.
According to the geometry of the cavity the modes gather in frequency degenerate families, because the frequency of the modes depends on the combination $q = 2p + |l|$; so that each frequency degenerate family is composed by $q + 1$ modes. In order to simplify the model we start considering a ring laser in which the atomic line is tuned to one of the families, and suppose that all other families are far enough from resonance to experience loss larger than gain, so that only the tuned family will contribute to the emission.

Under this approximation, we have expanded the slowly varying envelope of the electric field on the Gauss-Laguerre basis, only considering the contributions of the modes belonging to the active family $2p+|l|=q$

$$F(\rho,\varphi,t)=\sum_{p,l}^{2p+|l|=q} f_{pl}(t)A_{pl}(\rho,\varphi) \quad , \tag{2}$$

This ansatz is then introduced in the Maxwell Bloch equations including diffraction (see Ref. 1) which govern the dynamics of this system. In the mean field limit and after adiabatic elimination of the atomic variables the dynamical equations for the modal amplitudes f_{pl} read

$$\frac{df_{pl}}{d\tau}=-f_{pl}+2C\int_0^{2\pi}d\varphi\int_0^{\infty}d\rho\,\rho\,e^{-2\rho^2/\psi^2}.$$

$$\cdot A_{pl}(\rho,\varphi)\sum_{p',l'}^{2p'+|l'|=q}\frac{A_{p'l'}(\rho,\varphi)\,f_{p'l'}(\tau)}{1+\left|\sum_{p'',l''}^{2p''+|l''|=q}A_{p''l''}(\rho,\varphi)\,f_{p''l''}(\tau)\right|^2} \quad , \tag{3}$$

where $\tau=k\,t$, k = cavity damping constant, C = pump parameter ψ = pump width / beam width.

The results of the steady state and linear stability analysis are presented in the case when the family $2p+|l|=2$ is active: Figs. 1a,b,c,d show the four different intensity patterns which are stable. These patterns are in general a linear combination of modes; we demonstrated that the coefficients of this combination minimize a generalized free energy whose form is presented in Ref. 2.

By varying the pump parameter we discovered transitions from the state of Fig. 1a to the configuration of Fig. 1b, and from the state of Fig. 1d to the configuration of Fig. 1c . Both transitions correspond to a phenomenon of spontaneous breaking of the cylindrical symmetry of the transverse intensity pattern. We point out that all of these configurations have been observed experimentally at P.T.B. in Braunschweig, in close correspondance with the theoretical predictions (3).

Some of the different configurations arise in overlapping regions of the parameter space, thus realizing the phenomenon of *spatial multistability* . Here we have the coexistence of stationary solutions, which differ in their spatial structure, much more than in the total output intensity. If we consider the field pattern, taking into account phase relations, we discover cases of tristability for $2p+|l|=2$, up to eight coexisting patterns for $2p+|l|=3$.

The stationary patterns display points in which both the real and the imaginary part of the electric field vanish simultaneously, and such that, by covering a closed loop around one of these points, the phase of the electric field changes by a positive or negative integer multiple of 2π. The existence of these *phase singularity points* is evident both in the theory and in the experimental observations /4/. The behaviour of the electric field around each of these points

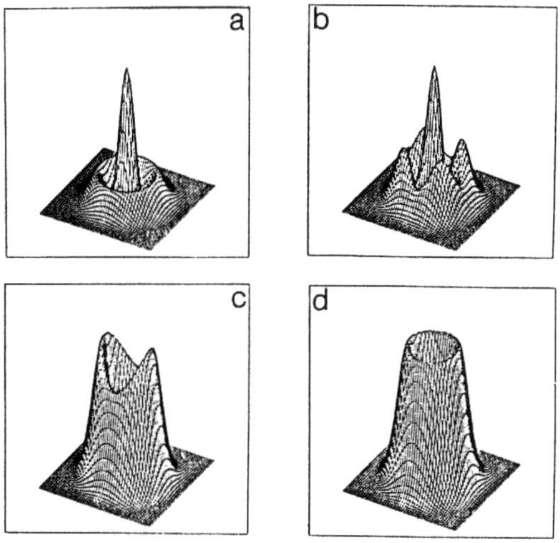

Fig.1 : Intensity profile of the transverse e.m. field for the four stable steady states in the case $2p + |l| = 2$.

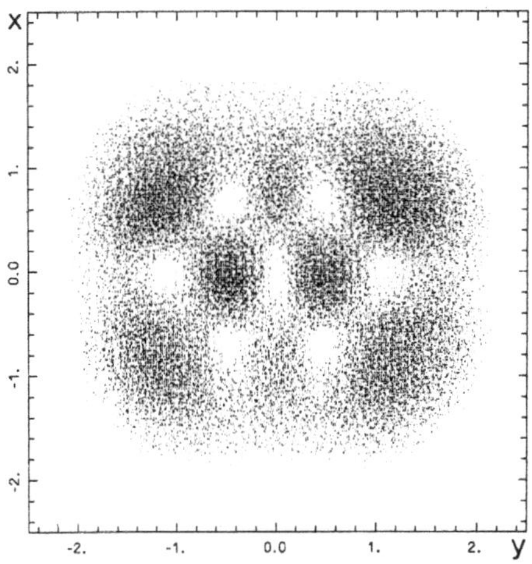

Fig.2 : Intensity of the transverse field for a stable steady state with seven phase singularities in the case $2p + |l| = 3$. The intensity is greater, the darker is the plot.

exhibits many similarities to that of the optical vortices, recently discovered /6/ in the framework of the model of Ref. 5. In our problem, the singularities arise from the combination of a few modes instead of several modes. In our case the singularities occur at fixed points, arranged in the form of regular arrays ("phase singularity crystals").

Fig. 2 shows the intensity profile of a stable stationary pattern for $2p + |l| = 3$ where an array of 7 phase singularities is evident. Finally, Fig. 3 shows the behaviour of the gradient of the phase of the electric field for the configuration of Fig. 2; the vortex structure of the field is evident, and in fact the gradient of

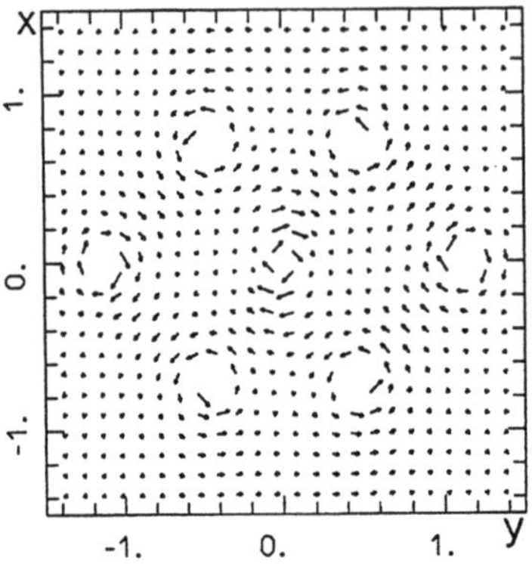

Fig.3 : Plot of the gradient of the field phase for the pattern of Fig. 2.

the field phase plays the same role of the velocity field in hydrodynamics /2/. Another interesting feature that our model displays is the capability of operation as a pattern recognition device. Let us suppose that the system is in the region of the control parameter space where m stable steady state patterns coexist (call them $N_i(\rho,\varphi), i = 1...m$). An ambiguous (i.e., not perfectly coincident with any of the m 'objects' $N_i(\rho,\varphi)$ in 'memory') input pattern $A(\rho,\varphi)$ is offered to the system for recognition. Let us define a 'similarity parameter' between the input pattern A and a memorized pattern N_i in the following way

$$S_{A,N_i} = \left| \int_0^{2\pi} d\varphi \int_0^{\infty} d\rho\, \rho\, N_i^*(\rho,\varphi)\, A(\rho,\varphi) \right| \quad .$$

Practically, the input pattern is injected as an external driving signal while the laser is kept below threshold. Then we let the external field fade away and switch the laser above threshold so that the input field leaves a 'seed' for the initial condition of the lasing process.

Our numerical simulations performed on the model clearly show that the emitted pattern of the laser at the steady state will correspond to that pattern $N_{\bar{i}}(\rho,\varphi)$ among the m coexisting patterns which shows the highest value of the similarity parameter to the offered pattern $A(\rho,\varphi)$.

This research was carried out in the framework of the ESPRIT Basic Research Action on 'Transverse Optical Patterns' (TOPP).

References

1. L.A. Lugiato, F. Prati, L.M. Narducci and G.L. Oppo, Opt. Comm. **69**, 387 (1989)

2. M. Brambilla, F. Battipede, L. A. Lugiato and F. Prati, this volume.

3. C. Tamm and C. O. Weiss, Feature issue on *"Transverse Effects in Nonlinear Optical Systems"* N. B. Abraham and W. J. Firth eds. , Journ. Opt. Soc. Am. B , June 1990

4. M. Brambilla, F. Battipede, L. A. Lugiato, V. Penna, F. Prati, C. Tamm and C. O. Weiss, *"Transverse Laser Patterns : I. Phase Singularity Crystals"*, Phys. Rev. A , to appear.
5. L.A. Lugiato, C. Oldano and L.M. Narducci, Journ. Opt. Soc. Am. B **5** 879 (1988)
6. P. Coullet, L. Gil and F. Rocca, Opt. Comm. **73**, 403 (1989)

Order–Disorder Transition in the Local Dynamics of a TEA CO_2 Laser

J.M. Guerra, F. Encinas, and I. Pastor

Departamento de Optica, Facultad de Ciencias Físicas,
Universidad Complutense, Ciudad Universitaria, E-28040 Madrid, Spain

An experimental study has been carried out on the evolution of the near field local intensity dynamics in a TEA (Transversely Excited Atmospheric Pressure) CO_2 laser. The diameter of the intracavity diaphragm is the control parameter. A transition from periodic or quasiperiodic to chaotic behavior was observed. Chaotic oscillations appear when no inner diaphragm was used. The correlation between oscillations taken at different points of the laser pattern was measured. Fast decay autocorrelations and low cross-correlations between synchronic pairs of signals were found. The dynamics described is a case of spatio-temporal chaos.

1. Introduction

Laser instabilities have a great practical importance because of their influence in many laser applications. Also, from a scientific point of view, their study has led to a deeper insight in laser dynamics. Recently attention has been focused on the dynamics of laser transverse patterns [1-6].

In this communication, we report the study of local instabilities observed in a TEA CO_2 laser.

We call these instabilities 'local', because they are observed with the greatest amplitude when the laser near field was detected in a small enough area ($\simeq 1\ mm^2$).

Our communication includes two experiments, one dealing with the evolution of the instabilities when a variable aperture is included inside the resonator ; in the second experiment we correlate the signals measured at different points of the laser pattern.

2. Measurement System

A 8 J, 3.5 μs pulse duration home-made TEA CO_2 laser [7] was used. The near field laser pattern dimensions were 25 x 20 mm^2.

The laser pulse has a short (50 ns width) high power gain-switch peak followed by a long quasistationary collisional

Springer Proceedings in Physics, Vol. 55
Nonlinear Dynamics and Quantum Phenomena in Optical Systems
Editors: R. Vilaseca, R. Corbalán

transfer tail. The light pulses in this laser were shot to shot accurately reproducible in pulse shape and energy, when we simultaneously measured the complete laser pattern. The laser cavity is made up of one 10 m radius full reflector and one plane 60% reflectivity output mirror. The cavity length is 90 cm and its Fresnel number is ≃ 11.

The laser intensity in the quasistationary tail is detected by a fast (1 ns rise time) photon drag linked to a 500 MHz bandwidth transient digitizer inside a Faraday cage. The pulses were digitized to 9 bits at a maximum of 100 Gsamples/s and the 512 points from each signal stored in a computer for further analysis.

If I(t) are the raw data, we construct $\mathfrak{I}(t)$ = (I(t) - I)/I where I stands for time average. In the following $\mathfrak{I}(t)$ will be referred to as the signal or the fluctuation intensity.

3. Experimental Results

Apertures with different diameters were placed inside the resonator to change the transverse structure of the pattern (Fig. 1). Laser intensity was always measured in this experiment, selecting a low area (3 mm^2) around the spot's center for observation. In order to ensure good statistics, 40 signals were taken for each value of the opening diameter. The pulses were registered for a total time of 100 ns. Apertures of 8,10,12 and 14 mm were used.

With an 8 mm diameter, the laser pattern shows a TEMoo near field intensity distribution. The observed temporal behavior of intensity is nearly periodic (Fig. 2a). The averaged Fourier spectrum shows sharp pure axial mode beating peaks, with decreasing amplitude, and the phase portrait corresponds closely to a limit cycle (Figs. 2b and 3a).

In the remaining apertures the laser pattern seems to be a mixture of a growing number of low order cylindrical transverse modes. Fluctuations in the measured intensity are of increasing irregularity. The average Fourier spectra show a structure of transversal mode beating sidebands, outflanking each axial mode beating peak. On the low frequency side of the spectra, the same transversal mode beating modulation peaks and one slower one also appeared (Fig. 4).

Without the inner aperture, the laser pattern had a completely different structure (Fig. 1) due to the boundary conditions imposed by the confining surface of the laser electrodes. One array of nearly equispaced bright fringes form the complete pattern, suggesting the superposition of high order rectangular transverse modes.

The intensity fluctuations are completely irregular in aspect. The average Fourier spectrum appeared as a broad,

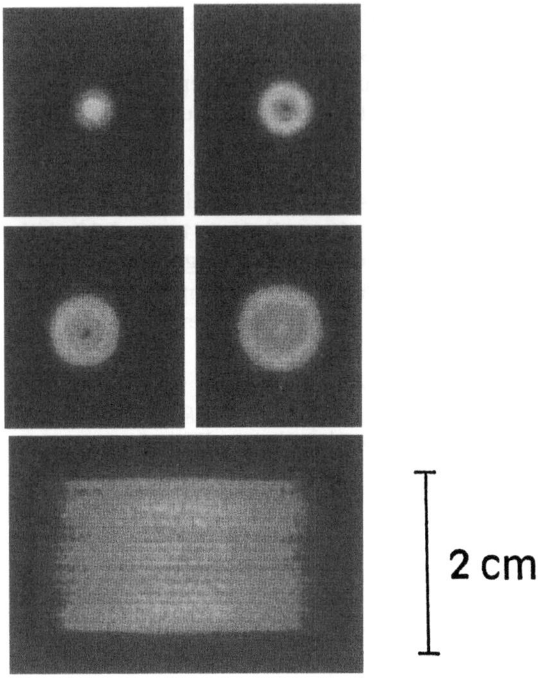

Fig.1 Laser near field burn patterns. From top to bottom and from left to right ϕ = 8,10,12,14 mm and without diaphragm.

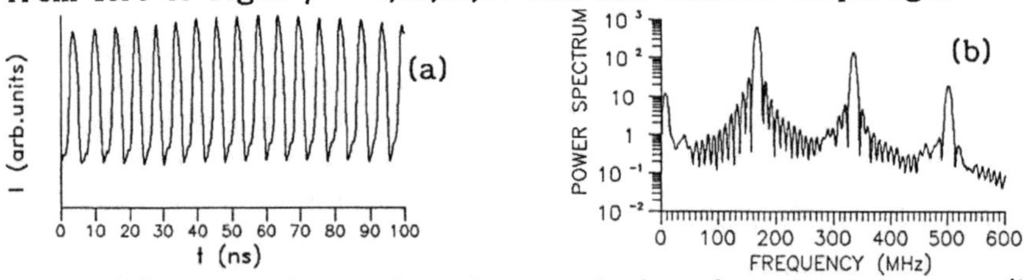

Fig.2 (a) Laser instensity time evolution for ϕ = 8 mm. (b) Averaged Fourier spectrum.

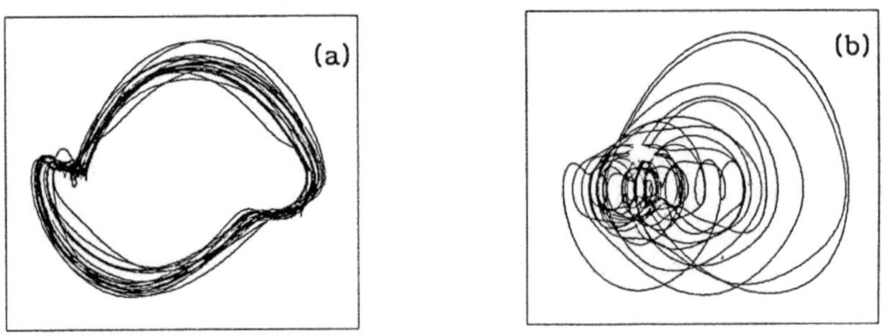

Fig.3 (a) Phase portrait for ϕ = 8 mm. (b) the same as (a) but without diaphragm.

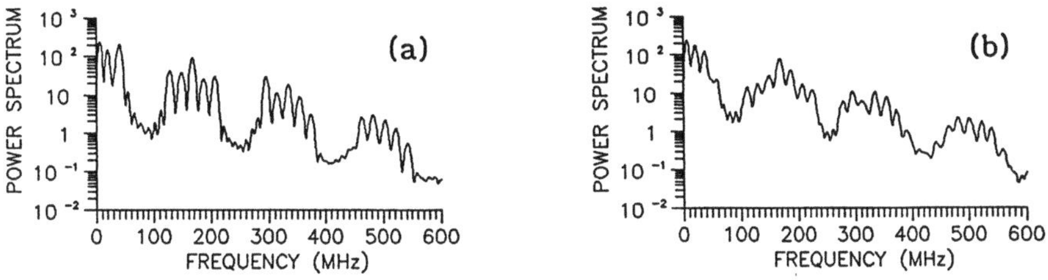

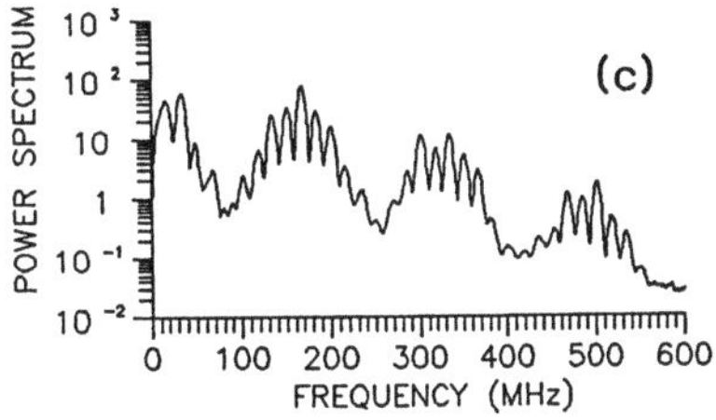

Fig.4 Averaged Fourier spectra for (a) ϕ = 10 mm, (b) ϕ = 12 mm and (c) ϕ = 14 mm.

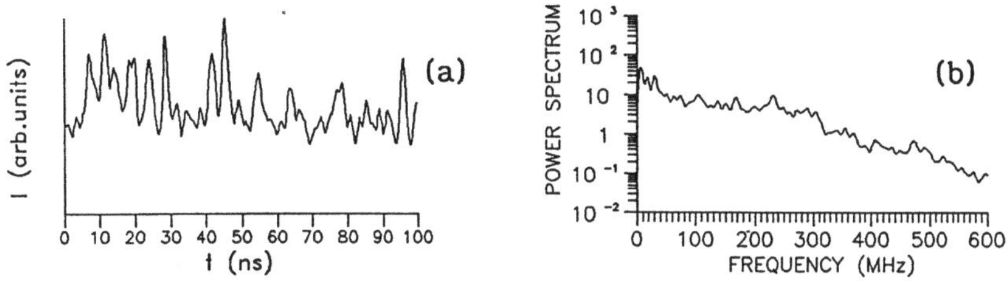

Fig.5 The same as figure 2, but without inner diaphragm.

nearly unstructured continuum (Figs. 5a,b). The individual pulse Fourier spectra on the contrary are completely structured in individual sharp peaks. But the place and the amplitude of the peaks are not reproducible from shot to shot. The phase portrait is reminiscent of a strange attractor (Fig. 3b).

The Grassberger–Procaccia algorithm applied to these signals shows good slope convergence to a dimension of 2.7 (Fig. 6).

In the correlation experiment, we simultaneously measured the intensity time evolution from two different small areas (3 mm^2) on the near field laser pattern (Fig. 7). These areas were taken 6 mm apart at several points of the laser pattern. The individual Fourier spectra corresponding to each pair of simultaneously registered signals have a completely different spectral composition both in frequency and amplitude. In fact, it is not possible to know by inspection of the signals and their Fourier spectra, whether they pertain to the same shot or to different shots (Fig. 8).

Autocorrelations of both signals have a fast decay of the envelope, as can be expected from their irregular aspect. The cross correlation between each pair of signals were normalized to the geometric mean of the two autocorrelations for zero time

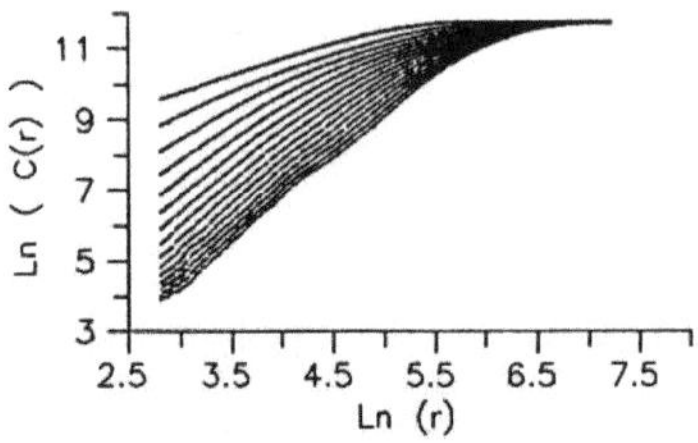

Fig.6 Grassberger-Procaccia algorithm applied to a signal obtained without diaphragm. Embedding dimensions go from 1 (top) to 15 (bottom). A linear fit in the range 3.08-3.76 of ln(r) gives slope saturation to ≈ 2.7 with correlation coefficient > 0.999 for embedding dimensions ≥ 10.

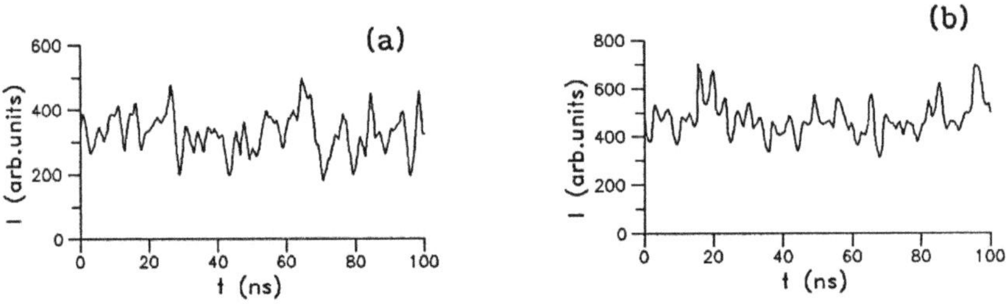

Fig.7 Laser intensities from two different points on the beam cross section separated by 6 mm.

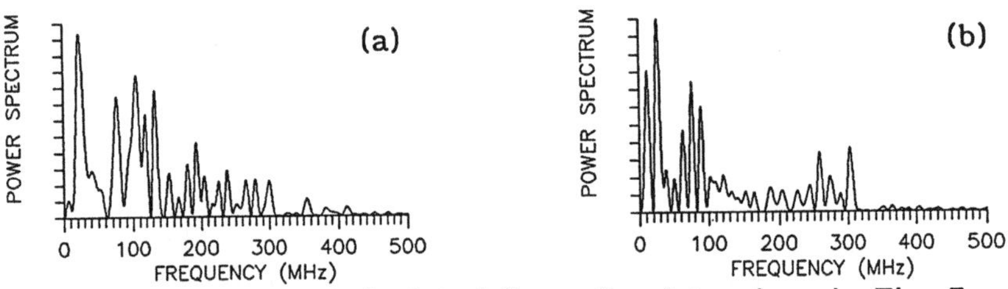

Fig.8 Power spectra calculated from the data given in Fig. 5

delay. It was observed (Fig. 9) that the cross correlation was significantly lower than either of the autocorrelations around zero time delay. The envelope of the cross correlation does not show any recognizable structure. That is, the signals taken from two different points on the laser near field pattern are uncorrelated. This result in itself supports the local character that we attribute to the studied instabilities.

When the whole laser pattern is detected, a residual, completely regular (over the dc component) mode beating oscillation remains. It seems that this residual regular axial mode beating was hidden below the greater irregular fluctuations observed in the local mode.

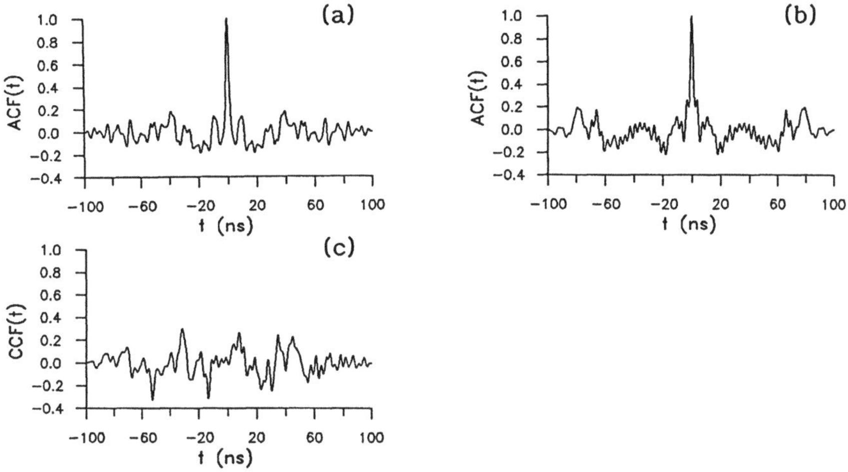

Fig.9 (a),(b) Autocorrelation functions calculated from the intensity data. (c) Cross correlation function from the intensity data.

4. Discussion and Conclusions

The persistence of phase coupling between axial modes indicates the existence of two levels of nonlinear phase couplings between the field distributions. In the first level, the axial modes with the same transverse spatial distribution (i.e. the same transverse mode) are phase coupled, as is usual in homogeneously broadened systems. In the second level, the groups of modes with different transverse structures become coupled.

The first level of coupling remains for all experimental conditions, while the second disappears when the inner diaphragm is removed. When the number of transverse structures is big enough the transverse phase coupling is lost and the oscillation becomes irregular. That seems to be the physical origin of the transition from order to chaos. Moreover the good slope convergence and low correlation dimension ($\simeq$ 2.7) of the applied Grassberger-Procaccia algorithm, without intracavity aperture, provides evidence of a local chaotic deterministic behavior.

The lack of correlation at several points of the laser pattern is to a certain point equivalent to the result reported in [5].

In an ordinary beating between transverse modes with discrete frequencies, the correlation would show a periodic envelope. But such a feature is not observed in the cross correlation envelope. Furthermore, the lack of cross correlation has been observed in several pairs of detecting areas placed in different zones of the laser pattern.

It is also worthwhile to point out the different Fourier spectra corresponding to signals synchronically measured at two points of the laser pattern. A recent theoretical approach to the transverse effects in laser patterns (see L.A. Lugiato et al. [6]) predicts complete correlation between the fluctuations on symmetric points of the laser pattern.

As far as we know, no complete theoretical explanation exists for the described results. In fact, the physical basis of the described laser behavior remains unclear.

References

[1] G.L. Lippi, N.B.Abraham, G.P. Puccioni, F.T. Arecchi, and J.R. Trédicce, Phys. Rev. A, 35 (9), 3978 (1987)

[2] D.J. Biswas, Vas Dev, and U.K. Cratterjee, Phys. Rev. A, 38 (1), 555 (1988)

[3] L.A. Lugiato, F. Prati, L.M. Narducci, P. Ru, J.R. Tredicce, and D.K. Bandy, Phys. Rev. A, 37 (10), 3847 (1988)

[4] J.R. Tredicce, E.J. Quel, A.M. Ghazzawi, C. Green, M.A. Pernigo, L.M. Narducci, and L.A. Lugiato, Phys. Rev. Letters, 62 (11), 1274 (1989)

[5] W. Klische, C.O. Weiss. and B. Wellegehausen, Phys. Rev. A, 39 (2), 919 (1989)

[6] L.A. Lugiato, G.L. Oppo, J.R. Tredicce, L.M. Narducci, and M.A. Pernigo, J. Opt. Soc. Am. B, 7 (6), 1019 (1990)

[7] F. Encinas Sanz and J.M. Guerra Perez, to be published in Meas. Sci. Technol.

Part VI

Influence of Noise

Noise and Transient Dynamics in Semiconductor Lasers

P. Spano, A. Mecozzi, A. Sapia, and A. D'Ottavi
Fondazione Ugo Bordoni,
Via B. Castiglione 59, I-00142 Roma, Italy

Abstract. The noise properties and dynamical behavior of semiconductor lasers can be described by a set of coupled rate equations for photon and carrier densities. The main distinctive features of semiconductor lasers arise from two terms, present in these equations, which derive directly from the band structure of semiconductors, namely the linewidth enhancement factor and the gain saturation. The origin of these two terms and their effects on the noise properties of semiconductor lasers and in particular on the emission lineshape are discussed. Moreover, the dynamics of semiconductor lasers driven by current pulses from below to above threshold is also discussed and compared with experimental results. In such dynamical conditions the equation describing the evolution of the electrical field is similar to that describing a physical system continuously driven from a stable to an unstable equilibrium point. The differences between the behavior of single and multimode lasers are underlined and discussed. Finally, experimental results showing transient excitation of a side mode in lasers having a monomode spectrum in steady state conditions are reported and compared with a theoretical model.

1. Introduction

In the last few decades, the study of semiconductor lasers physics has attracted great interest because of the extensive use of these devices in practical applications and the fundamental differences of semiconductor lasers relative to other laser systems not based on a band structure of the electronic energies. Some of the results of the band structure of semiconductor based devices are ease of modulation [1], high modulation speed [2], small physical dimensions and high achievable gain [1]. Other less striking but more fundamental features characterizing these devices are connected with the presence of two terms in the rate equations which describe both the noise and the dynamical behavior of these lasers. The first term is the *linewidth enhancement factor* α [3]; it takes into account the nonzero value of the real part of the complex susceptibility at the laser frequency in a system composed by two electronic energy bands. The main effects of this term are connected with the phase noise [4] (it is responsible for the broadening of the emission linewidth and the distortion of the lineshape) and with the dynamical change of frequency (frequency chirping) owing to intensity modulation [5]. The

Nonlinear Dynamics and Quantum Phenomena in Optical Systems
Editors: R. Vilaseca, R. Corbalán

second term is connected to the reduction of the gain at fixed excited carrier density when the optical power inside the laser cavity increases [6]. This term is responsible for the damping of the oscillatory phenomena typical of class B lasers [7], in which the rate equation for the carrier density cannot be adiabatically eliminated.

Another peculiar property of laser diodes is related to the system's response to a sudden change of the external pump which, in these devices, is the electrical current flowing through the laser. A sudden change of the injection current from an initial value below, to a final one above threshold, does not immediately affect the population inversion but only causes a gradual increase of the excited carrier density [1]. Therefore, the dynamics of a system which is continuously moved through a bifurcation point (the laser threshold) can be easily observed. The study of this phenomenon, in which the initial fluctuations play an essential role in driving the system towards the new stable state, can be framed in the more general context of the decay from unstable states [8,9,10]. The transient behavior of several physical systems, for example explosive chemical reactions [11], population dynamics [12] and superfluorescence pulse dynamics [13], can be framed in this very general class, but the investigation of this phenomenon in semiconductor lasers has the advantage of an easy experimental variation of the control parameters [14], generally not allowed in other physical systems. The study of the statistical properties of semiconductor laser dynamics can be in fact performed by varying the numbers of lasing modes [15], which corresponds to a change of the number of ways the system has to decay from the unstable state, as well as the rate of increase of the inversion [16]. Furthermore semiconductor lasers allow an easy observation of a new phenomenon which is related to a non-zero probability of excitation, during the laser switching, of a mode which is highly suppressed in stationary conditions [17]. This phenomenon can be considered as a different effect of the decay from an unstable state. In strictly single mode lasers, the effect of fluctuations is to randomize the time the laser needs to reach the final stationary state. In this case of quasi single mode lasers, fluctuations can also drive the system along different paths which represent large deviations from the most probable evolution.

2. Basic Relationships

The dynamics and the noise properties of semiconductor lasers can be described by using a standard approach based on the rate equations. Semiconductor lasers belong to general class B lasers [7]; the three equations necessary to describe the behavior of a generic laser can be reduced to two, after adiabatic elimination of polarization. For a single-mode laser, they become

$$\frac{d}{dt}E = \left[i\,\omega_M + \frac{1}{2}\left(G - 1/\tau_p\right)\right]E + F_E(t)\,, \tag{1a}$$

$$\frac{d}{dt}n = -\frac{n}{\tau_s} - G\,|E|^2 + C + F_n(t)\,, \tag{1b}$$

where E is the (complex) electric component of the EM field normalized in such a way that $|E|^2$ is the photon density inside the laser cavity, ω_M is the resonance frequency of the M-th cavity mode, G and τ_p are the gain and the cavity losses, n is the excited carrier density, τ_s is the carrier lifetime and C is the external injected current. $F_E(t)$ and $F_n(t)$ are Langevin noise terms accounting for the stochastic nature of spontaneous emission. These terms have zero mean and second order correlation functions [18]

$$\langle F_E(t)^* F_E(t') \rangle = \frac{R}{V}\, \delta(t - t') \tag{2a}$$

$$\langle F_E(t) F_E(t') \rangle = \langle F_E(t)^* F_E(t')^* \rangle = 0 \tag{2b}$$

$$\langle F_n(t) F_n(t') \rangle = \frac{1}{V}\left[2(R|E|^2 + S) + C - \frac{n}{\tau_s} - G|E|^2 \right] \delta(t - t') \tag{2c}$$

$$\langle F_E(t) F_n(t') \rangle = -\frac{R}{V} E\, \delta(t - t') \tag{2d}$$

where the brackets stand for ensemble average, R is the spontaneous emission rate inside the lasing mode, S is the rate of the total radiative recombinations and V is the active volume. Eqs. (1) define a set of stochastic differential equations which have to be interpreted in the Ito sense. In Eq. (1b) the spontaneous lifetime is generally a function of the carrier density because of bimolecular and Auger recombination processes [1].

Eqs. (1a) and (1b) are general; Eq (1a), however, can be rewritten in a more useful form which underlines the main peculiarities of semiconductor lasers. Consider a laser having a cavity with a Fabry-Perot structure. In this case, one can write

$$\omega_M = \frac{M\pi c}{\eta(n,\omega_M)L}, \tag{3}$$

where c is the light velocity in vacuum, η is the refractive index of the lasing medium which, in semiconductor lasers, is a function of the carrier density and of ω_M itself and L is the cavity length. After some manipulations, the following result is obtained:

$$\omega_M \approx \omega_M(n_t) + \frac{d\omega_M}{dn}(n - n_t) = \omega_t + \frac{\alpha G_n}{2}(n - n_t), \tag{4}$$

where $\omega_t = \omega_M(n_t)$ is the oscillation frequency of the laser corresponding to the transparency value of carrier density n_t,

$$\alpha = -2\, \frac{\omega_t\, v_g \left(\frac{d\eta}{dn}\right)_{n=n_t}}{c\, G_n} \tag{5}$$

is the linewidth enhancement factor and $G_n = dG/dn$ is the differential gain.

In most lasers the value of α is zero. α can assume a nonzero value, even if very small, only in the case of detuned lasers in which the lasing frequency does not correspond to the maximum of the gain vs. frequency curve. In semiconductor lasers, on the contrary, the band structure of the semiconductor gives rise to a value of α which is different from zero, as shown in section 3.1; typical values are in the range between 3 and 8 [19].

Another peculiarity of laser diodes is the dependence of gain on the field intensity at a fixed value of carrier density. There is not a general agreement on the specific way to introduce nonlinear gain saturation [20]. A widely used expression is [21]

$$G = \frac{G_n(n - n_t)}{1 + \varepsilon |E|^2}, \tag{6}$$

where ε is the nonlinear gain coefficient. This expression has been recently confirmed by a theory based on a density matrix approach [22].

Introducing both α and ε into Eqs. (1) one obtains the more explicit form

$$\frac{d}{dt}E = \left[i\omega_t + i\frac{\alpha G_n}{2}(n - n_t) + \frac{1}{2}\frac{G_n(n - n_t)}{1 + \varepsilon|E|^2} - \frac{1}{2\tau_p} \right] E + F_E(t), \tag{7a}$$

$$\frac{d}{dt}n = -\frac{n}{\tau_s} - \frac{G_n(n - n_t)}{1 + \varepsilon|E|^2}|E|^2 + C + F_n(t). \tag{7b}$$

The rate equations (7) will be the starting points for the treatment of both stationary and dynamical properties of semiconductor lasers. Some typical dynamical behaviors of laser diodes can be directly explained on the basis of these equations. For example, the response of a laser diode to a stepwise change of the injected current from a value C_0 below threshold, to one C_1 above (see Fig. 1) is due to a large difference between the time scales involved in semiconductor laser dynamics: the cavity photon lifetime τ_p (of the order of 2 ps) and the mean lifetime of the excited carriers τ_s (of the order of 1 ns). In the first stage of the evolution which follows the current switch the depletion due to stimulated emission can be neglected in Eq. (7b), so that the deterministic solution of Eq. (7b) is

$$n(t) = C_1\tau_s + \left[n(0) - C_1\tau_s\right] \exp\left(-t/\tau_s\right). \tag{8}$$

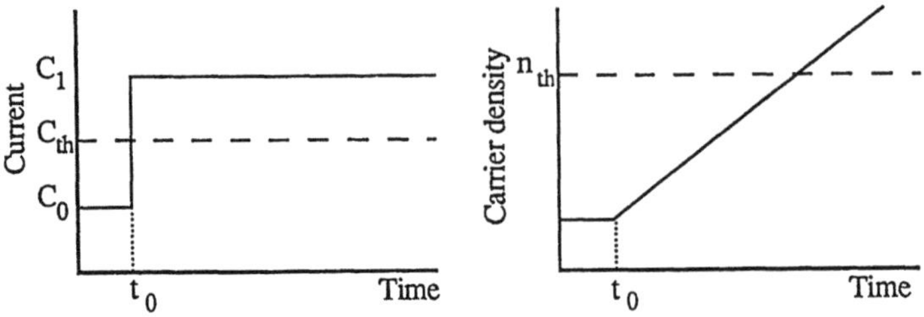

Fig. 1 Plot of inversion vs. time for a semiconductor laser modulated by a current step.

This equation gives an increase of the carrier density with a time constant τ_s. Only after the carrier density exceeds the threshold value, but is still increasing, the photon density grows and couples with the carrier density through the gain depletion term $G\,|E|^2$. This gives a response of the laser to a stepwise change of the pump parameter, different from that of other lasers, where the gain follows the sudden change of the pump parameter.

3. Origin of α and ε

At this point, a question naturally arises: what is the physical mechanism that makes the linewidth enhancement factor and the gain saturation coefficient much larger in laser diodes than in other lasers? Is the band structure of the semiconductor responsible for this behavior?

3.1 Origin of the Linewidth Enhancement Factor

As far as α is concerned, a simple model accounting for the large value of this parameter in semiconductor lasers was proposed by K. Vahala et al. [23]. The basic idea of their explanation is that the conduction-valence band system of a semiconductor laser can be approximated by an ensemble of collisionally broadened two level systems with different transition energies. For such a system

$$\alpha = -\frac{d\chi_R/dn}{d\chi_I/dn} \tag{9}$$

where χ_R and χ_I are the real and imaginary part of the complex susceptibility. Using the rate equation approximation, it can be shown that the definition (9) is equivalent to that given in Eq. (5). For a particular couple of levels, the real and imaginary part of the susceptibility are linked by the Kramers-Kronig relationships

$$\hat{\chi}_R = A\left[P_c(n,\omega) - P_v(n,\omega)\right] \frac{\omega_L - \omega}{(\omega_L - \omega)^2 + 1/T_2^2}, \tag{10a}$$

$$\hat{\chi}_I = A\left[P_c(n,\omega) - P_v(n,\omega)\right] \frac{1/T_2}{(\omega_L - \omega)^2 + 1/T_2^2}, \tag{10b}$$

where ω is the transition line center frequency, ω_L is the lasing frequency, T_2 is the collisional broadening time, P_c and P_v the conduction and valence band occupation probability and, finally, A is proportional to the dipole matrix element.

Eqs. (10) show that a system in which the laser transition occurs between two atomic levels is characterized by a lasing frequency $\omega_L = \omega$ which maximizes the gain $\hat{\chi}_I$, unless the laser is forced to work in a detuned condition. This implies a zero value of $\hat{\chi}_R$ and consequently of α. In semiconductor lasers, on the contrary, the total susceptibility is the sum of $\hat{\chi}_R$ and $\hat{\chi}_I$ over all states

$$\chi_R = \int_{-\infty}^{\infty} \rho(\omega)\, A\left[P_c(n,\omega) - P_v(n,\omega)\right] \frac{\omega_L - \omega}{(\omega_L - \omega)^2 + 1/T_2^2}\, d\omega\,, \tag{11a}$$

$$\chi_I = \int_{-\infty}^{\infty} \rho(\omega)\, A\left[P_c(n,\omega) - P_v(n,\omega)\right] \frac{1/T_2}{(\omega_L - \omega)^2 + 1/T_2^2}\, d\omega\,, \tag{11b}$$

where $\rho(\omega)$ is the effective density of states.

At this point the formal expression of α can be obtained

$$\alpha = \frac{\displaystyle\int_{-\infty}^{\infty} \rho(\omega) \frac{d}{dn}\left[P_c(n,\omega) - P_v(n,\omega)\right] \frac{\omega_L - \omega}{(\omega_L - \omega)^2 + 1/T_2^2}\, d\omega}{\displaystyle\int_{-\infty}^{\infty} \rho(\omega) \frac{d}{dn}\left[P_c(n,\omega) - P_v(n,\omega)\right] \frac{1/T_2}{(\omega_L - \omega)^2 + 1/T_2^2}\, d\omega}\,. \tag{12}$$

In semiconductor lasers with a FP cavity the lasing frequency is determined by the position of the χ_I maximum. In Ref.[23] it has been shown that, performing the exact calculation with parabolic band approximation, the function

$$\rho(\omega) \frac{d}{dn}\left[P_c(n,\omega) - P_v(n,\omega)\right] \tag{13}$$

is asymmetrical around ω_L, so that the contributions for $(\omega_L-\omega)>0$ and $(\omega_L-\omega)<0$ are differently weighted in the integral at the numerator of Eq. (12) and the net result is a non-vanishing value of α. Moreover α is shown to be a function of the threshold carrier density and the laser frequency and increases for lower oscillation frequencies.

3.2 Origin of the Nonlinear Gain

The physical mechanisms which affect the value of gain saturation coefficient are less known than those affecting the linewidth enhancement factor. Different mechanisms have been proposed to justify the non-zero value of ε in the gain expression (6)[24, 25, 26], but only two of them seem to be effective, namely spectral hole burning [25] and carrier heating [26], although an open question about their relative importance still remains.

In Fig. 2, a sketch representing the idea on which spectral hole burning is based is reported. For high field intensity inside the cavity, the finite intraband carrier scattering time (~100 fs) becomes comparable or larger than

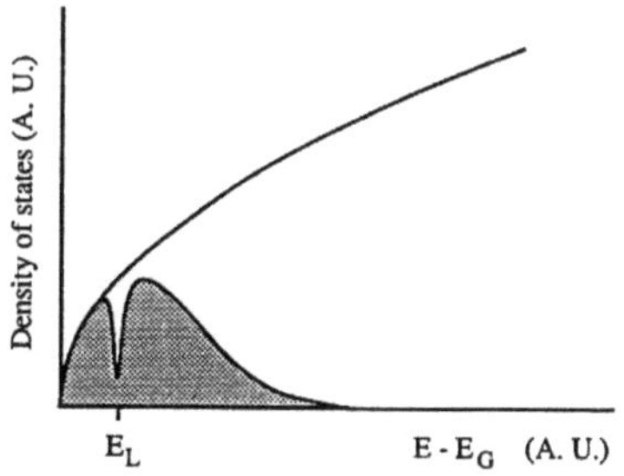

Fig 2. Pictorial representation of relative carrier density distribution (shadowed region) in the case of spectral hole burning.

the stimulated emission time, so that a hole in the carrier distribution around the laser line appears. This hole decreases the gain at the laser frequency even if the total carrier density remains almost unchanged. Many attempts were made in the past to directly observe the spectral hole burning, but only recently has this mechanism been experimentally verified [27] and a carrier scattering time of ≈ 70 fs was measured in InGaAsP devices.

The carrier heating mechanism is based on the idea that the optical field present in a cavity generates an increase of carrier temperature. This affects the occupation probability of the states and hence gain, because of the decrease of carrier density available at the lasing frequency. This mechanism was suggested to explain the results of a famous experiment [26] in which two short optical pulses (~100 fs) at the same wavelength are sent through a GaAlAs optical amplifier (a semiconductor laser with the end facets antireflection coated). One of the two pulses (pump) has an intensity high enough to induce gain saturation, while the other one (probe) has much lower intensity. They are orthogonally polarized so that a polarizer at the output of the amplifier can select the transmitted probe pulse. The experiment was performed by measuring the transmission coefficient of the probe pulse when the wavelength is moved from positive to negative gain regions (see Fig. 3). If spectral hole burning were responsible for gain saturation, when the pump pulse is present, a decrease of transmission coefficient in the gain region, an increase in the absorption region and no variation at the transparency wavelength should be observed The results, however, show that a decrease

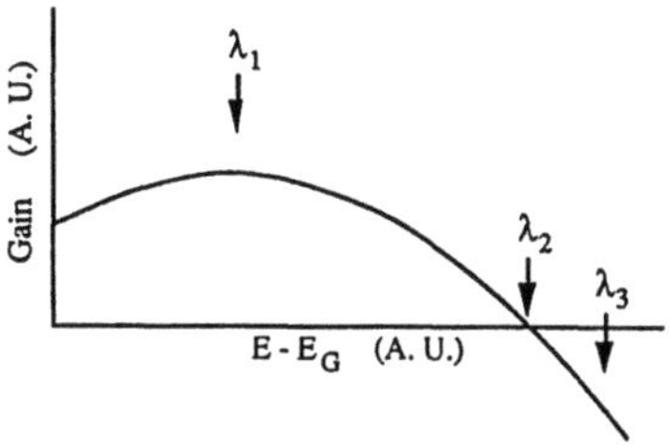

Fig. 3 Schematic of optical gain in laser diodes. The wavelength regions in which the experiment was performed are indicated.

of the transmission is always present for delays between the probe and pump pulses below ≈ 100 fs. This fact may be explained by supposing that a mechanism of carrier heating is responsible for the observed results. The same experiment repeated in InGaAsP gave more complex results than those in GaAlAs, which suggest that spectral hole burning plays an important role in the determination of gain compression characteristics of this material [28].

4. Main Effects of α and ε

The main effect of the linewidth enhancement factor is the broadening of the emission linewidth with respect to the usual Schawlow-Townes expression. When semiconductor lasers are used in optical communication systems, low values of linewidth can be very important. In order to understand why, one should remember that the particular modulation scheme of the semiconductor laser depends on the physical quantity in which the information is codified. This quantity can be either the intensity when traditional systems are used [29], or the field phase, frequency or amplitude if coherent systems are employed [30]. When intensity modulation is used in digital transmission systems, a fast response of the output power to a sudden change of the injected current and a very low value of time jitter between one pulse and another are essential requirements of the device, while the linewidth value is less important. In coherent systems, the detection is obtained through a beating of the signal with a strong local field, usually generated by another semiconductor laser (the local oscillator); thus a high degree of spectral purity is needed in order to improve the system's performance and this results in narrow emission lines or, more precisely, low phase noise for both the source and the local oscillator. Methods to reduce the linewidth enhancement factor and, hence, the linewidth have then been sought. Among the different techniques proposed in literature, two are more attractive. The first one is based on the use of distributed feedback (DFB) or distributed Bragg reflector (DBR) lasers [31]. In these devices, the feedback mechanism is obtained by a grating inside or at the end of the cavity and the emission frequency is determined by the grating period and can be shifted with respect to the maximum of the gain curve. In particular it can be moved toward higher values of frequency, so effectively reducing the value of α. The second method is based on the use of active layers having dimensions so small that quantum size effects appreciably affect the density of states $\rho(\omega)$ in Eq. (12) This can be obtained in quantum well and quantum wire lasers in which values of α reduced by a factor of approximately 2 and 4 with respect to the standard bulk devices have been predicted [32] and observed [33]. In quantum dot lasers the band structure typical of semiconductors is removed and the energy diagram of the allowed states resembles that of an atomic system, so that the value of α should approach zero in these devices [32].

Another important effect of the linewidth enhancement factor is frequency chirping [5], that is the unavoidable frequency modulation associated with intensity modulation. This effect is usually detrimental, because it generates a distortion of the pulse under intensity modulation due to fiber dispersion. However, this effect can be exploited in coherent systems because a simple frequency modulation can be obtained by a direct modulation of the injected current.

As far as the nonlinear gain coefficient is concerned, this parameter affects the modulation characteristics of semiconductor lasers by reducing the maximum attainable modulation speed and by damping effectively the resonance peaks [34] due to coupling between the field and the carrier density.

4.1 Linewidth of Semiconductor Lasers

The effect of the stochastic nature of the spontaneous emission events in lasers is to broaden their emission line with respect to the ideal Dirac delta function. Shawlow and Townes found that the emission line has a Lorentzian shape, with full width at half-maximum equal to [35]

$$\Delta\nu_L = \frac{h\omega_L \Delta\nu_c^2}{2P} \frac{N_2}{N_2 - N_1} \tag{14}$$

where h is the Planck constant, $\Delta\nu_c$ is the linewidth of the passive resonator of the laser, N_2 and N_1 the number of electrons in the upper and lower state of the laser and P is the emitted power. When this expression is applied, for instance, to a helium-neon laser emitting a power of 1 mW and having $\Delta\nu_c$ of the order of 10^5 s^{-1} one obtains $\Delta\nu_L \approx 10^{-3}$ s^{-1}. In these devices other extrinsic sources of broadening, like the Doppler effect and thermal fluctuations of the cavity length, give a broadening of the line which far exceeds this value so that the effect of the spontaneous emission cannot be observed. In semiconductor lasers, the short cavity and the low value of the mirror reflectivity result in $\Delta\nu_c \approx 2\ 10^{10}$ s^{-1} and $\Delta\nu_L \approx 10^6$ s^{-1}, so the linewidth is mainly determined by spontaneous emission and the effect of quantum noise can be measured directly.

The first accurate measurement of the emission linewidth of semiconductor lasers was performed by Fleming and Mooradian [36]. They found a linear dependence of the linewidth with the inverse of the output power as predicted by Eq. (14), but the magnitude of $\Delta\nu_L$ was about 50 times larger than expected. Moreover, measurements of the emission lineshape did not give the expected Lorentzian shape because of the presence of two side peaks at a distance from the central emission frequency ranging from about 0.5 to 5 GHz [37, 38]. The amplitude and the position of the side peaks depend on the output power of the laser as shown in Fig. 4. The same discrepancies were found when the power spectrum of the frequency noise was measured. The low frequency content of this spectrum, which is proportional to the linewidth, was much higher than expected and, in the high frequency portion, a peak not predicted by the Shawlow and Townes theory arose at a frequency equal to the frequency difference between the side and

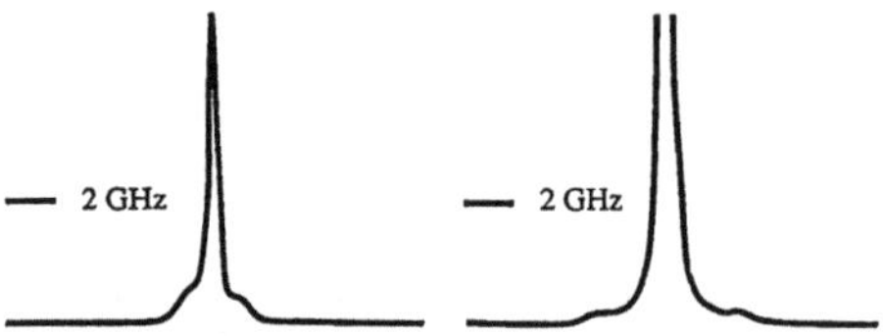

Fig. 4 Emission lineshape obtained for a DFB laser with a ridge structure. The curve on the left has been obtained for a current of 55 mA; the curve on the right for a current of 75 mA.

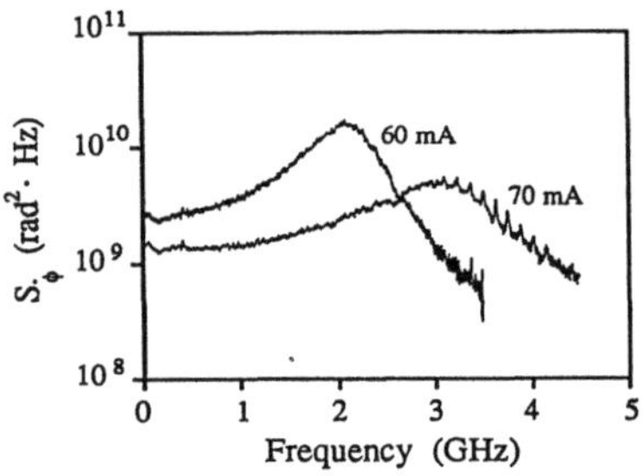

Fig. 5 Power spectral density of instantaneous frequency obtained for the same laser of Fig. 4.

central peaks of the emission line [39]. This is depicted in Fig. 5, where the spectrum measured for the same laser of Fig. 4 is reported at two different output powers.

A simple intuitive model to explain the anomalous broadening of the emission line was given by Henry [3]. He showed that this phenomenon is due to coupling between intensity and phase of the field induced by the linewidth enhancement factor. This model, however, does not account for the presence of temporal correlations of the phase. Models able to explain the shape of both the emission line and the measured frequency noise spectra were performed [4, 40, 41, 42]. A slightly modified version of these theories accounting for the nonlinear gain saturation term ε will be reported here.

The starting point is the set of Eqs. (7), which can be written in terms of intensity and phase of the field by making the transformation

$$E(t) = \sqrt{I(t)} \exp \left\{ i \left[\omega_0 t + \phi(t) \right] \right\} . \tag{15}$$

By making use of the Ito rules for stochastic calculus, Eqs. (7) are equivalent to the following equations with the new variables:

$$\frac{d}{dt} I = \left[\frac{G_n(n - n_t)}{1 + \varepsilon I} - \frac{1}{\tau_p} \right] I + \frac{R}{V} + F_I(t) , \tag{16a}$$

$$\frac{d}{dt} \phi = \left[\omega_t - \omega_0 + \frac{\alpha G_n}{2} (n - n_t) \right] + F_\phi(t) , \tag{16b}$$

$$\frac{d}{dt} n = - \frac{n}{\tau_s} - \frac{G_n(n - n_t)}{1+ \varepsilon I} I + C + F_n(t) . \tag{16c}$$

The new Langevin noise terms have the following correlation functions

$$\langle F_i(t) \rangle = 0 , \tag{17a}$$

$$\langle F_i(t) F_j(t) \rangle = D_{ij}\, \delta(t - t') , \tag{17b}$$

where the nonzero diffusion coefficients are given by [42, 43]

$$D_{II} = \frac{2RI}{V} \tag{18a}$$

$$D_{\phi\phi} = \frac{R}{2IV} \tag{18b}$$

$$D_{nn} = \frac{2(RI+ S) + C - n/\tau_s - GI}{V} \tag{18c}$$

$$D_{In} = - \frac{2RI}{V} \tag{18d}$$

Let I_0 and n_0 be the steady state values of I and n for the set of equations obtained by neglecting all the Langevin noise terms in Eqs. (16). Their expressions can be easily found by using Eqs. (16a) and (16c). The stationary emission frequency can be found by making use of Eq. (16b):

$$\omega_0 = \omega_t + \frac{\alpha G_n}{2} (n_0 - n_t). \tag{19}$$

To obtain the power spectra of the instantaneous frequency, $\dot{\phi}$, and of the fluctuations of I and n, $I = I_0 + \delta I$, $n = n_0 + \delta n$ and $\phi = \phi_0 + \delta\phi$, where ϕ_0 is an arbitrary reference value for the phase, have to be substituted in Eqs. (16). Taking into account only first order terms, one obtains

$$\frac{d}{dt} \delta I = - \Gamma_{II}\, \delta I + \Gamma_{In}\, \delta n + F_I(t), \tag{20a}$$

$$\frac{d}{dt} \delta\phi = \Gamma_{\phi n}\, \delta n + F_\phi(t), \tag{20b}$$

$$\frac{d}{dt} \delta n = - \Gamma_{nI}\, \delta I - \Gamma_{nn}\, \delta n + F_n(t), \tag{20c}$$

where

$$\Gamma_{II} = \frac{R}{I_0 V} + \frac{G_n(n_0 - n_t)}{(1+ \varepsilon I_0)^2} I_0 \varepsilon = \frac{\varepsilon I_0}{\tau_p(1+ \varepsilon I_0)} , \tag{21a}$$

$$\Gamma_{In} = \frac{G_n I_0}{1+ \varepsilon I_0} , \tag{21b}$$

$$\Gamma_{nI} = \frac{1}{\tau_p} - \frac{R}{I_0 V} - \frac{G_n(n_0 - n_t)}{(1+\varepsilon I_0)^2} I_0 \varepsilon \approx \frac{1}{\tau_p(1+\varepsilon I_0)}, \tag{21c}$$

$$\Gamma_{nn} = \frac{1}{\tau_s} + \frac{G_n I_0}{1+\varepsilon I_0}, \tag{21d}$$

$$\Gamma_{\phi n} = \frac{\alpha G_n}{2}, \tag{21e}$$

and the correlation properties of the Langevin noise term are still given by Eqs. (18) with I replaced by its stationary value I_0. As a consequence of this substitution, the diffusion coefficient of the carriers becomes the usual $D_{nn} = 2(RI_0 + S)/V$.

Defining the power spectrum of the ergodic process $\delta A(t)$ as the Fourier transform of the two time correlation function

$$S_{\delta A}(\Omega) = \int_{-\infty}^{\infty} \exp(-i\Omega\tau) \langle \delta A(t) \delta A(t+\tau) \rangle \, d\tau \tag{22}$$

and making use of standard techniques, one obtains

$$\begin{aligned} S_{\dot\phi}(\Omega) &= D_{\phi\phi} + S_{\delta n}(\Omega)\, \Gamma^2_{\phi n} \\ &= D_{\phi\phi} + \frac{1}{\Delta(\Omega)} \left[(\Omega^2 + \Gamma^2_{II})\, D_{nn} + \Gamma^2_{nI}\, D_{II} - 2\, \Gamma_{II} \Gamma_{nI}\, D_{In} \right] \Gamma^2_{\phi n}, \end{aligned} \tag{23}$$

where

$$\Delta(\Omega) = (\Omega^2 - \Omega^2_R)^2 + \Omega^2 \Gamma^2_R, \tag{24}$$

being

$$\Omega_R = (\Gamma_{II} \Gamma_{nn} + \Gamma_{nI}\, \Gamma_{In})^{1/2} \tag{25}$$

and

$$\Gamma_R = \Gamma_{nn} + \Gamma_{II} \tag{26}$$

the resonance peak frequency and the damping constant of the relaxation oscillations, respectively.

Eq. (23) shows that the frequency-noise power-spectrum consists of two parts. One, due to the random phase of the spontaneously emitted photons, is independent of frequency and is the term accounted for in the Shawlow-Townes theory. The other term is proportional to the power spectrum of the fluctuations of the carrier density through the proportionality factor α^2 and is due to the coupling between carrier density fluctuations and frequency fluctuations induced by α. This term is frequency dependent and shows a peak near the relaxation oscillation frequency Ω_R. It is worthwhile to note

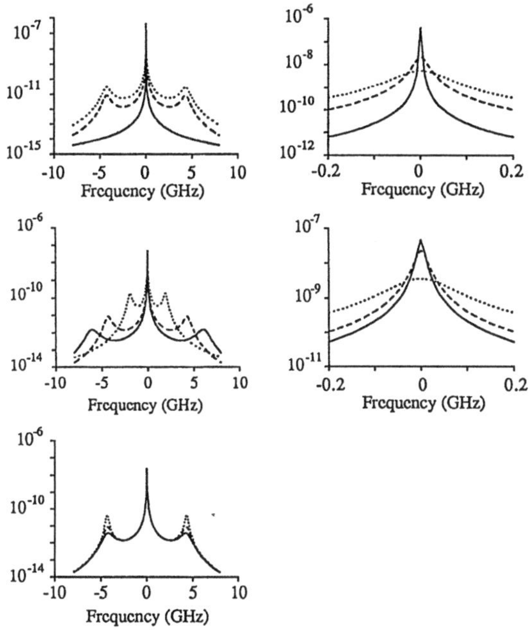

Fig. 6 Emission lineshape of a single mode laser evaluated through Eq.(23) using the relations reported in [45]. The figures on the right are magnifications of the central portion of the figures on the left. The top plots are obtained for $I_0V=5x10^3$, $\varepsilon/V=1.67x10^{-7}$ and $\alpha=0$ (solid), $\alpha=4$ (dashed), $\alpha=8$ (dotted). The central plots are obtained for $\alpha=4$, $\varepsilon/V=1.67x10^{-7}$ and $I_0V=10^4$ (solid), $I_0V=5x10^3$ (dashed), $I_0V=10^3$ (dotted). The bottom plots are obtained for $I_0V=5x10^3$, $\alpha=4$ and $\varepsilon/V=0.3x10^{-7}$ (solid), $\varepsilon/V=1.67x\ 10^{-7}$ (dashed), $\varepsilon=0$ (dotted).

that, for large values of the intensity I_0, the damping of the fluctuations is affected by the nonlinear gain coefficient ε.

The emission lineshape of a laser is defined as

$$S_E(\omega) = \int_{-\infty}^{\infty} \exp(-i\omega\tau)\, \langle E(t+\tau)E^*(t)\rangle \, d\tau \ . \tag{27}$$

In Fig. 6 the emission lineshape, obtained through Eqs. (23) and (27) using the relations reported in [44], is reported for different values of laser parameters. The dramatic effect of α as well as the effect of ε are clear.

When the frequency noise spectrum is flat, the emission line is a lorentzian and its linewidth $\Delta\nu_L$ can be evaluated by using the expression [45]

$$\Delta\nu_L = \frac{1}{2\pi}\, S_{\dot\phi}(0). \tag{28}$$

In semiconductor lasers the emission line does not have a Lorentzian shape due to the presence of two relaxation peaks at frequencies equal to $\omega_0 \pm \Omega_R$. The main lobe of the emission line, however, is Lorentzian to a good degree of approximation when typical parameters of semiconductor lasers are considered, so that Eq. (28) can be used to give a good approximation of the linewidth of semiconductor lasers. By substituting in Eq. (28) the expression of $S_{\dot{\phi}}(0)$ given by Eq. (23) one gets

$$\Delta\nu_L = \frac{R}{4\pi I_0 V}\,(1+\alpha_{eff}^2)\,, \tag{29}$$

where

$$\alpha_{eff} = \alpha\,(1+\frac{\varepsilon}{G_n \tau_s}). \tag{30}$$

To obtain Eq. (29), the terms proportional to D_{nn} and D_{In} have been neglected because they are much smaller than the term proportional to D_{II}.

Eq. (29) gives a linear dependence of linewidth $\Delta\nu_L$ with the inverse of the total photon density inside the cavity I_0. This is the same result of Shawlow and Townes if an enhancement factor $(1+\alpha_{eff}^2)$ is included in their expression. By virtue of the linear relationship between the photon density inside the cavity and the total output power

$$P = \frac{I_0 V}{\hbar\omega_0}\,\sigma, \tag{31}$$

being σ the mirror losses, the linewidth is also proportional to the inverse of the output power P.

Recently, a saturation and/or rebroadening of the emission lineshape at high power have been observed in DFB and in DBR lasers, even for very narrow linewidth lasers [46, 47]. In Fig. 7, a typical sketch of the linewidth versus the inverse of the output power is reported. Sometimes, the rebroadening of the linewidth is linked to an increase of low frequency content of the frequency noise spectrum. Different theories have been proposed to explain this phenomenon, but agreement on the most effective mechanism has not been reached. A theory proposes that 1/f noise, always observed in the frequency noise spectrum at very low frequencies, contributes to the linewidth [48]. When the output power increases, the amount of frequency noise due to spontaneous emission, accounted for in Eq. (23), decreases, while the 1/f noise contribution does not change. This causes a saturation of linewidth when spontaneous emission contribution to frequency noise becomes comparable with the 1/f noise picked up by the detection system. This mechanism, however, cannot explain rebroadening and, in some cases, also the saturation of the linewidth. Another mechanism, effective in some devices, is connected with the presence of a non-lasing

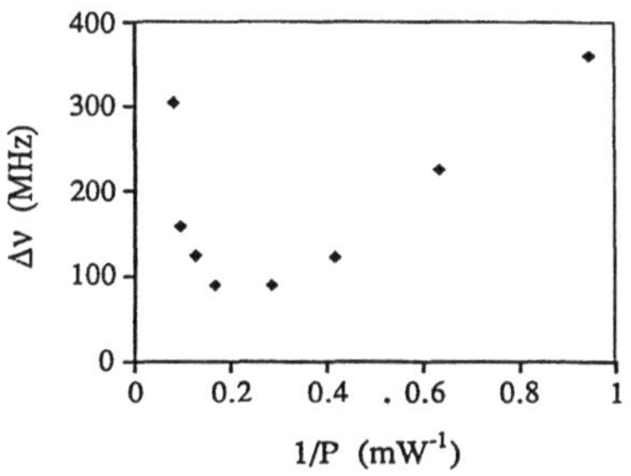

Fig. 7 Emission linewidth vs. the inverse of the output power for a DFB laser. The laser has a ridge structure.

mode in the spectrum of the laser [49, 50]. It has been shown that, in this case, the side mode induces fluctuations on the carrier density which affect the phase noise of the main mode and, hence, its linewidth. Although a ratio of the order of 10^3 between the intensities of main and side modes is sufficient to give rebroadening of the line, this mechanism cannot explain saturation or rebroadening in lasers having much higher values of this ratio. A third mechanism can be deduced from an inspection of Eqs. (29) and (30). When εI_0 is of the order of unity, α_{eff} becomes a function of I_0 so that the linear relation between $\Delta\nu_L$ and I_0 is no longer maintained. In particular, the increase of α_{eff} with I_0 results in an increase of the linewidth [51]. Recently, however, linewidth measurements performed on lasers which have very high values of photon density inside the cavity (obtained by means of a very high value of the cavity quality factor Q) did not show measurable rebroadening; whereas an enhanced effect of the nonlinear gain coefficient is expected in this case [52]. Furthermore, recent theoretical analyses [53] have shown that the effect of the nonlinear gain coefficient in DFB lasers is negligible with respect to other effects related to spatial hole burning, which is the inhomogeneity of gain and refractive index along the cavity induced by spatial distribution of the field. This effect has been shown to produce a power-dependent longitudinal variation of the linewidth enhancement factor along the cavity and a value of σ in Eq. (31) which also depends on power. The effect of spatial hole burning on linewidth seems to dominate over the effect of the nonlinear gain coefficient, but it cannot explain rebroadening of linewidth, but only an asymptotic nonzero value at $1/P = 0$ [54], which was actually observed [47].

4.2 Frequency Chirping

The effect of the linewidth enhancement factor is to couple intensity and frequency of the field through the carrier density, so that a dynamical evolution of the intensity reflects on a correspondent dynamical change of the frequency. When a laser operates at steady state, only spontaneous emission fluctuations affect the intensity and the coupling induces a broadening of the

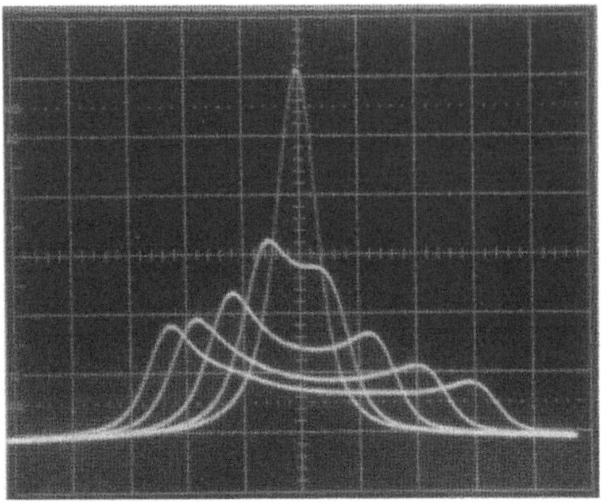

Fig. 8 Time averaged power spectra of a 1.5 μm laser under a 2 MHz sinusoidal modulation. The bias current is 10 mA above threshold, while the amplitudes of the modulation current are 0, 2, 4, 6, 8 mA. The spectrum broadens when the modulation current increases.

emission line. However, when the driving current of the laser is modulated the carrier modulation induces both an intensity and a frequency modulation which cannot be disjoint. This is shown in Fig. 8 where the emission spectrum of a single mode laser is reported for different values of the modulation current. The frequency response can be analytically evaluated in the small signal limit when a sinusoidal modulation of the kind

$$C(t) = C' + \delta C \exp(-i\Omega t), \qquad \delta C \ll C' \tag{32}$$

is considered, C' being an arbitrary value well above the threshold current. Performing a linearization of the rate equations (16) similar to that reported in the previous section and neglecting all the Langevin noise terms, one obtains

$$\frac{\delta\dot{\phi}}{\delta C} = \Gamma_{\phi n}\frac{\delta n}{\delta C} = \Gamma_{\phi n}\,\frac{i\Omega - \Gamma_{II}}{\Omega^2 - \Omega_R^2 + i\Omega\Gamma_R}, \tag{33}$$

δn being the carrier change induced by the current modulation. This equation shows that the frequency modulation induced by the current modulation is proportional to $\Gamma_{\phi n}$ and, hence, to the linewidth enhancement factor α.

4.3 Modulation Bandwidth

The maximum attainable modulation rate is of great importance for semiconductor laser applications. The large transmission bandwidth allowed by single mode fibers makes the speed of the end components of transmission

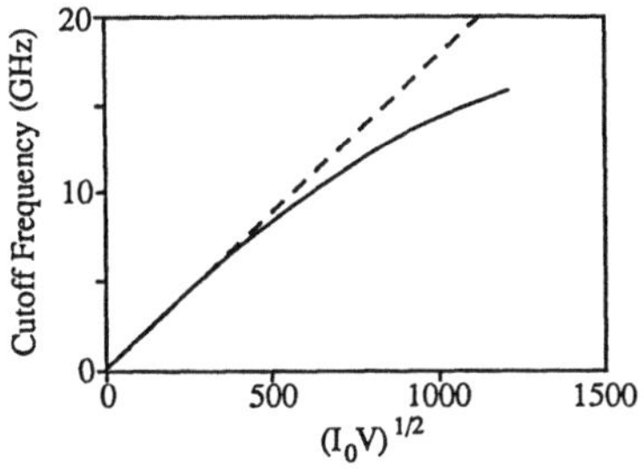

Fig. 9 Cutoff frequency as a function of the square root of the photon number inside the cavity. Dashed line refers to $\varepsilon = 0$, solid line to $\varepsilon/V = 0.7 \times 10^{-7}$.

systems (semiconductor lasers and photodetectors) the limiting factors for the maximum attainable transmission rate. Extremely high values of modulation bandwidth have been obtained for semiconductor lasers [55] (22 GHz for devices working at room temperature). The limiting parameter for higher modulation bandwidth are the parasitic elements of the equivalent electrical circuit of semiconductor lasers and, mostly, the nonlinear gain coefficient ε. The limiting effect of ε can be inferred from the intensity modulation response obtained from the solution of the rate equations. In the small signal limit, by using the same current modulation reported in Eq. (32), one obtains

$$\frac{\delta I}{\delta C} = \frac{-\Gamma_{nI}}{\Omega^2 - \Omega_R^2 + i\Omega\Gamma_R}. \tag{34}$$

This equation accounts only for the intrinsic response of the laser and disregards all the extrinsic ones such as thermal and parasitic effects, which modifies the real modulation response of the laser. An inspection of this equation shows that the effect of ε, which is contained in the term Γ_R, is twofold. The first effect is the damping of the resonance at Ω_R, as already underlined in section 4.1; the second effect is the reduction of the modulation bandwidth. This reduction can be quantified by calculating the cutoff frequency, that is the frequency at which the modulation response (34) reduces to one half:

$$\Omega_{3dB}^2 = (\Omega_R^2 - \Gamma_R^2/2) + \left[(\Omega_R^2 - \Gamma_R^2/2)^2 + \Omega_R^4\right]^{1/2}. \tag{35}$$

The detrimental effect of the gain saturation coefficient ε is evident from Fig. 9, where the cutoff frequency is reported as a function of the square root of the output power of the laser for two different values of ε.

5. Transient Dynamics

The low value of quality factor of semiconductor laser cavities causes the spontaneous emission noise to be the dominant factor in the broadening of the emission line. Another typical phenomenon in which spontaneous emission fluctuations play an important role is the transient dynamics which follows a sudden switch of the laser gain from below to above threshold [56]. In the initial state below threshold, the probability distribution of the complex field inside the laser cavity has a zero mean, whereas in the final one above threshold, the field has a non-zero mean value. The transition from the initial state in which the field is incoherent to the final coherent state is driven by spontaneous emission fluctuations, which are amplified during transient dynamics. This implies that the delay between the transition of the laser gain from negative to positive values and the attainment for the first time of a fixed value of the field intensity (first passage time) undergoes large fluctuations [56].

The first experiments concerning the time-dependent photon statistics during the laser switch-on were performed in 1967 by Arecchi et al. [57] and in 1970 by Meltzer and Mandel [58], both on Q-switched He-Ne lasers. Later, the results of these experiments were fully explained by resorting to the theory of the decay from an unstable equilibrium point [59] and to the quasi-deterministic theory [60]. In Q-switched lasers, the switching time of the laser gain from below to above threshold is much shorter than all the characteristic times involved in laser dynamics. Only recently has the problem of lasers in which the cavity loss (or the gain) is slowly varied been faced [61],[62],[63]. Semiconductor lasers are usually modulated in this second way, because the minority carrier lifetime is much longer than the photon lifetime. A sudden increase of the current from below to above threshold in these devices, in fact, induces a quasi linear increase of the gain, which lasts until the amplitude of the field affects the gain, reducing it to the value corresponding to threshold. The modulation from below to above threshold of semiconductor lasers, which is usually referred to as gain-switching, has received considerable attention because very short pulse lengths can be easily achieved in this way [64]. Fluctuations of the first passage time can, however, limit the practical use of laser diodes in these driving conditions. For this reason a study able to show the limitations for the use of this modulation technique has been recently embarked on [65]. From this study a particular property of laser systems has been recognized. It consists in the excitation, during the transient, of a side mode which, in steady state conditions, is highly suppressed [66],[67]. This property is rather evident in gain switched lasers and, in the particular case of semiconductor lasers, can have detrimental practical effects. The physical mechanism responsible for this behavior is the growth, during the transient, of the side mode optical gain beyond its threshold value. Although the gain of this mode is always lower than that of the main mode, random fluctuations due to spontaneous emission can trigger the lasing action to start on the side mode itself [17]. The random nature of this phenomenon makes its effects particularly detrimental in digital transmission systems because the side mode propagates at a different speed with respect to the main mode due to the transmission line dispersion.

5.1 Fluctuations of the Turn-On Instant

After the switching of the injected current from a value $C_0 < C_{th}$ to a value $C_1 > C_{th}$, C_{th} being the threshold current, there is a delay in the optical response because the injected currier number has to reach a level just beyond the threshold before the optical pulse emission. Up to this level, there is a negligible lasing emission so that the term $G\,|E|^2$ in Eq. (7b) can be neglected. As a consequence, Eqs. (7a) and (7b) decouple and Eq. (7a) can be easily solved when the Langevin noise term $F_n(t)$ is neglected. This is a reasonable approximation because the Langevin noise term on the carriers gives rise only to small fluctuations of the gain which have small effects during the transient compared with the fluctuations induced by the Langevin noise term on the photons, as they are amplified by the stimulated emission process. The result of the integration of Eq. (7a) can be inserted into Eq. (7b),yielding

$$I(t) = |E(t)|^2 = |\eta_1 + \eta_2(t)|^2 \exp\left\{\int_{t_{th}}^{t}\left(G(n(t')) - \frac{1}{\tau_p}\right)dt'\right\}, \tag{36}$$

where

$$\eta_1 = E_0 \exp\left\{\int_0^{t_{th}}\left[i\Delta\omega(n(t')) + \frac{1}{2}\left(G(n(t')) - \frac{1}{\tau_p}\right)\right]dt'\right\} \tag{37}$$

accounts for the deterministic evolution starting from the initial state E_0,

$$\eta_2(t) = \int_0^t F_I(t')\sqrt{R/V} \times \exp\left\{-\int_{t_{th}}^{t'}\left[i\Delta\omega(n(t'')) + \frac{1}{2}\left(G(n(t'')) - \frac{1}{\tau_p}\right)\right]dt''\right\}dt' \tag{38}$$

accounts for the contribution of any spontaneous emission event taking place after the electrical switching,

$$t_{th} = \tau_s \ln\left[(C_1 - C_0)/(C_1 - C_{th})\right] \tag{39}$$

is the time at which the carrier density reaches the threshold value $n_{th} = n_t + (G_n\tau_p)^{-1}$ and, finally,

$$C_{th} = n_{th}/\tau_s \tag{40}$$

is the threshold current.

In the previous equations, n(t), given by Eq. (8), is the deterministic solution of Eq. (7b) obtained by neglecting the depletion due to stimulated emission, while $\Delta\omega = \omega_1 + \alpha G_n(n-n_t)/2$. At a fixed time t, both η_1 and η_2 are complex zero mean independent Gaussian variables. Eq. (8) can be rewritten as

$$n(t) = n_{th} + \tau_s(C_1 - C_{th}) \left\{ 1 - \exp\left[- (t - t_{th})/\tau_s \right] \right\}, \tag{41}$$

and substituting this expression into Eq. (37) one obtains

$$\langle|\eta_1|^2\rangle = \langle|E_0|^2\rangle \exp \left\{ \tau_s (C_1 - C_{th}) \left[t_{th} + \tau_s \left(1 - \exp(t_{th}/\tau_s) \right) \right] \right\}. \tag{42}$$

To obtain analytical results for $\langle|\eta_2(t)|^2\rangle$, one should notice that the main contribution to the exponential in Eq. (38) is given by the times around $t = t_{th}$ when the expression inside the integral is zero. This justifies in Eq. (38) the approximation of the carrier density n(t) with its expansion around t_{th}, that is

$$n(t) \approx n_{th} + (C_1 - C_{th})(t - t_{th}), \tag{43}$$

thus giving

$$\langle|\eta_2(t)|^2\rangle = \int_0^t \sqrt{R/V} \exp \left\{ - \frac{1}{2} (C_1 - C_{th})(t - t_{th})^2 \right\} dt. \tag{44}$$

The exponential appearing in Eq. (42) is a quantity vanishingly small when the threshold time is large enough, that is, when the bias C_0 is sufficiently below the threshold current C_{th} (see Eq. (39)). One can deduce that the influence of the initial distribution of the field E_0 appearing in Eq. (42) is negligible when compared to the contribution of the spontaneous emission events accounted for in Eq. (44). The physical meaning of this result is that the statistical properties of the laser switching do not depend on the initial probability distribution of the field, because the photons initially present inside the cavity are absorbed before the laser attains the threshold gain. The statistical properties of the switching depend, on the other hand, on the spontaneous emission events taking place after the current switching and the most effective ones are those taking place around the threshold time (the variance of $\eta_2(t)$ given by Eq. (44) is the integral of a Gaussian centered around the threshold time). Photons spontaneously emitted before are not effective because they have a high probability of being absorbed, while those emitted after the threshold time are less effective because they add to a field already large.

Let us define now the first passage time $\bar{t}$ as the time at which the laser reaches for the first time a given value $\bar{I}$ of the photon density. The statistical properties of the first passage time can be found by making use of the expansion (43) in the integral in Eq. (36). After some algebra, one gets

$$\langle \bar{t} \rangle = t_{th} + \left(\frac{2}{G_n C_f}\right)^{1/2} \left\{ \sqrt{\ln\mu - \psi(1)} + O\left[(\ln\mu - \psi(1))^{-1/2}\right] \right\}, \quad (45)$$

$$\langle (\bar{t} - \langle \bar{t} \rangle)^2 \rangle = \frac{2}{G_n C_f}$$

$$\times \left\{ \frac{\psi'(1)}{4(\ln\mu - \psi(1))} + \frac{\psi''(1)}{8(\ln\mu - \psi(1))^2} + O\left[(\ln\mu - \psi(1))^{-3}\right] \right\}, \quad (46)$$

where $C_f = C_1 - C_{th}$, the $\psi^{(k)}(z)$ are the k-th derivatives of the digamma function and

$$\mu = \bar{I}\left(\frac{R}{V}\left(\frac{\pi}{2G_n C_f}\right)^{1/2}\left\{\mathrm{erf}\left[\left(\frac{G_n C_f}{2}\right)^{1/2} t_{th}\right] + 1\right\} + \langle |\eta_1|^2 \rangle\right)^{-1}. \quad (47)$$

In the last equation, the term related to the initial conditions, $\langle |\eta_1|^2 \rangle$, is usually much smaller than the other term, so that the first two moments of the first passage time are independent of the initial field distribution, which is related to the initial biasing current. This result is confirmed by computer simulations of the complete set of Eqs. (7) which show that this is true up to biasing currents of the order of 0.95 C_{th} [68]. The only relevant parameter is the difference between the injected and the threshold currents C_f, which is in turn proportional to the final output power in the on state. The variance of the first passage time distribution given by Eq. (46) is usually referred to as time jitter of semiconductor laser gain switching.

Eqs. (45)-(47) hold for single-mode semiconductor lasers, that is for DFB or DBR lasers. Semiconductor lasers with a Fabry-Perot resonator have in general many modes with a comparable value of gain and, consequently, many of them can be excited during the transient. In a simplified laser model in which M identical modes are considered, the same calculations performed for single-mode lasers can be repeated and the final results are equations identical to Eqs. (45)-(47) apart from the substitution of the $\psi^{(k)}(1)$ with $\psi^{(k)}(M)$ [15]. The effect is to lower the time jitter value of a factor of the order of M, being $\psi^{(k)}(M) \approx 1/M^k$. This result can be easily explained by considering that, when M increases, the number of different ways the system has to escape from the initial unstable state increases as well. A more accurate theory accounting for a realistic gain distribution among the different modes gives results in qualitative agreement with those obtained with the above described simplified model.

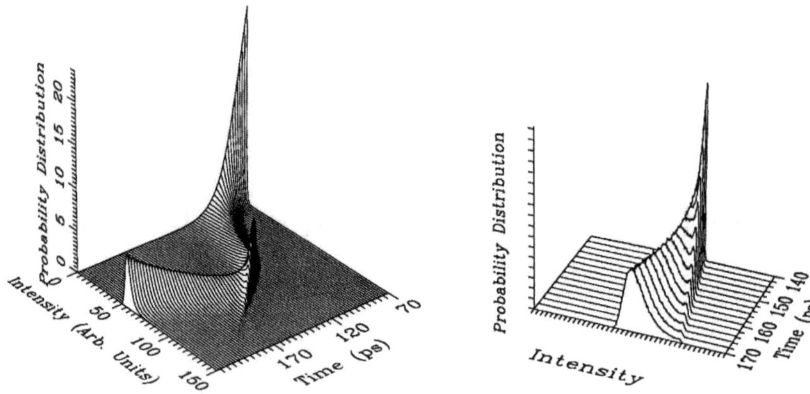

Fig. 10 Probability distribution of the intensity vs time obtained by simulation. On the right side an enlargement showing the double-peaked region is shown.

The slow response of the carrier density to an applied current step in single-mode laser diodes reflects also on the distribution of the switching trajectories in the time-intensity plane [69, 70,71]. It has already been shown that spontaneous emission gives rise to a spreading of the switching trajectories during the optical switching. In Q-switching, the trajectories spread out at the beginning of the leading edge of the optical pulse emission but they differ from one another only in a rigid temporal translation by a random amount. In semiconductor lasers, on the other hand, each trajectory attains a maximum of the intensity strictly dependent on the random delay between the current step and the emission of the optical pulse. This is due to the fact that the population inversion is still growing when the photon population builds up so that the later the switching the larger the values attained by the population inversion and, consequently, by the photon density during the first overshoot peak. As a consequence, the switching trajectories cluster under a line at a nonzero angle off the time axis. This peculiar behavior is shown in Fig. 10, where the probability distribution of the output intensity of the laser is reported as a function of time. This result has been obtained by a computer simulation of the set of Eq. (7) performed by means of a second order Runge-Kutta algorithm [72]. In the same figure, an enlargement for times slightly exceeding the maximum of the first overshoot of the output power is shown. In this region the probability distribution of the intensity shows two peaks. The peak at high values of laser intensity arises from those trajectories of the ensemble switching late. Such trajectories are in fact near the maximum of their overshoot peak, here they have a horizontal tangent which is responsible for the peak in the distribution. At the same time the most of the other trajectories, which are decreasing, give rise to the second peak of the probability distribution at lower values of intensity. The double-peaked probability distribution resembles that found by

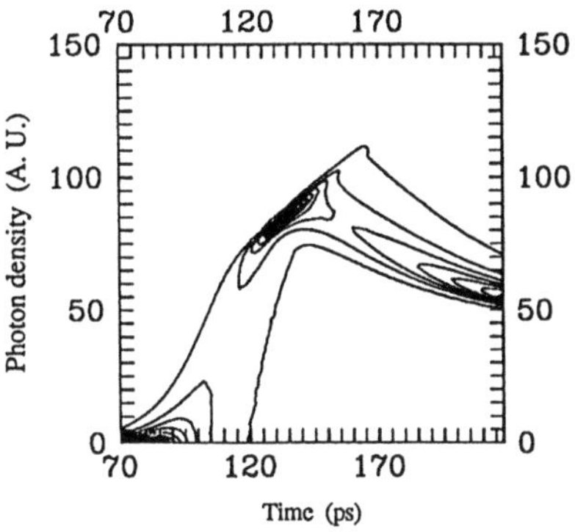

Fig. 11 The same as in Fig. 10, but as a two dimensional plot of level curves.

Nicolis ét al. [73] in the kinetic of exothermic chemical reactions and by Broggi and Lugiato [74] in optical bistability. The same quantity as in Fig.10 is reported in Fig.11 as a two dimensional plot of level curves.

At this point the principal theoretical results obtained for semiconductor laser gain switching can be summarized as follows:

1. The first passage time statistics and, in particular, its second order moment (time jitter) are independent of the biasing conditions.
2. Time jitter is lower in multimode than in monomode lasers.
3 A peculiar structure of the distribution of the trajectories in the time-intensity plane has been predicted.
4. The probability distribution of the intensity has been shown to have two peaks for times slightly exceeding the first overshoot of oscillation.

Two different experimental approaches have been proposed for measuring the time jitter in semiconductor lasers. The main problem one must solve when measuring the statistical properties of semiconductor laser switch-on is connected with the times involved in the phenomenon, which can be shorter than the response time of the best electronic equipments. For this reason in addition to methods based on the standard measurement in the time domain [75, 16], a different approach, based on a frequency domain measurement, has been proposed [76]. It has been shown, however, that this last method can be used only in the case of mode-locked lasers in which the different optical pulses are correlated and time jitter is due to slow fluctuations of laser parameters [77]. A typical set-up which can be used to perform measurements in the time domain is reported in Fig.12 [16]. The laser under test is biased below threshold and driven by a step current pulse with rise time of around 100 ps. The optical pulses emitted by the laser are detected by a fast photodiode with a rise time of 45 ps. Its output signal is displayed by a sampling oscilloscope equipped with a head having a rise time less than 25 ps and triggered by a part of the electrical pulse exciting the laser.

The need for a sampling oscilloscope arises from the short times involved in semiconductor laser dynamics. The use of such an oscilloscope, which gives a single dot for each realization of the process, allows one to perform a statistical analysis of the ensemble of the trajectories even if the single

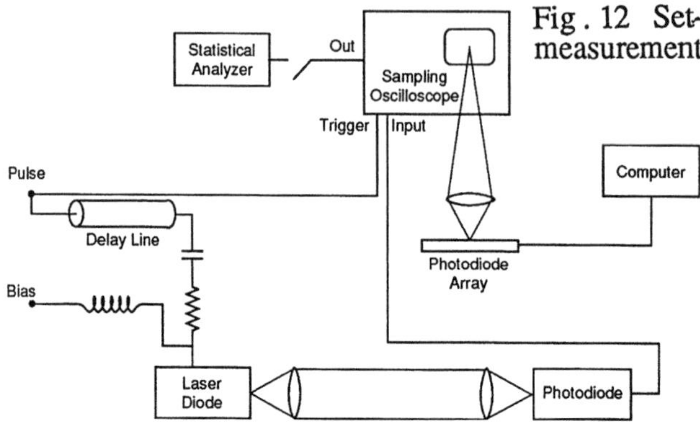

Fig . 12 Set-up for the time jitter measurements.

a b

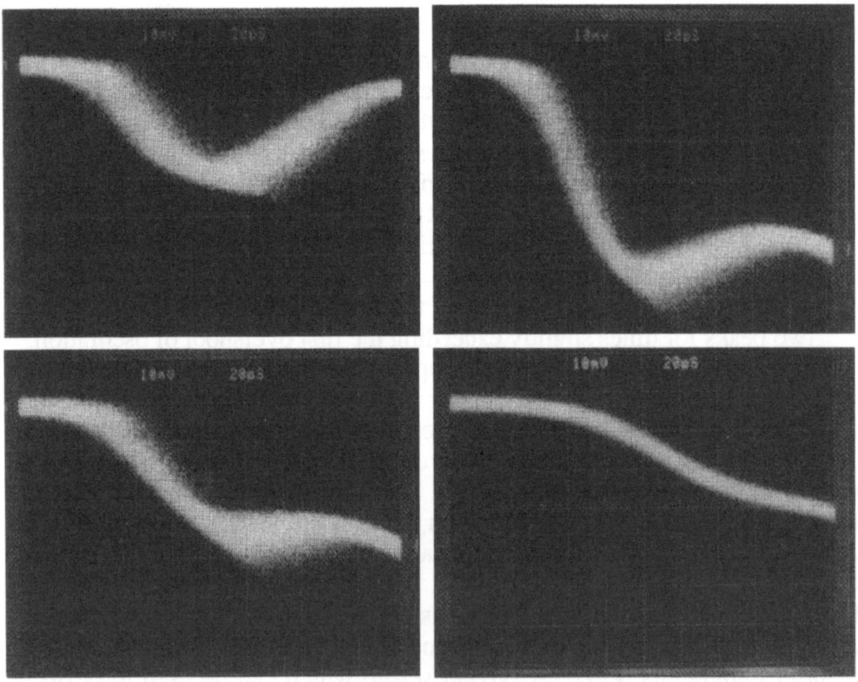

c d

Fig. 13 Traces of the laser switch on displayed by the sampling oscilloscope. The laser is a DFB laser with a threshold current of 28.9 mA. The driving conditions are pulse current 60 mA and bias current (a) 1mA, (b) 21 mA, (c) 28.1 mA, and (d) 29 mA. The horizontal scale is 20 ps/div.

trajectory is not experimentally accessible. With this experimental configuration a horizontal scanning of the display allows one to evaluate the variance of the first passage time while a vertical scanning gives the variance of the intensity noise. In Fig.13 some pictures of the oscilloscope screen are shown. These pictures represent the leading edge of the optical pulse emitted

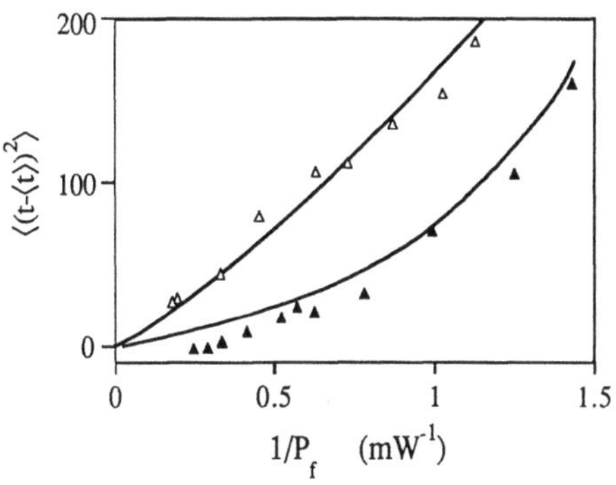

Fig. 14 Experimental results and theoretical fits of time jitter vs the inverse of P_f. Triangles refer to a multimode laser, open triangles to a single mode laser.

by a single mode laser (DFB) under different driving conditions. The picture d) refers to a bias above threshold. This last picture, which does not display any appreciable time jitter, verifies the theoretical prediction of a strong reduction of time jitter for bias above threshold [67] and shows that time jitter due to electronic apparatus is negligible. The amount of time jitter can be obtained by both acquiring the oscilloscope trace by a computer or by imaging the oscilloscope display on a photodiode array. In both cases the time jitter value can be obtained only after the variance of the uncorrelated noise contributions due to the vertical oscilloscope amplifier and electronic equipment are subtracted by the measured value [14]. The intensity distribution during the optical switch-on can be evaluated by resorting to a statistical analyzer which acquires the output from the oscilloscope at a fixed time delay with respect to the electrical switch-on.

The time jitter values, measured with the above technique, are reported in Fig. 14 as a function of the inverse of the value of the stationary final output power P_f in the on state. The experimental data have been obtained by separately varying the bias current and the amplitude of the current step, but only a dependence on the driving conditions through the value of P_f has been observed in all the tested devices. The experimental data reported in the figure refer to both a single (open triangles) and a multimode (triangles) laser. This figure shows that the the time jitter value decreases for increasing values of P_f in both cases, even if in a quite different fashion. In single mode lasers, in fact, this dependence is quasi linear as stated by Eq.(46) (one should remember that P_f is proportional to C_f) while it is strongly nonlinear in the multimode laser. This nonlinear behavior has to be attributed to the dependence of the number of the excited modes on the driving conditions. When the value of P_f increases, in fact, the overshoot of the carrier density beyond the threshold value is enhanced so that the number of modes suffering a positive gain also increases, so reducing the time jitter value. This can be experimentally observed using an apparatus similar to that reported in Fig.15 [15]. The number of modes for different values of P_f has been inserted in

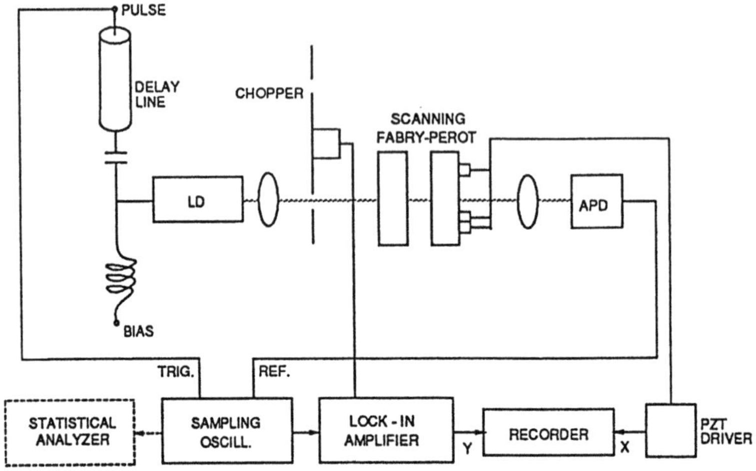

Fig. 15 Set-up for the measurement of the dynamical spectrum. The statistical analyzer is used to measure the statistical distribution of the intensity of the different modes. In this case, the chopper and the lock-in amplifier are removed.

Eq.(46). The result is shown in Fig. 14 together with the theoretical curve obtained for the single mode case. The agreement is quite good for both cases. The curves obtained by using the complete theory for multimode lasers and computer simulations of the multimode version of eqs. (1) reported in Ref. [15], are very close to that shown in Fig.14.

The statistical distribution of switching trajectories in the time intensity plane can be evaluated through an inspection of Fig.13. Here the number of sampling dots per unit area is proportional to the probability that a trajectory crosses it. The increased density of dots in curves a), b), c) for times a bit shorter than those corresponding to the attainment of the maximum intensity indicates that all the trajectories cluster in that region, however far apart they are at the beginning of the pulse. Moreover all the trajectories stay under a straight-line envelope making a nonzero angle with the horizontal axis. Unfortunately, a quantitative comparison between theory and experiment is impossible because the deconvolution of the measured statistical distributions of the intensity with the noise affecting the measurement is hard to perform, especially in the most meaningful region where the trajectories cluster. Nevertheless, some theoretical results can be confirmed. One of them is the presence of the double peaked probability distribution of the intensity for times slightly exceeding the time at which most of the trajectories reach their maximum value. The result of such a measurement, performed in the driving conditions of 13 a), is reported in Fig. 16.

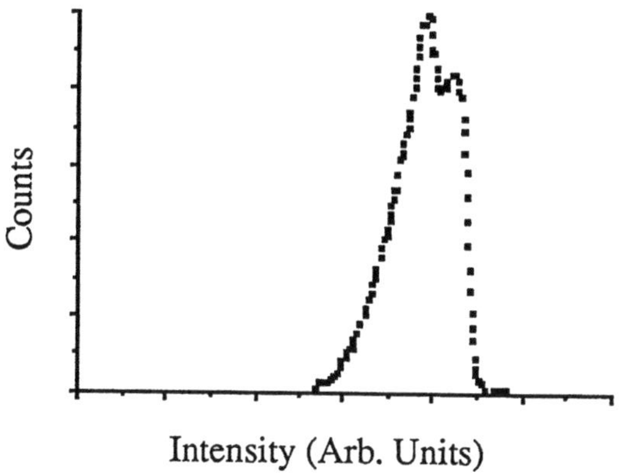

Fig. 16 Probability distribution on a linear scale of the intensity measured for time slightly exceeding the maximum of the first overshoot peak of the intensity.

5.2 Excitation of a Side-Mode During the Transient

In the previous section, the effect of spontaneous emission in spreading the switching trajectories at the leading edge of the optical pulse emission has been considered. The characterization of this phenomenon has been performed by defining the first passage time and by studying its statistical properties. In strictly single mode lasers, this is practically the only *macroscopic* effect of spontaneous emission during the transient. In multimode lasers, on the other hand, spontaneous emission events give rise also to a dynamical excitation of any mode having gain larger than zero at the moment of the photon build up. It has already been shown that the sharing of the laser intensity among different modes has the effect to reduce the time jitter of the laser switch on.

A different effect of spontaneous emission arises in single-mode lasers having side modes highly suppressed in stationary conditions. In these lasers, if the difference between main and side mode losses is low enough, immediately before the optical pulse emission the side mode suffers a gain larger than zero though much lower than that of the main mode. As a consequence, random spontaneous emission events can trigger the lasing action to start on the side mode. The realizations in which the side mode is dynamically excited produce anomalously long tails in the probability distribution of both main and side mode intensities. These tails are typical of very unlikely events corresponding to large deviations from the most probable evolution.

From a theoretical point of view, this phenomenon is different from the usual ones which can be framed in the context of the decay from unstable states. Up to now, the theoretical analysis of the decay from unstable states have only studied the statistical properties of the time the system needs to reach the new stable point after the switching. Here, on the other hand, one is rather interested in the statistics of the direction of leaving the initially unstable point.

The starting point of the analysis is the multimodal version of Eqs. (7) which, under the hypothesis of homogeneous broadening, reads

$$\frac{d}{dt}E_j = \left[i\omega_{tj} + i\frac{\alpha G_n}{2}(n - n_t) + \frac{1}{2}\frac{G_n(n - n_t)}{1+\varepsilon I} - \frac{1}{2\tau_{pj}} \right] E_j + F_{E_j}(t), \quad (48a)$$

$$\frac{d}{dt}n = -\frac{n}{\tau_s} - \sum_j \frac{G_n(n - n_t)}{1+\varepsilon I}|E_j|^2 + C + F_n(t), \quad (48b)$$

where $I = \Sigma\,|E_j|^2 = \Sigma\,I_j$ and j is the mode index. The analysis will be restricted to the case in which only two modes are present. In this case, j assumes the value 0 for the main and 1 for the side mode, and this implies $\tau_{p0} > \tau_{p1}$.

To find some analytical estimation of the probability of the side mode excitation during the laser switch on, the analysis is restricted to a limited region of the phase space $D = \{I \le I_T\}$. The intensity I_T is chosen low enough to ensure that inside D the carrier depletion due to stimulated emission can be neglected in Eq. (48). A linearized theory cannot give the probability density function of the side mode at a fixed time because the tails of the probability density extend at any time up to values of the intensity which lie outside D. To avoid this problem, one should look for an analytical expression of the probability density function $\phi(I_1)$ defined in such a way that $\phi(I_1)dI_1$ gives the probability that the side mode intensity lies from I_1 to I_1+dI_1 when the total intensity is I_T. An integration of $\phi(I_1)$ from 0 to $I_T/2$ gives the (conditional) probability that the side mode is larger than the main mode when the total intensity is I_T. To obtain an analytical expression for $\phi(I_1)$ the Fokker-Planck equation corresponding to the linearized version of the set of Eqs. (48) can be written down. Neglecting the Langevin noise term $F_n(t)$ in the equation for the carrier density, the Fokker-Planck equation can be solved and an analytical expression of $\phi(I_1)$ found [17]. The obtained analytical expression is in excellent agreement with the results of computer simulations of the complete set of Eqs. (48). In Fig. 17 the probability that the side mode is larger than the main mode when the total intensity is one half the stationary value of the main mode intensity is reported as a function of the gain difference between the two modes. The different curves refer to different values of the injected current in the on state; the bias current does not affect the results for the same reason as discussed in the previous section. One can see that this probability is larger than 10^{-9} for difference of gain as large as 10-15 cm^{-1}.

The apparatus used to perform the measurement of the dynamical spectrum of single mode lasers is the same as used to perform the measurements of the dynamical spectrum of multimode lasers and reported in Fig. 15. Here, however, one needs to measure not only the dynamical

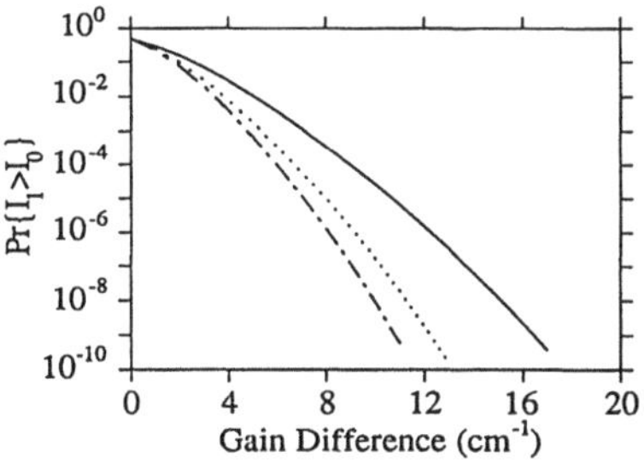

Fig. 17 Probability that the side mode is larger than the main mode, when the total intensity is one half the stationary value of the intensity in the on state,vs the gain difference between the two modes. The three curves refer to different values of C_f: solid line $C_f = 2C_{th}$, dotted $C_f = C_{th}$, dot-dashed $C_f = 0.75\ C_{th}$.

spectrum averaged over many different realizations of the switch-on process, as in the previous experiment, but also the statistical distribution of the main and side mode intensities. For this reason the chopper and the lock-in amplifier, which allow the signal-to-noise ratio to be increased in the measurement of the time dependent average spectrum, must be eliminated when measuring the statistical distribution of the intensity. Many different DFB lasers were tested for various values of the current of the final on state C_1. This has been done by varying the bias current from 0 to threshold while keeping the pulse height fixed and equal to 50 mA. Only in one sample was the dynamical excitation of a side mode observed. In this sample the gain difference between the main and the most intense side mode is 4 cm^{-1} and the threshold current is 18 mA. The side mode suppression ratio (SMSR) defined as the ratio between side and main mode powers in CW conditions is as high as $4x10^{-4}$ at an output power of 0.5 mW and $4x10^{-5}$ at 8 mW. The dynamical spectra averaged over many different realizations are shown in Fig.18 corresponding to different time delays t_d with respect to the turn-on instant (defined as the instant at which the optical intensity reaches 10% of its stationary value in the on state). Fig.18 refers to C_0=0.01 mA and 12 mA respectively. The broadening of the optical spectrum is due to frequency chirping occurring during the integration time of the photodiode (≈30 ps), while its spectral structure is due to relaxation oscillations of the laser. The statistical distribution of intensity at the side-mode wavelength, obtained for $C_0 =$ 12 mA and $C_1 - C_0 =$ 50 mA, is shown in Fig. 19 for two values of t_d. The small dots refer to the instrumental noise measured at zero output intensity. In spite of the large values of SMSR in the dynamical spectra averaged over different realizations, a tail of the side mode intensity distribution towards the larger values of the intensity is evident for both values of t_d. This underlines the nature of dynamical side mode excitation that shows up as rare but intense events which, though being unnoticed in the average spectrum, have a nonzero probability.

A quantitative agreement between theory and the experimental results is hard to establish because of the noise of the experimental apparatus that leads

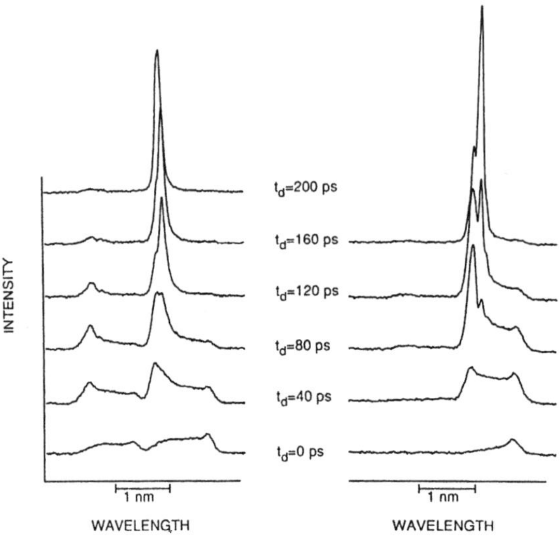

Fig. 18 Dynamical spectrum averaged over many different realizations, measured at different time delays t_d. The threshold current of the laser is 18 mA, and the current step is 50 mA. The left spectra are obtained with a bias current of 0.01 mA, while the right ones with a bias of 10 mA.

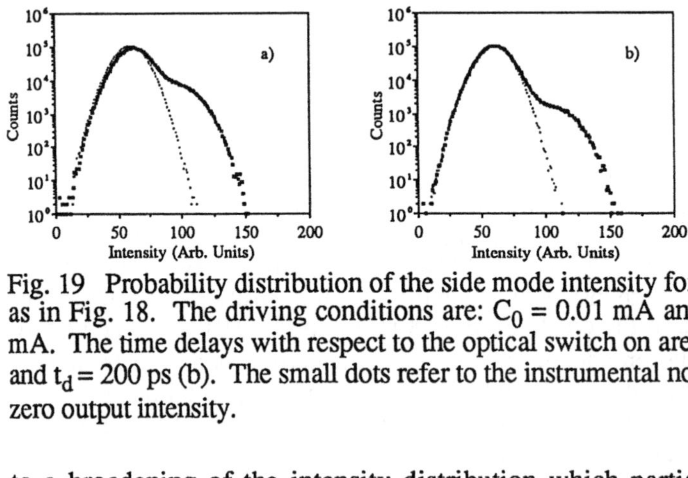

Fig. 19 Probability distribution of the side mode intensity for the same laser as in Fig. 18. The driving conditions are: C_0 = 0.01 mA and C_1 - C_0 = 50 mA. The time delays with respect to the optical switch on are t_d = 100 ps (a), and t_d = 200 ps (b). The small dots refer to the instrumental noise measured at zero output intensity.

to a broadening of the intensity distribution which partially masks the structure of the investigated phenomenon. One can, however, extract the following conclusions:

1) The average amplitude of the side mode increases with an increase in the pulse amplitude.
2) The value of SMSR increases for increasing t_d. This increase is due to the gain difference between the two modes which, for increasing values of t_d, tends to promote the main mode even in those realizations in which it starts later than the side mode.
3) SMSR increases by increasing C_0 for a constant value of C_1.

This last experimental observation, which is in contradiction with the theoretical results, can be explained by taking into account a different effect due an inhomogeneous distribution of the carrier density, which modifies the longitudinal properties of the grating during switching and induces a larger gain on the side mode.

6. Conclusions

In this paper, the effect of spontaneous emission noise on both stationary and dynamical properties of semiconductor lasers has been reviewed. It has been shown that the actual broadening of the semiconductor laser emission line cannot be explained on the basis of a model accounting only for the direct effect of the random phase of the spontaneously emitted photons on the intracavity field. In semiconductor lasers spontaneous emission noise has, in fact, also an indirect effect on the linewidth because it induces intensity fluctuations which give rise to fluctuations of the carrier density randomly modulating the refractive index and, hence, the phase of the field. It has also been shown that the strong coupling between carrier density and refractive index, peculiar to semiconductor lasers, can explain many characteristic features of this kind of laser, like, for example, the non-Lorentzian lineshape, the non-flatness of the frequency noise spectrum and the frequency chirping.

As far as the dynamical behavior is concerned, when a semiconductor laser is suddenly switched from below to above threshold by means of a current step, spontaneous emission noise has a twofold effect. First, it induces in both single and multimode lasers a spreading of the switching trajectories at the leading edge of the optical pulse emission. Second, in nearly single mode lasers it causes a random excitation of a side mode which is highly suppressed in stationary conditions. The long tails of the side mode probability distribution resulting from this second effect are the signature of very unlikely events which correspond to large deviations from the most probable evolution.

This work has been carried out at the Fondazione Ugo Bordoni in the framework of the agreement between the Fondazione Ugo Bordoni and the Italian P. T. Administration.

References

1 G. P. Agrawal and N. K. Dutta, *Long wavelength semiconductor lasers* Van Nostrand Reinhold Company, New York, (1986).
2 K. Y. Lau and A. Yariv, Semiconductors and Semimetals **22**, 69 (1985).
3 C. H. Henry, IEEE J. Quantum Electron. QE-**18**, 259 (1982).
4 P. Spano, S. Piazzolla, and M. Tamburrini, IEEE J. Quantum Electron. QE-**19**, 1195 (1983).

5 K. Kishino, S. Aoki, and Y. Suematsu, IEEE J. Quantum Electron. QE-**18**, 343 (1982).
6 M. Yamada and Y. Suematsu, IEEE J. Quantum Electron. QE-**15**, 743 (1979).
7 F. T. Arecchi, G. L. Lippi, G. P. Puccioni, and J. R. Tredicce, Opt. Comm. **51**, 308 (1985).
8 M. Suzuki, Phys. Lett. **67A**, 339 (1978).
9 F. de Pasquale and P. Tombesi, Phys. Lett. **72A**, 7 (1979).
10 F. Haake, J. W. Haus, and R. Glauber, Phys. Rew. A **23**, 3255 (1981).
11 F. de Pasquale and A. Mecozzi, Phys. Rev. A **31**, 2454 (1985).
12 F. de Pasquale, P. Tartaglia, and P. Tombesi, Nuovo Cimento **69B**, 228 (1982).
13 F. Haake, J. W. Haus, H. King, G. Schröder, and R. Glauber, Phys. Rev. A **23**, 1322 (1981).
14 P. Spano, A. D'Ottavi, A. Mecozzi, B. Daino, and S. Piazzolla, IEEE J. Quantum Electron. QE-**25**, 1440 (1989).
15 A. D'Ottavi, A. Mecozzi, P. Spano, and S. Piazzolla, Appl. Phys. Lett. **53**, 2362 (1988).
16 P. Spano, A. D'Ottavi, A. Mecozzi, and S. Piazzolla, Appl. Phys. Lett. **52**, 2203 (1988).
17 A. Mecozzi, A. Sapia, P. Spano, and G. P. Agrawal "Transient multimode dynamics in nearly single mode lasers," IEEE J. Quantum Electron, in press.
18 H. Haug, Phys. Rev. **184**, 338 (1969).
19 M. Osinski and J. Buus, IEEE J. Quantum Electron QE-**23**, 9 (1987).
20 G. P. Agrawal, J. Appl. Phys. **63**, 1232 (1988).
21 J. E. Bowers, B. R. Hemenway, A. H. Gnauck, and D. P. Wilt, J. Quantum Electron. QE-**22**, 833 (1986).
22 A. Tomita and A. Suzuki, Digest of "12th IEEE International Semiconductor Lasers Conference," paper L-24, Davos (CH) Sept 9-14, (1990).
23 K. Vahala, L. C. Chiu, S. Margalit, and A. Yariv, Appl. Phys. Lett. **42**, 631 (1983).
24 C. B. Su, Electron Lett. **24**, 370 (1988).
25 G. P. Agrawal, IEEE J. Quantum Electron. QE-**23**, 860 (1987).
26 M. P. Kesler and E. P. Ippen, Appl. Phys. Lett. **51**, 1765 (1987).
27 R. Frankenberger and R. Shimpe, Digest of "12th IEEE International Semiconductor Lasers Conference," paper G-5, Davos (CH) Sept 9-14, (1990).
28 K. L. Hall, J. Mark, E. P. Ippen, and G. Eisenstein, Appl. Phys. Lett. **56**, 1740 (1990).
29 P. S. Henry, IEEE J. Quantum Electron. QE-**21**, 1862 (1985).
30 R. A. Linke, A. H. Gnauck, IEEE J. Lightwave Technol. LT-**6**, 1750 (1988).
31 S. Ogita, M. Yano, and H. Imai, Electron Lett. **22**, 581 (1986).

32 Y. Arakawa, A. Yariv, IEEE J. Quantum Electron. QE-**22**, 1887 (1986).
33 Y. Arakawa, K. Vahala, A. Yariv, and K. Lau, Appl. Phys. Lett. **48**, 384 (1986).
34 C. B. Su, V. A. Lanzisera, IEEE J. Quantum Electron. QE-**22**, 1568 (1986).
35 A. Yariv, *Quantum Electronics 3rd ed.* John Wiley & Sons, New York (1988).
36 M. W. Fleming and A. Mooradian, Appl. Phys. Lett. **38**, 511 (1981).
37 J. M. Osterwalder and B. J. Rickett, Proc. IEEE **67**, 1671 (1979).
38 B. Daino, P. Spano, M. Tamburrini, and S. Piazzolla, IEEE J. Quantum Electron. QE-**19**, 266 (1983).
39 S. Piazzolla, P. Spano, and M. Tamburrini, Appl. Phys. Lett. **41**, 695 (1982).
40 R. Shimpe and W. Hart, Electron Lett. **19**, 136 (1983).
41 K. Vahala and A. Yariv, IEEE J. Quantum Electron. QE-**19**, 1102 (1983).
42 C. H. Henry, IEEE J. Quantum Electron. QE-**19**, 1391 (1983).
43 M. Lax, Phys. Rev. **157**, 213 (1967).
44 S. Piazzolla, P. Spano, Opt. Commun., **51**, 278 (1984).
45 P. Spano, S. Piazzolla, and M. Tamburrini, IEEE J. Quantum Electron. QE-**20**, 350 (1984).
46 H. Yasaka, M. Fukuda, and T. Ikegami, Electron. Lett. **24**, 760 (1988).
47 M. Okai, T. Tsuchiya, K. Uomi, N. Chinone, and T. Harada, "12th IEEE International Semiconductor Lasers Conference," paper O-1, Davos (CH) Sept 9-14, (1990).
48 K. Kikuchi, Electron Lett. **24**, 1001 (1988).
49 U. Krüger and K. Petermann, IEEE J. Quantum Electron. QE-**24**, 2355 (1988).
50 A. Sapia, P. Spano, and A. Mecozzi, Appl. Phys. Lett. **55**, 2582 (1989).
51 G. Morthier, P. Vanwikelberge, F. Buytaert, and R. Baets, IEE Proceedings **137** J, 30 (1990).
52 H. Burkhard, H. Hillmer, S. Hansmann, A. Mecozzi, A. Sapia, and P. Spano, Digest of "Optical Coherent Transmission in Multipoint Applications: Systems and Devices," Rome (Italy), January 29-31, (1991).
53 X. Pan, H. Olesen, and B. Tromborg, Electron. Lett. **26**, 1074 (1990).
54 J. E. E. Withway, G. H. B. Thompson, A. J. Collar, and C. J. Armistead, IEEE J. Quantum Electron. QE-**25**, 1261 (1989).
55 R. Olshansky, W. Powazinik, P. H. Hill, V. Lanzisera, and R. B. Lauer, Electron. Lett. **23**, 839 (1987).
56 F. T. Arecchi, R. Meucci, and J. A. Roversi, Europhys. Lett. **8**, 225 (1989).
57 F. T. Arecchi, V. Degiorgio, and B. Querzola, Phys. Rev. Lett. **19**, 1168 (1967).
58 D. Meltzer and L. Mandel, Phys. Rev. Lett. **25**, 1151 (1970).
59 F. Haake, J. W. Haus, and R. Glauber, Phys. Rev. A **23**, 3255 (1981).

60 F. de Pasquale, P. Tartaglia, and P. Tombesi, Phys. Rev. A **25**, 466 (1982).
61 P. Mandel and T. Erneux, Phys. Rev. Lett. **53**, 1818 (1984).
62 G. Broggi, A. Colombo, L. A. Lugiato, and P. Mandel, Phys. Rev. A **33**, 3635 (1986).
63 M. C. Torrent and M. San Miguel, Phys. Rev. A **38**, 245 (1988).
64 K. Y. Lau, Appl. Phys. Lett. **52**, 257 (1988).
65 M. M. Choy, P. L. Liu, and S. Sasaki, Appl. Phys. Lett. **52**, 1762 (1988).
66 P. L. Liu and M. M. Choy, IEEE J. Quantum Electron. QE-**25**, 1767 (1989).
67 S. E. Miller, IEEE J. Quantum Electron. QE-**25**, 1771 (1989).
68 A. Mecozzi, P. Spano, and A. D'Ottavi, Appl. Phys. Lett. **55**, 769 (1989).
69 S. Balle, P. Colet, and M. San Miguel, "Statistics for the transient response of single-mode semiconductor laser gain-switching," unpublished.
70 P. Spano, A. Mecozzi, and A. Sapia, Phys. Rev. Lett. **64**, 3003 (1990).
71 A. Mecozzi, P. Spano, and A. Sapia, Opt. Lett., **15**, 1067 (1990).
72 W. Rümelin, SIAM J. Numer. Anal. **19**, 604 (1982).
73 F. Baras, G. Nicolis, M. Malek Mansur, and J. W. Turner, J. Stat. Phys. **32**, 1 (1983).
74 G. Broggi and L. Lugiato, Phys. Rev. A **29**, 2949 (1984).
75 M. M. Choy, P. L. Liu, P. W. Shumate, T. P. Lee, and S. Tsuji, Appl. Phys. Lett. **47**, 448 (1985).
76 A. J. Taylor, J. M. Wiesenfeld, G. Eisenstein and R. S. Tucker, Appl. Phys. Lett., **49**, 681 (1986).
77 E. H. Böttcher and D. Bimberg, Appl. Phys. Lett., **54**, 1971 (1989).

Relation Between Passage Time and Maximum Intensity in Semiconductor Laser Gain Switching

S. Balle, P. Colet, and M. San Miguel

Departament de Física, Universitat de les Illes Balears, E-07071 Palma de Mallorca, Spain

The statistics of laser switch-on processes have been considered in a variety of systems and situations. An historical perspective on the problems associated with transient fluctuations during the switch-on is given in Ref. 1. There are in general two different statistical problems to be considered. One is the statistics of the switch-on time at which laser emission is observed. This is described by the method of Passage Times (PT)and it can be studied during a linear regime of laser amplification. The second problem refers to the large statistical fluctuations during the nonlinear regime. The two problems have been extensively considered for lasers in which the laser intensity grows monotonically towards its steady state value after they are switched on. More recent work refers to lasers displaying relaxation oscillations, as for example CO_2 or semiconductor lasers.

Theoretical calculations and experiments have been reported for the PT statistics of semiconductor[2] and CO_2 lasers[3]. The nonlinear regime of intensity fluctuations during relaxation oscillations has also been recently considered for semiconductor lasers[4]. An interesting new problem which appears in the nonlinear regime is the characterization of the statistics of the laser intensity at its maximum value occurring in the first relaxation oscillation[5]. This maximum value shows a rather large dispersion. We address here this problem for single mode gain switched semiconductor lasers, establishing a linear relationship between the maximum intensity and the Passage Time for each individual switch-on event.

The dynamics of a single mode semiconductor laser is well described by the Langevin formulation of the rate equations for the slowly varying complex amplitude E(s) of the optical field and the minority carrier number N[2, 4]

$$d_s E = \frac{1}{2}\left[g(1+i\alpha)\,N - \gamma\right]E + (\beta N)^{1/2}\,\xi(s) \tag{2.1}$$

$$d_s N = -\gamma_e N + C - gN|E|^2 \tag{2.2}$$

where α is the linewidth enhancement factor; g gives the gain rate per carrier; γ, the inverse cavity lifetime; γ_e, the carrier-recombination rate; $C \equiv \gamma_e N_0$, the injection current; β, the spontaneous emission rate per carrier. The random spontaneous emission process is modeled by a noise term $\xi(s)$ taken as a complex Gaussian white noise of zero mean value and correlation $\langle \xi(s)\,\xi^*(s')\rangle = 2\,\delta(s-s')$. In these equations gain saturation effects are neglected. Also note that the results for the statistics of $I \equiv |E|^2$ and N are independent of the value of α. In the following we set $\alpha = 0$.

The stochastic rate equations are conveniently written in dimensionless form normalizing E and N to their steady-state values $z = E/I_{st}^{1/2}$, $n = N/N_{st}$. Introducing a

Nonlinear Dynamics and Quantum Phenomena in Optical Systems
Editors: R. Vilaseca, R. Corbalán

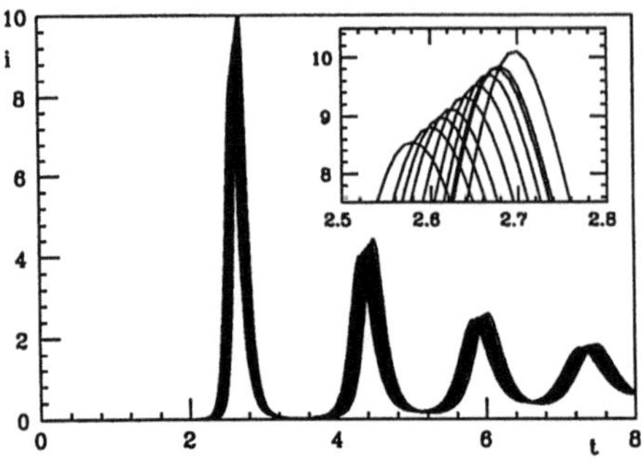

Fig. 1. Switching events as obtained from numerical simulation of (2.3) and (2.4) for λ=20. The insert describes the structure of the first intensity maximum.

dimensionless time $t=(\gamma\gamma_e)^{1/2}s$,

$$d_t z = \frac{z}{2\varepsilon}(n-1) + \sqrt{\eta n}\, q(t) \tag{2.3}$$

$$d_t n = -\varepsilon[(1+\lambda|z|^2)n - \lambda - 1] \tag{2.4}$$

where

$$\varepsilon \equiv (\gamma_e/\gamma)^{1/2}, \quad \lambda \equiv (N_0/N_{st}) - 1, \quad \eta \equiv (\delta/\lambda) \equiv \beta/[\varepsilon^2(\gamma\gamma_e)^{1/2}\lambda] \tag{2.5}$$

and $q(t)$ has the same correlation as $\xi(t)$. The parameter λ gives the operating point of the laser. For a typical semiconductor laser[2, 4] $\gamma_e^{-1} \approx 10^{-9}$ sec, $\gamma^{-1} \approx 10^{-12}$ sec, $\beta \approx 10^3$ sec^{-1}, $\varepsilon = 0.036$, $\delta = 1.54 \times 10^{-4}$. Laser gain switching is modeled by the solution of (2.3)-(2.4) with initial conditions $n(t=0) = 0$, $z(t=0) = 0$ and $\lambda > 0$.

The superposition of 10 switching events for $i(t) \equiv |z(t)|^2$ is plotted in Fig. 1 as obtained from the numerical integration of (2.3) - (2.4) with 10 different sequences of random numbers. Characteristic relaxation oscillations are displayed. An important point is that the intensity trajectories are not just shifted versions of the same trajectory with a different rising time. In fact, different rising times imply different heights of the maximum intensity. The later the amplification of the intensity occurs, the larger the carrier number grows and as a consequence the larger is the peak in the laser light intensity. Dispersion of 2% in the PT (jitter) is transformed in 10% dispersion in the maximum value of the intensity i_{max}. We are here interested in the relation between the statistics of the PT and i_{max}.

The PT statistics can be calculated in the linear regime in which the saturation term $\lambda|z|^2 n$ in (2.4) can be neglected. In this regime

$$n(t) = (1+\lambda)(1-e^{-\varepsilon t}) \approx (1+\lambda)\varepsilon t \tag{2.6}$$

where the last approximation holds for times of interest $\varepsilon t << 1$. Replacing (2.6) in (2.3) we are led to a situation considered in Ref. 6 for the turn-on of a laser of class A (He-Ne, Ar, etc.) by slowly sweeping the gain or loss parameter through threshold. Here we have a sweeping rate $(1+\lambda)/2$ and the slow sweeping condition corresponds to the requirement that the final value of the amplification factor $(n-1)/2\varepsilon$ in (2.3), as obtained from (2.6), be much larger than the sweeping rate. This is guaranteed whenever $\varepsilon^2\lambda^{-1} \ll 1$. A similar analysis to the one in Ref. 6 leads to the following explicit results for the mean PT, $\langle T \rangle$ and its variance[5]:

$$\langle T \rangle - \bar{t} = \sqrt{2}\,\lambda^{-1/2} t_r^{1/2} - \frac{1}{\sqrt{2}} (t_r \lambda)^{-1/2} \psi(1) \tag{2.7}$$

$$\langle (\Delta T)^2 \rangle = \frac{1}{2} (\lambda t_r)^{-1} \psi'(1) \tag{2.8}$$

where $\bar{t}$ is the threshold time defined by $n(\bar{t}) = 1$ and

$$t_r = \ln \frac{i}{\langle |h|^2 \rangle} \quad , \quad \langle |h|^2 \rangle = 2\eta\varepsilon + \eta (2\pi/\lambda)^{1/2} \tag{2.9}$$

For typical values of the parameters, the second term in (2.7) can be neglected. The variance of the PT distribution is a small quantity due to the factor t_r^{-1} in (2.8) and it becomes smaller with larger values of λ. A precise test of these results on the PT statistics has been performed by numerical simulations of (2.3)-(2.4) by averaging over 10^4 trajectories. The agreement is remarkable for a broad range of values of λ of interest and different values of i (Ref. 5). These results for the PT statistics are essentially the same as those reported in Ref. 2 where good agreement with experimental results is also found.

Different trajectories corresponding to different switch-on events have different intensity maximum, as seen in Fig. 1. It turns out that the statistics of this transient dynamics in the nonlinear regime can be described by the same idea successfully used for lasers which do not exhibit relaxation oscillations[7]: The time dependent ensemble average obtained from a simulation of (2.3)-(2.4) is well reproduced by the average over a random initial condition $z(t=\bar{t})$ of the deterministic solution of (2.3)-(2.4) with $n(t=\bar{t})=1$. It can be seen that the intensity fluctuations show large transient anomalous fluctuations. These fluctuations follow the relaxation oscillations and have an interesting local minimum at the time that the mean intensity reaches a maximum in the oscillations. The calculation of the statistics of i_{max} can then be addressed by a deterministic approximation to the dynamics in the vicinity of the time T_{max} in which i(t) reaches i_{max}. Different trajectories can be identified by a random initial time for such approximation. Close to $t = T_{max}$ saturation effects dominate, so that (2.4) can be approximated as

$$d_t n = -\varepsilon \lambda |z|^2 n \tag{3.1}$$

which together with the deterministic limit ($\eta=0$) of (2.3) can be solved to give

$$i(t) = i(t_0) + \frac{n(t_0) - n + \ln n - \ln n(t_0)}{\varepsilon^2 \lambda} \tag{3.2}$$

where t_0 is an initial time. This approximation, which is valid around $t = T_{max}$, can be extrapolated down to the PT which is taken as the initial time $t_0 = T$ for each

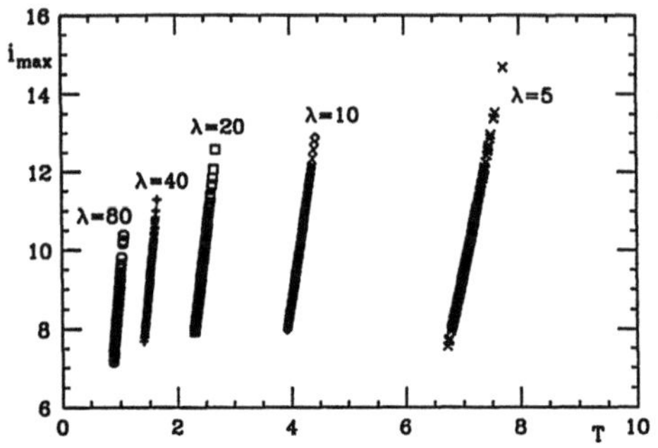

Fig. 2. i_{max} vs the Passage Time T for each trajectory and different values of λ.

trajectory. This PT is also the upper limit of validity of linear theory, so that $n(t_0{=}T)$ is given by (2.6). With this choice of t_0 and recalling that $n(t{=}T_{max}){=}1$, (3.2) gives for $t = T_{max}$

$$i_{max} = i_0 + \frac{(1+\lambda)(1-e^{-\varepsilon T}) - 1 - \ln\{(1+\lambda)(1-e^{-\varepsilon T})\}}{\varepsilon^2 \lambda}. \qquad (3.3)$$

This equation relates the maximum value of the intensity i_{max} to the PT for each individual trajectory identified by a different value of T. The result (2.8) implies a small dispersion in the values of T, so that an expansion of (3.3) around $T{=}\langle T\rangle$ is appropriate:

$$i_{max} = i^0_{max} + a(\lambda)(T - \langle T \rangle) \qquad (3.4)$$

.where

$$a(\lambda) = \frac{1}{\varepsilon\lambda}\left[(1+\lambda)\, e^{-\varepsilon\langle T\rangle} - \left(e^{\varepsilon\langle T\rangle} - 1\right)^{-1}\right] \qquad (3.5)$$

The validity of the linear relation implied by (3.4) is strikingly displayed in the results of numerical simulations of the rate equations shown in Fig. 2 for different operational points of the laser. The approximation becomes better as the operating point is higher above threshold, since this reduces $\langle(\Delta T)^2\rangle$. It also requires a judicious choice of the reference intensity for the PT: It has to be large enough for (3.3) to hold down to T and small enough so that (2.6) still holds. A comparison of the slope $a(\lambda)$ obtained from (3.5) and for the best linear fit to the data in Fig. 2. is shown in Table I for a reference intensity $i = 0.5$.

The main consequence of (3.4) is that the knowledge of the dispersion of the PT distribution is enough to know the dispersion of i_{max}. The dispersion in PT values is amplified by the factor $a(\lambda)$ to give the dispersion of maximum values of the intensity. This amplification is larger the higher is the operating point of the laser above threshold.

Table I

λ	a (λ) (best fit)	a (λ) (theory)
5	6.99	6.68
10	9.29	9.04
20	11.80	11.68
40	14.61	14.43
80	17.22	17.09

Acknowledgment

Financial support from Comisión Interministerial de Ciencia y Tecnología (CICYT) Project TIC 90/080 is acknowledged.

References

1. F. T. Arecchi in *Noise in Nonlinear Dynamical Systems*, edited by F. Moss, L. Lugiato and W. Schleich (Cambridge University Press, 1990), pp. 263-269.
2. A. Mecozzi, S. Piazzolla, A. D'Ottavi and P. Spano, Phys. Rev. A 38, 3136 (1988); Appl. Phys. Lett. 55, 769 (1989); P. Spano, A. D'Ottavi, A. Mecozzi, B. Daino and S. Piazzolla, IEEE, J. Quantum Electron. 25, 1440 (1989); A. Mecozzi, P. Spano, A. D'Ottavi, and S. Piazzolla, Appl. Phys. Lett. 55, 769 (1989).
3. F. T. Arecchi, R. Meucci and J. A. Roversi, Europhys. Lett. 8, 225 (1989); F. T. Arecchi, W. Gadomski, R. Meucci and, J. A. Roversi, Phys. Rev. A 39, 4004 (1989). M. Ciofini, R. Meucci and F. T. Arecchi, Phys. Rev. A 42, 482 (1990).
4. P. Spano, A. Mecozzi, and A. Sapia, Phys. Rev. Lett. 64, 3003 (1990); A. Mecozzi, P. Spano and A. Sapia, Opt. Lett. 15, 1067 (1990); See also the paper by P. Spano in this book.
5. S. Balle, P. Colet and M. San Miguel, Phys. Rev. A (to appear).
6. M. C. Torrent and M. San Miguel, Phys. Rev. A 38, 245 (1988); M. C. Torrent, F. Sagués and M. San Miguel, Phys. Rev. A 40, 6662 (1989).
7. F. De Pasquale and P. Tombesi, Phys. Lett. A 72, 7 (1979); F. De Pasquale, P. Tartaglia, and P. Tombesi, Physica A 99, 581 (1979); Z. Phys. B. 43, 353 (1981); Phys. Rev. A 25, 466 (1982).
8. See the paper by P. Spano in this book; G. P. Agrawal, J. Appl. Phys. 63, 1232 (1988).

Transition from Deterministic to Stochastic Behaviour in Bistable Systems

C. Boden, I. Roloff, F. Mitschke, and H. Welling

Institut für Quantenoptik, Universität Hannover,
Welfengarten 1, W-3000 Hannover 1, Fed. Rep. of Germany

The average switching time and its variance in a noisy bistable system are experimentally and numerically investigated and compared to an analytic prediction.

The dynamics of a one-dimensional bistable system can be described by $dx/dt = F(x,y)$, with x, y being the output and input quantities, respectively. The process can be visualised as the over-damped motion of a particle in a potential $V(x,y) = -\int F(x,y)\,dx$. To describe the upswitching behaviour, it is a common procedure to locally describe the potential near the marginal point by $V(x,\beta) = -ax^3 - \beta x$ $(a > 0)$, where $\beta = y - y_{th}$ denotes the distance from threshold. In the absence of noise, switching can only occur for an input value β greater than zero.

The influence of additive noise dramatically changes the behaviour of the system. There is now switching even below threshold because the potential well can be surpassed. The dynamics is given by $dx/dt = F(x,\beta) + \mu(t)$, where we take $\mu(t)$ to be a white noise signal of power ε with a Gaussian amplitude distribution. The average switching time $\langle\tau\rangle$ then depends on two parameters: β and ε .

Certain scaling laws for $\langle\tau\rangle$ have been known for quite a while such as: $\langle\tau\rangle \sim \beta^{-1/2}$ for $\varepsilon=0$, $\langle\tau\rangle \sim \varepsilon^{-1/3}$ for $\beta=0$ and $\langle\tau\rangle \sim \exp(|\beta|^{3/2}/\varepsilon)$ for $|\beta| >> \varepsilon$, $\beta < 0$ (Kramers escape rate). However, a complete picture has been missing up to now.

Recently a theory was published /1/ that describes the switching behaviour for a whole parameter region as an analytic function of both parameters β and ε. The authors derived the analytic formula (1) for the switching time near marginality:

$$\langle\tau\rangle = \Phi(k)\,(a^2\varepsilon)^{-1/3} + D(x_1, x_2). \tag{1}$$

Here, $D(x_1,x_2)$ is a model dependent constant that accounts only for minor corrections. The essential behaviour of $\langle\tau\rangle$ is determined by the first term that contains the function:

$$\Phi(k) = \sum_{n=0}^{\infty} (-1)^n \frac{B_n}{n!} k^n \tag{2}$$

Springer Proceedings in Physics, Vol. 55
Nonlinear Dynamics and Quantum Phenomena in Optical Systems
Editors: R. Vilaseca, R. Corbalán

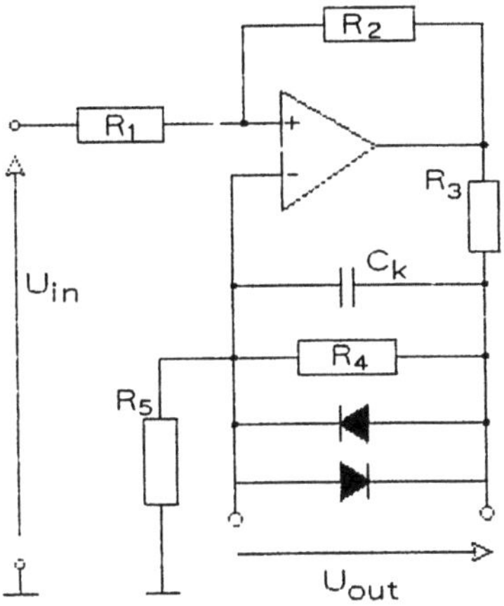

Evaluation of the variables in Eq.(5) from circuit components

$$X = \frac{U_{out}}{U_T}$$

$$Y = \frac{U_{in}}{U_T} \cdot \frac{R_2 R_D}{R_2 R_5 - R_1 R_3}$$

$$\tau = \frac{t}{\tau_0} = \frac{t}{R_D C_k}$$

$$C = \left(\frac{R_1}{R_2 R_5 - R_1 R_3} - \frac{1}{R_4} \right) R_D$$

$$R_D = \frac{U_T}{2 I_0}$$

$$\varepsilon = \left(\frac{U_{rms}}{U_T} \cdot \frac{R_2 R_D}{R_2 R_5 - R_1 R_3} \right)^2$$

Fig. 1: Schmitt trigger circuit, U_T and I_0 are characteristic for the pair of diodes and were measured under operating conditions.

where $$B_n = \frac{1}{3}\left[\frac{\pi}{3}\right]^{1/2} 2^{(2n+1)/3}\, \Gamma\left[\frac{2n+1}{6}\right] \tag{3}$$

and $$k = (\beta / a)(a / \varepsilon)^{2/3}. \tag{4}$$

k combines all relevant parameters of the problem: a and β for the potential, and ε for the noise. Its importance can also be seen from the observation that below threshold, $k = 3/(4^{2/3})(\Delta V/\varepsilon)^{2/3}$, which says that k measures the height of the potential barrier in terms of the noise power. The vicinity of marginality can thus be defined as the regime where $|k| \lesssim 1$.

We tested this theory through experiments with a bistable electronic Schmitt trigger circuit shown in Fig. 1. Its nonlinearity is provided by the nonlinear characteristics of a pair of diodes and its dynamics is described by Eq. (5):

$$dX/dt = Y + CX - \sinh X. \tag{5}$$

All the variables of the equation can be evaluated from circuit components. Eq. 5 implies a potential $V(X,Y) = \cosh X - {}^1\!/_2 CX^2 - YX$. After transforming to the potential near marginality and neglecting higher orders we obtain a and β for the Schmitt trigger: $a = (C^2-1)^{1/2}/6$ and $\beta = y - y_{th}$ with the threshold $y_{th} = (C^2-1)^{1/2} - C \cdot \operatorname{arcosh} C$.

Noise from a pseudo-random generator was added to the input voltage. All parameters in Eq. 1 are now fixed by either circuit component values, the input voltage, or the input noise

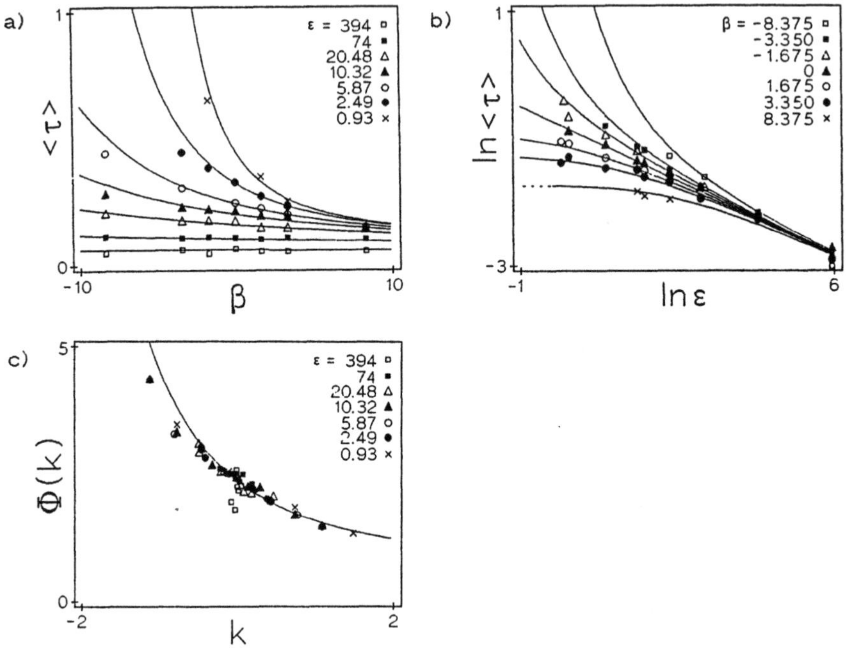

Fig. 2: Experimental results from Schmitt trigger: C=76

a) average switching time ⟨τ⟩ versus distance β from threshold

b) ⟨τ⟩ versus noise power ε

c) rescaled data and the universal scaling function Φ(k)

power so that we should not need a single fit parameter to compare experiment and theory.

The experiment was controlled by a computer. 1000 measurements of τ were taken for each parameter set to obtain $\langle\tau\rangle$. The results of our measurements are displayed in Fig. 2. Fig. 2a shows the average switching time versus distance from threshold. The curves are the corresponding analytic results from the theory. We first expected $D(x_1,x_2) = 0$ for all curves but it turned out that a minor correction was required. We found an explanation in the intrinsic noise in our system, which is 1/f rather than white. This conclusion is supported through numerical simulations with low-pass filtered noise. Consideration of 1/f noise should be necessary in any experiment when the intrinsic noise level is approached.

Fig. 2b shows $\langle\tau\rangle$ versus noise power ε. There is good agreement between experimental data points and the analytic curves.

We turn to the universal scaling function $\Phi(k)$ of Eq. (2). Each data point is specified by three parameters ε, β and the measured switching time $\langle\tau\rangle$. From these values we can cal-

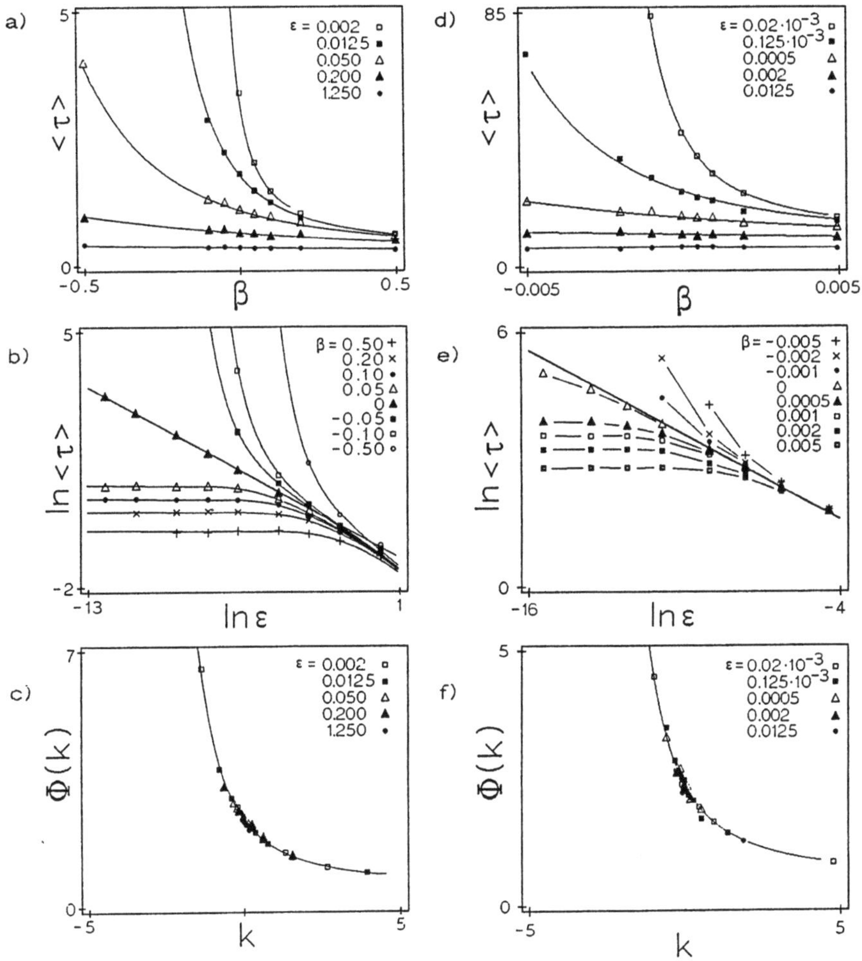

Fig. 3: a)–c) numerical simulation of the Schmitt trigger: C=76
d)–f) simulation of the Bonifacio–Lugiato model: C=20, a=2.82

culate k for every point with Eq. (4) and Φ(k) from Eq. (1). Fig. 2c shows the result: almost all data points fall on one curve and also match the analytic curve from the theory.

Deviations occur for $k<-1$, and for noise amplitudes larger than the width of the hysteresis. Either case is beyond the limits of validity of the theory.

We also performed numerical simulations for the Schmitt trigger (see Fig. 3a –c) , and for the Bonifacio–Lugiato model of bistability (see Fig. 3d –f) . The agreement of analytics and simulation is very good in both cases.

This experimental test of an analytic formula for the average switching time in noisy bistable systems gives a good agreement within the limits of validity of the theory so that this description now convincingly covers the whole transition region from the deterministic behaviour to stochastically dominated behaviour.

/1/ P. Colet, M. San Miguel, J. Casademunt, J. M. Sancho, Phys. Rev. A 39, 149 (1989)

Phase and Amplitude Correlations Induced by the Switch-On Chirp of a Detuned Laser

S. Ciuchi[1], *F. de Pasquale*[2], *M. San Miguel*[3], *and N.B. Abraham*[4]

[1]Dipartimento di Fisica, Università di Roma "La Sapienza", Piazzale Aldo Moro 2, I-00185 Roma, Italy
[2]Dipartimento di Fisica, Università de L'Aquila, I-67100 L'Aquila, Italy
[3]Departament de Física, Universitat de les Illes Balears, E-07071 Palma de Mallorca, Spain
[4]Department of Physics, Bryn Mawr College, Bryn Mawr, PA 19010, USA

Abstract. We study the amplitude-phase correlations during the switch-on process for a simple model of a single mode detuned laser. Deterministic frequency chirps induced by detuning during the switch-on [1] delay and significantly strengthen the transient correlation of phase and intensity fluctuations when the detuning is sufficiently large compared to the strength of the noise. In this case the peak in the correlations is related to the anomalous intensity fluctuations characteristic of transient switching.

1. The Model

We consider the simple model for a single mode laser in which the material variables can be adiabatically eliminated. We have taken the third-order approximation of the saturation and in the adiabatic elimination of the polarization we have transferred the noise, due to spontaneous emission, from the polarization to the amplitude equation. The equation for the slowly varying field amplitude is

$$dE/dt=(1+i\Theta)(\alpha E(t)-\beta|E(t)|^2)E(t)+\xi(t). \tag{1a}$$

The spontaneous emission noise is well described by a gaussian white noise $\xi(t)$ whose correlations are given by

$$<\xi(t)\xi^*(t')> = 2\varepsilon\delta(t-t'). \tag{1b}$$

The degree of laser excitation is governed by α and the threshold is given by $\alpha=0$. Equation (1a) is written in a rotating reference frame, which gives a constant amplitude for the lasing steady state solution (*on-state*). The detuning of the laser cavity from the atomic resonance is given by $\alpha\Theta$, as indicated by the fact that the non-lasing solution (*off-state*) has a frequency of $\alpha\Theta$ while the lasing solution has a frequency of zero. We consider switching processes, i.e. the situation in which there is an abrupt change in the linear coefficient of the net gain from a negative value $\alpha_0<0$ to a positive value α. We can separate this equation for the complex field amplitude into equations for the intensity and phase by defining

Springer Proceedings in Physics, Vol. 55
Nonlinear Dynamics and Quantum Phenomena in Optical Systems
Editors: R. Vilaseca, R. Corbalán

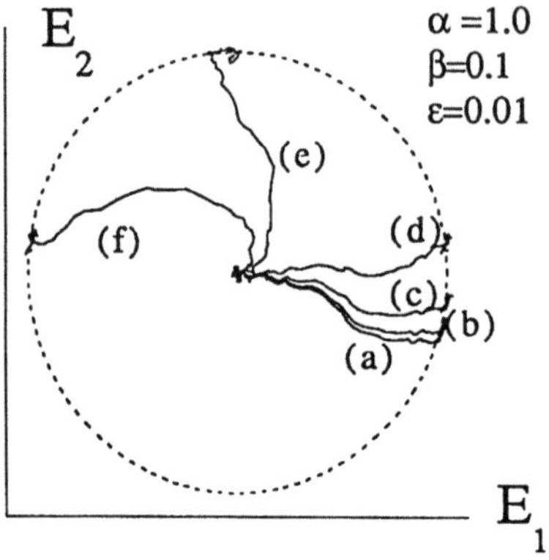

FIG.1 Evolution of the complex field for: (a) Θ=0 (b) Θ=0.01, (c) Θ=0.035, (d) Θ=0.1, (e) Θ=0.1, (f) Θ=1.0. Parameters shown in the inset have been used for all calculations in this paper.

$$E(t) = (I(t))^{1/2}\exp(i\phi(t)). \tag{2}$$

we have (using Ito calculus)

$$dI/dt = 2(\alpha I-\beta I^{2})+2\varepsilon+2(\varepsilon I)^{1/2}\xi_R(t) \tag{3a}$$

$$d\phi/dt = \Theta(\alpha- \beta I) + I^{-1/2}\varepsilon^{1/2}\xi_\phi(t), \tag{3b}$$

where the noise processes ξs have zero means and correlation functions given by

$$<\xi_I(t)\xi_I(t')> =\delta(t-t'), \qquad <\xi_\phi(t)\xi_\phi(t')> =\delta(t-t'), \qquad <\xi_\phi(t)\xi_I(t')> = 0. \tag{3c}$$

There are two terms on the right hand side of eq. (3b). The first, detuning dependent term is associated with the angular velocity of the phase (*instantaneous frequency*), the second, noise-dependent term is associated with the transient and final phase diffusion. The qualitative features of the motion of the field can be identified in fig. 1 which shows single realizations of the trajectories *using the same sequence of random numbers* for the noise during the transients. For small detunings the evolution occurs with almost constant phase once the trajectory leaves the vicinity of $|E| = 0$. In contrast, for larger detunings we see the trajectories spiral out away from the origin. As the final value of the intensity is approached, the phase rotation slows and in the final state there is only the ordinary phase diffusion indicated by the spread along the circumference of a circle that represents the set of all possible constant intensity "deterministic" equilibrium states.

2. The Quasi-Deterministic Theory Approximation

We have used the usual quasi-deterministic theory (QDT) developed to study the transient evolution of many non-linear systems [2-5]. Within this approximation the intensity process is approximated by replacing the initial condition in the deterministic solution of (1) by an effective time dependent process H(t)

$$I_{QDT}(t) = |H(t)|^2e^{2\alpha t}/[1+(\beta/\alpha)|H(t)|^2(e^{2\alpha t}-1)] \;, \tag{4}$$

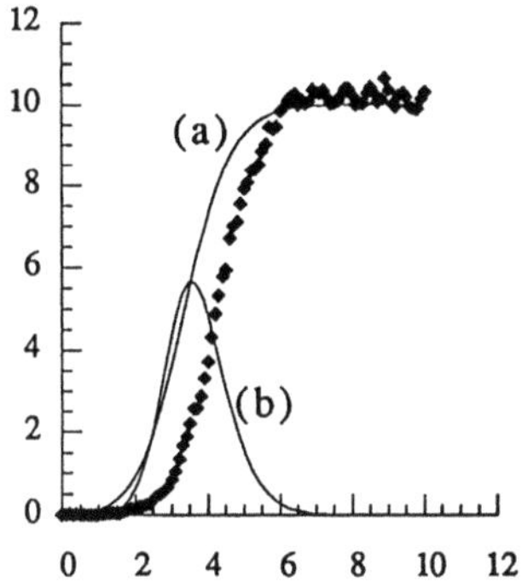

FIG.2 Mean intensity (a) and its variance (b) vs. time in QDT. The intensity of a single realization obtained from numerical solution of eq. (1) is shown (diamonds).

The *initial condition* $H(t)$ is a gaussian process that takes into account the effects of spontaneous emission events both before and after the switch-on has occurred

$$H(t)=\int_0^t dt' e^{-\alpha(1+i\Theta)t'}\xi(t')+E(0)\,. \tag{5}$$

For $t \gg 1/2\alpha$, $H(t)$ is replaced by a time-independent complex gaussian variable H, whose variance is

$$\langle |H|^2 \rangle = \varepsilon/2\alpha + \varepsilon/2\alpha_0\,. \tag{6}$$

Fluctuations of the time of departure from the initial unstable state imply fluctuations of the intensity among different switches. In fig. 2 is shown the behaviour of intensity and of intensity fluctuations as given by QDT. The peak of the variance of the intensity occurs at the turning point in the evolution of the average intensity. The amplitude of this peak does not scale with the the strength of the noise ε. For this reason these fluctuations are called *anomalous* [2-7].

Once the QDT has been used to determine the intensity, the equation of motion of the phase can be solved by a direct integration:

$$\phi(t) = \phi(0) + \Theta\left(\alpha t - \beta\int_0^t dt' I_{QDT}(t') \right) + \sqrt{\varepsilon}\int_0^t dt' [I_{QDT}(t')]^{-1/2}\xi_\phi(t'). \tag{7}$$

Equations (4) and (7) together with the assumption (6) form the basic set of equations of our approximation. We see that the intensity evolution under these assumptions is independent of the phase evolution. The evolution of the detuning dependent term in eq. (7) depends linearly on intensity and vanishes as the intensity reaches its equilibrium value ($I \simeq \alpha/\beta$). Its fluctuations are due to intensity fluctuations. The phase diffusion term contributes mainly to the phase evolution in the intial time regime in which the intensity is very low.

3. The Modulus-Phase Correlation Function

We have found it convenient to subtract the uncorrelated amplitude-phase part from the usual field-field correlation function, defining [8-9]

$$C(t_1,t_2) \equiv \langle E(t_1)E^*(t_2)\rangle - \langle (I(t_1)I(t_2))^{1/2}\rangle \langle e^{i(\phi(t_1)-\phi(t_2))}\rangle\quad . \tag{8}$$

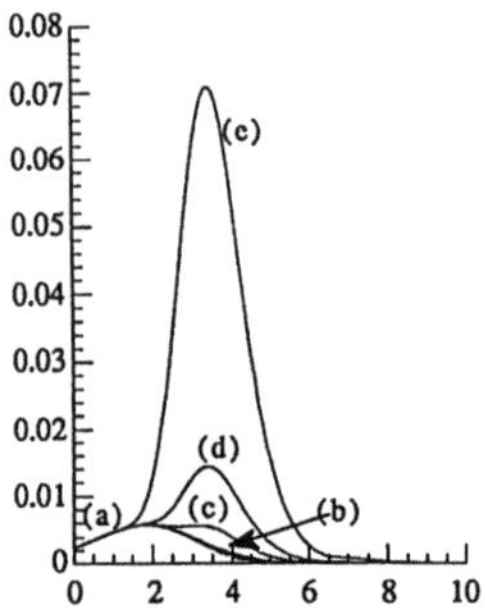

FIG.3 Modulus of C for t=0.25 for several values of detuning. (a) to (e) as in Fig.1.

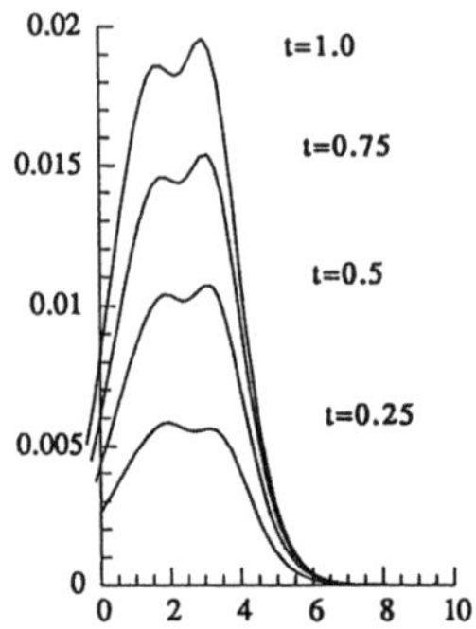

FIG.4 Modulus of C vs T for several values of time difference t.

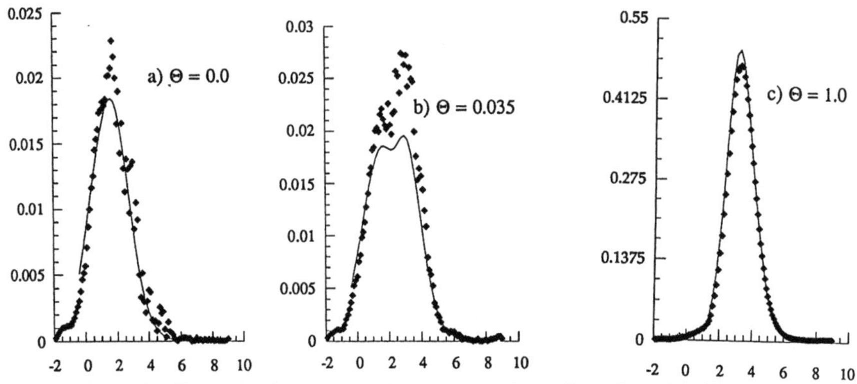

FIG.5 |C| vs T. Data obtained averaging 3072 trajectories of eq. (1) (diamonds) and QDT results (solid line). (a) $\Theta = 0$, (b) $\Theta = 0.035$, (c) $\Theta = 1$

Correlation functions for transients are functions of two times rather than one. $\langle .. \rangle$ means an average over the ensemble since time averages cannot be meaningfully taken during the transient. We introduce a notation for the arguments of the transient correlation function. Supposing $t_1>t_2$ we define $T = t_2$ and $t = t_1-t_2$. Let us consider the T dependence of C with t as a parameter. Separating into real and imaginary parts and considering $t<<T$ we get

$$\mathrm{Re}\, C(T,t) = \langle I(T)e^{-\varepsilon t/2I(T)}\rangle - \langle I(T)\rangle\langle e^{-\varepsilon t/2I(T)}\rangle \tag{9a}$$

$$\mathrm{Im}\, C(T,t) = -i\Theta\beta t(\langle I(T)^2\rangle - \langle I(T)\rangle^2). \tag{9b}$$

The real part of C takes into account fluctuations due to the phase-diffusion while the imaginary part of C takes into account the fluctuations of the instantaneous frequency (see eqs. (3b-7)) [8-9].

We have found a crossover in the behaviour of the norm of C vs T as detuning is increased. This is shown in fig. 3 for a fixed value of time difference t. To explain this behaviour we notice from eqs. (9) that Re C does not depend on the detuning while Im C increases in modulus as detuning increases. At a certain crossover value of the detuning ($\Theta_c \approx 0.035$) the imaginary part peak becomes greater than that of the real part. This value Θ_c increases almost linearly as the noise increases [8-9]. The behaviour of |C| for several values of t near Θ_c is shown in fig. 4 We see that the amplitude of |C| increases as t increases while the double peak structure is always present. For this reason we have chosen quite a large t (t=1) in numerical simulations of eq. (1). Results for |C| are shown in figs 5a,b,c compared with QDT calculations.

4. Summary

We have developed here an extension of the QDT formalism that takes into account phase fluctuations. The theory predicts a rather sharp crossover phenomenon in the modulus-phase cross correlation function. We have compared our analytical results with numerical simulation of the model equations and found a reasonable agreement. As experiments have recently used heterodyne techniques to measure the instantaneous phase of a laser signal [10], it should be possible to measure these effects and to study their relevance for transient laser processes. This may give rise to an important form of self-phase modulation in the generation of Q-switched pulses in detuned lasers.

References

[1]D.E. Chyba and N.B. Abraham, in *Coherence and Quantum Optics VI*, ed. J. Eberly, L. Mandel and E. Wolf (Plenum, NY 1990); D.E. Chyba and N.B. Abraham, Opt. Communications, submitted (1990).
[2]F.T. Arecchi, V. Degiorgio and B. Querzola, Phys. Rev. Lett. **19**, 1168 (1967).
[3]M.Suzuki, Adv. Chem. Phys. **46**, 195 (1981).
[4]F.Haake, J.Haus, R.Glauber, Phys. Rev. **23**, 3235 (1981).
[5]F.de Pasquale, P.Tartaglia, P.Tombesi, Phys. Lett. **72A**, 7 (1979).
[6]F.T. Arecchi, V. de Giorgio and B. Querzola, Phys. Rev. Lett. **19**, 1168 (1967).
[7]D. Meltzer and L. Mandel, Phys. Rev. Lett., **25**, 1151 (1970).
[8]S.Ciuchi, F.De Pasquale, M. San Miguel, N.B. Abraham, to be published in *Non - linear Dynamics in Optical Systems,* ed. by N.B. Abraham (OSA Proceedings, 1990)
[9]S.Ciuchi, F.De Pasquale, M. San Miguel, N.B. Abraham, Phys. Rev. A, submitted (1990).
[10]C.O.Weiss, N.B.Abraham and U.Hubner, Phys. Rev. Lett. **62**, 1587 (1988).

Noise and Symmetry Restoring in Transverse Laser Patterns

P. Colet[1], *M. San Miguel*[1], *M. Brambilla*[2], *and L.A. Lugiato*[2]

[1]Departament de Física, Universitat de les Illes Balears, E-07071 Palma de Mallorca, Spain
[2]Dipartimento di Fisica, Università di Milano, Via Celoria 16, I-20133 Milano, Italy

Transverse pattern formation in nonlinear optical systems has been receiving increased attention in recent years [1-9]. In this context we analyze in this paper the effects of noise in a model for spontaneous breaking of cylindrical symmetry in a nonlinear optical system [6,10]. The model is described in this book [10]. We consider the case 2p+l=2. The dynamics of the system is governed by a three mode model in which the normalized slowly-varying envelope of the electric field is $F(\rho,\varphi,t)=\Sigma_{i=1}^{3}A_i(\rho,\varphi)f_i(t)$, $f_i(t)$ are the complex mode amplitudes and $A_i(\rho,\varphi)$ are the modes of the cavity, A_1 corresponds to the mode A_{10} of [10] and A_2 and A_3 to the two modes A_{02}.

After adiabatic elimination of the atomic variables and after the cubic approximation, the dynamics of the model is defined by

$$\frac{d f_i}{d\tau} = - f_i + 2\,C\left[\sum_j M_{ij}\, f_j - \sum_{jkl} A_{ijkl}\, f_j\, f_k\, f_l^*\right] + \eta_i(\tau) \tag{1}$$

where τ is a dimensionless time-variable, C the pump parameter and M_{ij} and A_{ijkl} are known mode coupling coefficients. The random force $\eta_i(\tau)$ is a Gaussian white noise of zero mean and correlation $<\eta_i(\tau)\,\eta_j^*(\tau')>=\varepsilon\delta_{ij}\delta(\tau-\tau')$. The deterministic anal-ysis of (1) indicates that the stationary solution with cylindrical symmetry ($f_2=f_3=0$) becomes unstable at $C=C^I$. Beyond the instability there exists an infinite class of multimode stationary solutions in which f_1 coexists with $f_2\neq0$, $f_3\neq0$. These solu-tions exhibit four lateral peaks with an arbitrary orientation. Setting $f_j=|f_j|e^{i\beta_j}$, we can define the relative phases $\xi=(\beta_3-\beta_2)/4, \delta=(\beta_3+\beta_2)/4-\beta_1/2$. Then F is given by

$$F(\rho,\varphi,\tau)=\sqrt{\frac{2}{\pi}}\, e^{-\rho^2} e^{i\beta_1}\left\{(1-2\rho^2)|f_1|+2\rho^2 e^{2i\delta}\left(|f_2|\, e^{2i(\varphi-\xi)}+|f_3|\, e^{-2i(\varphi-\xi)}\right)\right\}. \tag{2}$$

A change of φ implies a change of ξ of the same amount. So that ξ is the angular variable which determines the global orientation of the pattern.

Noise is expected to restore the global symmetry of the problem by some mechanism in which the system randomly visits the different possible deterministic stationary solutions. Noise is also crucial during the transient dynamics leading to pattern formation starting from an unstable state.

The steady state properties can be obtained from the stationary solution of the Fokker-Planck equation associated with (1). It turns out that (1) admits a potential form such that the stationary solution is

Springer Proceedings in Physics, Vol. 55
Nonlinear Dynamics and Quantum Phenomena in Optical Systems
Editors: R. Vilaseca, R. Corbalán

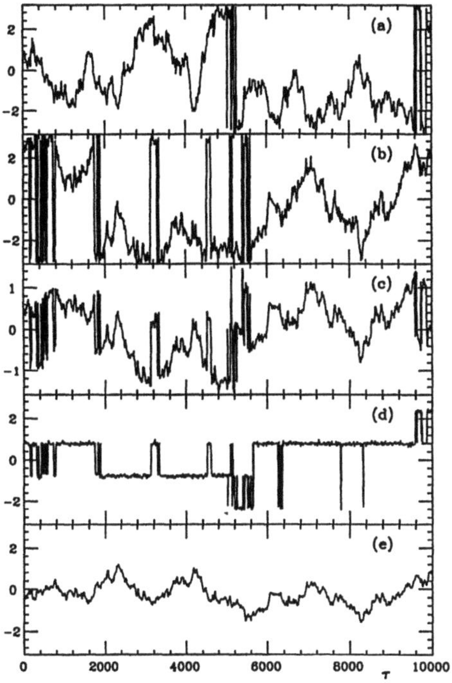

Fig.1. Stochastic evolution of mode phases and relative phases for a single realization. ε=0.001, C=1.1, k=1. a) β_2; b) β_3; c) ξ; d) δ; e) φ_1 (as defined in (4)).

$$P_{st}(f_1,f_2,f_3) = N \exp \left\{ -\frac{1}{\varepsilon} \int_0^{2\pi} d\varphi \int_0^{\infty} d\rho\, \rho \left[|F|^2 - C\,\chi(\rho) \left(2|F|^2 - |F|^4 \right) \right] \right\} \qquad (3)$$

where $\chi(\rho)$ gives a Gaussian transverse shape for the equilibrium population difference and N is a normalization constant. The distribution of the angular variable ξ is obtained from (4) by the corresponding change of variables. The stationary distribution is independent of the angular variable ξ. This result expresses the cylindrical symmetry restoring by noise. The mechanism of symmetry restoring is a phase diffusion process. Indeed, linearizing around the deterministic steady state solution, the equation for variable ξ obtained from (1) is a diffusion equation. The stationary distribution has maxima at the metastable states $\delta = \pm \pi/4, \pm 3\pi/4$.

The actual dynamics of the system is well described by the stochastic evolution of a given deterministic steady state solution. Such evolution is shown in Fig. 1 obtained from the numerical solution of (2) for a stochastic realization of the $\eta_i(t)$. The variable δ jumps randomly between the values $\pm\pi/4$, $\pm 3\pi/4$ which minimize the potential associated with (3). We find that ξ also has jumps of $\pi/2$ at the same times as δ. However on long time scales there is no preferred value of ξ as indicated by the fact that the potential associated with (3) is independent of ξ. In any metastable state with $\delta = \pm \pi/4, \pm 3\pi/4$ the variable ξ is seen to diffuse freely. The restoring of the spatial cylindrical symmetry is described by the evolution of the angle variable ξ. On the other hand the simultaneous changes of ξ and δ in $\pi/2$

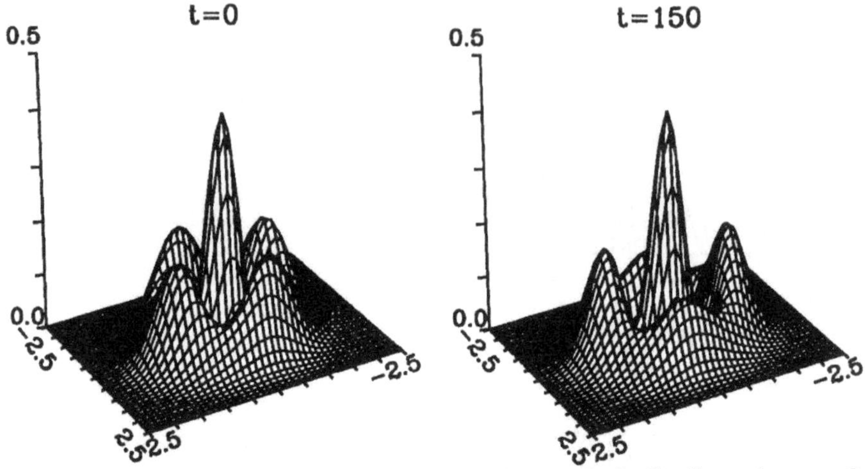

Fig.2.Pattern evolution due to noise in the steady state.Left: Configuration at t=0 corresponding to a deterministic steady state. Right: Configuration at time t=150. Same parameters as in Fig. 1

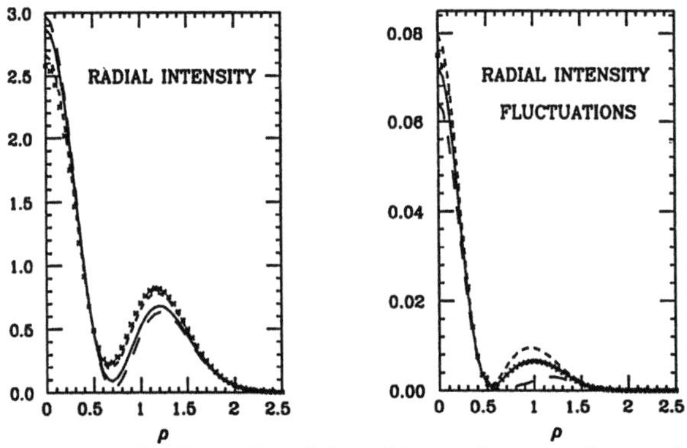

Fig.3. Radial intensity I(ρ) and its variance at times: (– –) t=10, (—) t=20, (- - -) t=40, (xxx) t=60. Averages are performed over 10000 realizations.

corresponding to changes in 2π in β_2 or β_3, leave the spatial orientation of the pattern (φ) unchanged.

The stochastic motion of the f_i leads to a random rotation and distortion of the pattern as seen in Fig. 2. The motion of the lateral peaks associated with symmetry restoring is here made explicit.

Transient dynamics has also been studied by a numerical solution of (2) with an unstable initial condition $f_2=f_3=0$. The ensemble average of the radial distribution of the total intensity $I(\rho)=\int d\varphi\, |F(\rho,\varphi)|^2$ and its variance are shown in Fig. 3. We recall that the first mode has maxima at $\rho=0$, $(3/2)^{1/2}$ with a minimum at $\rho=2^{-1/2}$, while modes 2 and 3 have a single maximum at $\rho=1$. Fig. 3 shows the gradual change in the position of the relative maximum at $\rho\neq 0$ under the growing influence of modes 2

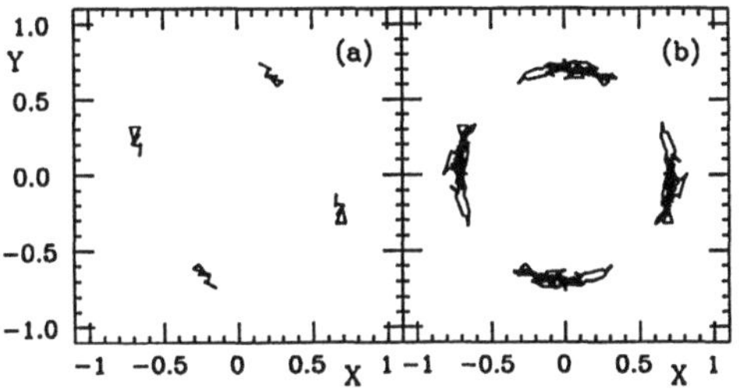

Fig. 4. Stochastic trajectories described by the phase singularities for different times: a) t=100; b) t=1000. For larger times the symmetry is completely restored.

and 3. Characteristic transient anomalous fluctuations represented by a peak of $(\Delta I)^2$ as a function of time are observed in the vicinity of the local extrema of mode amplitudes.

Phase singularities [10,11] occur at the points in the transverse plane (ρ_k,φ_k) where the electric field vanishes. The total variation of the field phase over a closed loop around one of them is equal to an integer multiple of 2π. In this model there exist four such points equally separated by angles of $\pi/2$. Their position is given by (k=0,1,2,3):

$$\varphi_k = \frac{i}{4} \ln \left(\frac{f_2 f_1^* - f_1 f_3^*}{f_1 f_2^* - f_3 f_1^*} \right) + k\frac{\pi}{2}, \quad 2\rho_k^2 = \frac{f_1}{f_1 - f_2 \exp(2i\varphi_k) - f_3 \exp(-2i\varphi_k)} . \tag{4}$$

The variables φ_k and ρ_k are time dependent stochastic quantities. In the metastable states mentioned above the four singularities are located over a circle of radius $\rho=2^{-1/2}$. The effect of noise is to produce a random motion of these points which results in an overall random motion of the spatial pattern. This is shown in Fig. 4 where the trajectories described by phase singularities at different times are plotted for the same stochastic realization considered in Figs. 1 and 2.

Acknowledgments: Work carried out by M.B. and L.A.L. in the framework of the ESPRIT Basic Research Action on "Transverse Optical Patterns" (TOPP). P.C. and M.S.M acknowledge financial support from DGCYT (PB 86-0534, Spain).

References

[1] Moloney J H 1987 in *Instabilities and Chaos in Quantum Optics* eds F T Arecchi and R Harrison (Berlin: Springer-Verlag) pp 139 ff
[2] Firth W J 1988 in *Instabilities and Chaos in Quantum Optics II* eds N B Abraham, F T Arecchi and L A Lugiato (New York: Plenum) pp 219 ff
[3] Lugiato L A and Lefeber R 1987 *Phys. Rev. Lett.* **50** 220
[4] Aguado M, Rodríguez R F and San Miguel M 1989 *Phys. Rev. A* **39** 5686
[5] Lugiato L A, Oldano C and Narducci L M 1988 *J. Opt. Soc. Am. B* **5** 879
[6] Lugiato L A, Prati F, Narducci L M and Oppo G L 1989 *Optics Comm.* **69** 387

[7] Tredicce J R, Quel E J, Ghazzawi A M, Green C, Pernigo M A, Narducci L M and Lugiato L A 1989 *Phys. Rev. Lett.* **62** 1274
[8] Coullet P, Gil L and Rocca F 1989 *Optics Comm.* **73** 403
[9] Feature issue on Transverse Effects in Nonlinear Optical Systems ed N B Abraham and W Firth *J. Opt. Soc. Am. B* June 1990.
[10] Battipede F, Brambilla M, Lugiato L A, Prati F and Penna V, this volume.
[11] Weiss C O, this volume.

Analysis of the Decay of Intensity Correlation Functions for the Gain-Noise Model of Dye Lasers

J.M. Noriega[1], *L. Pesquera*[2], *M.A. Rodríguez*[2], *J. Casademunt*[3], and *A. Hernández-Machado*[4]

[1]Departamento de Matemáticas, Universitat de Oviedo, E-33007 Oviedo, Spain
[2]Departamento de Física Moderna, Universitat de Cantabria, E-39005 Santander, Spain
[3]Department of Physics, University of Pittsburgh, PA 15260, USA
[4]Departamento de Estructura y Constituyentes de la Materia, Universitat de Barcelona, Diagonal, 647, E-08028 Barcelona, Spain

Abstract. The decay of intensity correlation functions, $C(t)$, for the colored gain–noise model of dye lasers is analyzed and compared with the relaxation of $C(t)$ for the loss–noise model. We use a method that provides explicit expressions of nonmarkovian correlation functions, approximating simultaneously short and long time behaviours. It is shown that when the correlation time of the noise increases differences between the gain and loss noise models tend to disappear. The decay of $C(t)$ can be described for both models by a time scale that approaches the deterministic time. However, in contrast with the loss–noise model, a larger time scale with a vanishing amplitude remains for the gain–noise model.

1. Introduction

It is by now well established that the anomalous statistical properties of dye lasers are due to pump parameter fluctuations with a finite correlation time (external colored noise). Experimental [1] and theoretical [2] work on dye laser fluctuations have been recently reviewed. The standard theoretical model of dye laser fluctuations can be formally obtained by replacing the loss parameter by a colored noise. We will refer to it as loss–noise model. This model seems to describe the experimental results [1] and in particular intensity correlation functions [3,4]. However, experimental evidence has been reported which identifies the pump laser as the source of the noise [3]. Then, it seems natural to consider a model incorporating fluctuations of the gain parameter [5–7]. It has been shown [5] that in the white noise limit the statistical properties predicted by loss and gain noise models are very different. In fact most of the qualitative predictions associated with colored noise in the loss–noise model can be recovered within a white gain–noise model [5]. Only the early time decay of $C(t)$ remains as a signature of colored noise [5,7].

Springer Proceedings in Physics, Vol. 55
Nonlinear Dynamics and Quantum Phenomena in Optical Systems
Editors: R. Vilaseca, R. Corbalán

Numerical simulations have shown [6] that differences between both models are masked by the correlation time of the noise, τ. The aim of this communication is to analyze the decay of $C(t)$ for the colored gain–noise model and to compare it with that corresponding to the loss–noise model. To make this analysis we use a method [8] that provides explicit expressions of nonmarkovian correlation functions. We find that as τ increases the decay of $C(t)$ for times larger than τ can be described for both models by a time scale that approaches the deterministic time. In this sense the differences between the gain and loss–noise models tend to disappear with τ. However, in contrast with the loss–noise model, a larger time scale with a vanishing amplitude is involved in the decay of $C(t)$ for the gain–noise model.

2. Gain–noise model

The gain–noise model for a single mode dye laser on resonance is defined in dimensionless units by the following equation for the intensity [5]

$$\dot{I} = v(I) + p(t)g(I) = I[-\alpha_2 + \alpha_1/(1 + I/\alpha_1)] + p(t)I/(1 + I/\alpha_1), \quad (1)$$

where we have neglected the spontaneous emission noise, since we consider situations above threshold [1,2]. Here α_1 and α_2 are the gain and loss parameters, respectively. The gain noise, $p(t)$, is taken as a Gaussian noise with zero mean and correlation: $\langle p(t)p(t')\rangle = \exp(-|t-t'|/\tau)/\tau$. The loss–noise model is obtained taking the limit $\alpha_1 \to \infty$ with $\alpha = \alpha_1 - \alpha_2$ fixed. In the following we analyze the decay of $C(t)$ for different values of the parameter α/α_1. This parameter gives information on the importance of saturation effects in the fluctuations. When α/α_1 is close to zero, saturation effects are small and we recover the loss–noise model.

3. The double–expansion method

Nonmarkovian correlation functions, $C(t) = \langle \delta I(t)\delta I(0)\rangle / \langle(\delta I)^2\rangle$, can be approximated to first order in τ by the following expression [8]

$$C(t) = C_M(t) + \tau\gamma_0[1 - \exp(-t/\tau)]C_M^L(t), \quad (2)$$

where $\gamma_0 = -\langle \delta I L^+ \delta I\rangle / \langle(\delta I)^2\rangle$ and C_M, C_M^L are correlation functions associated to a markovian process with Fokker–Planck operator L

$$C_M(t) = \langle \delta I(t)\delta I\rangle_M / \langle(\delta I)^2\rangle, \quad C_M^L = \langle \delta I(t)L^\dagger \delta I\rangle_M / \langle \delta I L^\dagger \delta I\rangle \quad (3a)$$

$$L = -\partial_I v(I) + \partial_I g(I)\partial_I[g(I) - \tau(v(I)g'(I) - g(I)v'(I))]. \quad (3b)$$

Eq.(2) implies the characteristic initial plateau of any nonmarkovian correlation function: $\dot{C}(0) = 0$. A more global characterization of $C(t)$ is given by the so-called relaxation moments, $T_k = \int_0^\infty dt\, C(t)\, t^k$. For $t \gg \tau$, $C(t)$ given by (2) corresponds to the correlation function of a markovian process defined by L with an effective initial distribution $(1-\tau L^\dagger)P_{st}(I)$ instead of $P_{st}(I)$. Then time scales larger than τ are associated to the operator L.

As the second step in our procedure we apply the double expansion method [9] to obtain the markovian correlation functions defined by eq.(3).The basic idea is to approximate simultaneously short and long time behaviours with the use of a superposition of $N = (n+m+1)/2$ exponentials. The amplitudes and time scales are such that n derivatives and m relaxation times are exactly reproduced. In the following we use conditions of one derivative and two relaxation moments of $C_M(t)$ and the relaxation time of $C_M^L(t)$. In this case we have

$$C(t) = [a_1 \exp(-\lambda_1 t) + (1-a_1)\exp(-\lambda_2 t)] \\ +\tau\gamma_0[1-\exp(-t/\tau)]\exp(-\gamma_0 t), \tag{4}$$

where the first derivative and the relaxation time to order τ of $C(t)$ are exact.

4. Intensity correlation functions

To analyze the decay of $C(t)$ we use (4). We have checked by performing numerical simulations that this approximation is accurate when τ is smaller than the evolution time of the deterministic linearized eq.(1) (deterministic time, t_d). As we have discussed above, the second term of (4) is important only for times not larger than τ. In the following we analyze the decay of $C_M(t)$, given by the first term of (4), for two values of α/α_1: 0 (loss-noise model) and 0.2 (gain-noise model). We compare both models as τ increases.

We show in Fig.1 a_1 for $\alpha/\alpha_1 = 0.2$. We see that, in contrast with the white noise case, when τ increases the dominant time scale is λ_2. As concerns the loss-noise model, a_1 grows with $\langle I\rangle$ and τ. Then the dominant time scale is λ_1. Fig.2 shows λ_2 rescaled with t_d for $\alpha/\alpha_1 = 0.2$. We see that λ_2 approaches t_d when τ increases. The same situation occurs for the dominant time scale λ_1 of the loss-noise model (see Fig.3). Therefore the differences between the gain and loss-noise models tend to disappear with τ. This is in agreement with simulation results [6]. As concerns the dominated time scale, we find that for the loss-noise model λ_2^{-1} is always smaller than t_d. On the contrary, λ_1^{-1} is larger than t_d for $\alpha/\alpha_1 = 0.2$. We conclude that as τ increases a difference remains between both models. The decay of $C(t)$ is slower for the gain-noise model than for the loss-noise model.

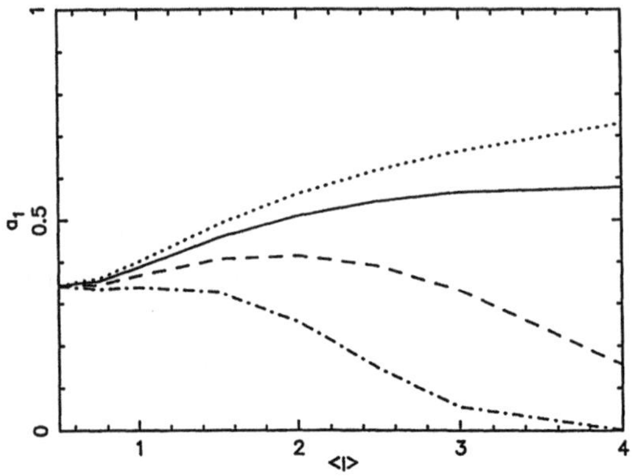

FIG. 1. a_1 vs. $\langle I \rangle$ for $\alpha/\alpha_1 = 0.2$ and different values of τ: · · · ·, 0; —, 0.1; – – –, 0.25; – · – · –, 0.5.

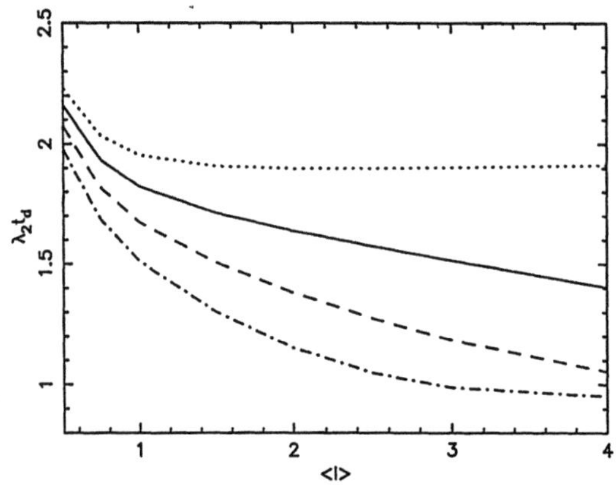

FIG. 2. $\lambda_2\, t_d$ vs. $\langle I \rangle$. Same parameters as in Fig. 1.

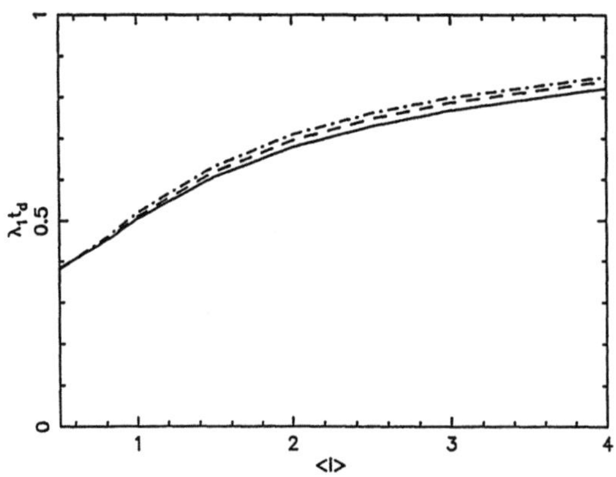

FIG. 3. $\lambda_1\, t_d$ vs. $\langle I \rangle$ for $\alpha/\alpha_1 = 0$ and τ: —, 0.1; – – –, 0.25; – · – ·, 0.5.

Financial support from Dirección General de Investigación Científica y Técnica (Spain), Project PB-87-0014 is acknowledged.

References

[1]R. Roy, A. W. Yu, and S. Zhu, in *Noise in nonlinear dynamical systems*, edited by F. Moss and P. V. E. McClintock (Cambridge University Press, Cambridge, England, 1989), Vol.3, p. 90.
[2]M. San Miguel, in *Instabilities and Chaos in Quantum Optics II*, edited by N. B. Abraham, F. T. Arecchi and L. A. Lugiato (Plenum, New York, 1988), p. 147.
[3]A.W. Yu, G. P. Agrawal, and R. Roy, Opt. Lett. **12**, 806 (1987).
[4]P. D. Lett and E. C. Gage, Phys. Rev. A **39**, 1193 (1989).
[5]M. Aguado, E. Hernández–García, and M. San Miguel, Phys. Rev. A **38**,5670 (1988).
[6]E. Hernández–García, R. Toral, and M. San Miguel, Phys. Rev. A (1990).
[7]J. M. Noriega, L. Pesquera, and M. A. Rodríguez (preprint).
[8]A. Hernández–Machado, J. Casademunt, M. A. Rodríguez, L. Pesquera, and J. M. Noriega (preprint).
[9]W. Nadler and K. Schulten, J. Chem. Phys. **82**, 151 (1985).

Anomalous Fluctuations for a Good Cavity Laser with Swept Losses

A. Valle, L. Pesquera, and M.A. Rodríguez

Departamento de Física Moderna, Universitat de Cantabria,
E-39005 Santander, Spain

Abstract. We study the statistical properties of the intensity, I, of a good cavity laser driven across the threshold region by a slow variation of the cavity losses. The quasideterministic theory is shown to be valid when the sweeping rate, v, is large with respect to the spontaneous emission noise. Anomalous fluctuations are characterized by the time, t_m, at which the maximum occurs and by the value of this maximum, σ_m. Our results show that t_m coincides with the time of maximum rate of growth of I. We find that t_m^{-1} and σ_m scale with $v^{1/2}$. We also show that relative fluctuations are given by a universal scaling function.

1. Introduction

It has been stated quite generally [1] that in a dynamical bifurcation when the control parameter is continuously swept through the instability point the bifurcation is delayed. Numerical integration has shown [2] that the delay persists in the presence of the spontaneous emission noise when the sweeping rate, v, is larger than the intensity of the noise, D. The dynamical bifurcation point has been characterized in terms of the mean first–passage time [3] and using the average intensity [4]. Experimental results for a Ar^+ laser show [5] that the time at which the intensity has a maximum slope, T, scales with $v^{-1/2}$. This scaling is obtained in the deterministic framework when nonlinearities are taken into account [6]. It is known that in the nonlinear regime there are anomalous large transient fluctuations [7]. A stochastic description is then desirable.

The aim of this communication is to study anomalous fluctuations for a good cavity laser with swept losses. We show that a generalization of the quasideterministic theory (QDT) [8] is valid when v is large with respect to D. Anomalous fluctuations are characterized by the time at which the maximum occurs, t_m, and by the value of this maximum, σ_m. Numerical simulations show that the maximum of the fluctuations disappears when $v \sim D$. In this situation the delay in the bifurcation is negligible [2]. When v is larger than D, t_m is found to coincide with T. We also find

Springer Proceedings in Physics, Vol. 55
Nonlinear Dynamics and Quantum Phenomena in Optical Systems
Editors: R. Vilaseca, R. Corbalán

that t_m^{-1} and σ_m scale with $v^{1/2}$. Finally, relative fluctuations are shown to be given by a universal scaling function.

2. Model and quasideterministic theory (QDT)

The electric field amplitude E in a tuned single–mode laser in the good cavity limit can be described by the equation

$$\dot{E} = aE - B \mid E \mid^2 E + \xi(t), \tag{1}$$

where $a = \Gamma - \kappa$ (Γ and κ are the gain and loss parameters, respectively) and B is proportional to Γ. The spontaneous–emission term $\xi(t)$ is taken as a Gaussian white noise of zero mean and correlation $\langle \xi^*(t)\xi(t')\rangle = 2D\delta(t-t')$. We are interested in a situation in which the laser is continuously driven from below to above threshold at a certain sweeping rate by changing the loss parameter [5]: $\kappa(t) = \kappa_0 - vt$. Then, we have $a(t) = a_0 + vt\,(a_0 < 0, v > 0)$, where $a_0 = \Gamma - \kappa_0$. The time at which $a(t)$ changes sign is denoted by $\bar{t}$, $\bar{t} = \mid a_0 \mid /v$. In a deterministic analysis [1] the system becomes dynamically unstable at $t^* = 2\bar{t}$. It has been shown [2] that for a linear variation of Γ the delay is negligible when $v \sim D$. We have checked by performing numerical simulations that this result also holds with swept losses. The delayed bifurcation can be analyzed with the use of a linear model [3,4]. However, in order to study anomalous fluctuations nonlinear terms must be taken into account. To perform this analysis we generalize the QDT [8] to the present case.

The first step of the QDT is to consider the linearized version of (1), whose solution for the intensity is

$$I(t) = \mid h(t) \mid^2 \exp[2 \textstyle\int_0^t dt'\, a(t')] = \mid h(t) \mid^2 \exp[v(t-\bar{t})^2 - v\bar{t}^2], \tag{2}$$

where h is a complex Gaussian process with variance [3]

$$\langle \mid h(t) \mid^2 \rangle = D / \mid a_0 \mid + 2D/\sqrt{v}\, \exp(v\bar{t}^2) \Big[\int_{-\sqrt{v}\,\bar{t}}^{\sqrt{v}(t-\bar{t})} du \, \exp(-u^2) \Big]. \tag{3}$$

Here we assume that $E(0)$ is distributed according to the Gaussian distribution below threshold, so that $\langle I(0) \rangle = D/ \mid a_0 \mid$. In the linear solution (3) $h(t)$ plays the role of an effective random initial condition. This process becomes a time–independent random variable h_∞ when $\sqrt{v}(t-\bar{t}) > 2$ (see (3)). In this time regime the QDT corresponds to the solution of (1) without noise but with the initial condition replaced by h_∞,

$$I(t) = \frac{\mid h_\infty \mid^2 \exp[v(t-\bar{t})^2 - v\bar{t}^2]}{1 + 2B \mid h_\infty \mid^2 \displaystyle\int_0^t ds \, \exp[v(s-\bar{t})^2 - v\bar{t}^2]}. \tag{4}$$

It is easy to see that (4) can be also deduced when a matching time t_0 exists, such that the nonlinear terms are negligible but $\sqrt{v}(t_0 - \bar{t}) > 2$. This corresponds to the conditions of validity of the QDT. In this way we find that (4) is valid when

$$\eta = \frac{BD}{|a_0|\sqrt{v}} \ll 10^{-2} \;\text{ if }\; \frac{\sqrt{v}}{|a_0|} \gtrsim 1; \;\; \eta \ll \exp(-a_0^2/v) \;\text{ if }\; \frac{\sqrt{v}}{|a_0|} \ll 1. \tag{5}$$

These conditions agree when $\sqrt{v} \gtrsim |\, a_0 \,|$ with the ones found in Ref.[9]. However, when $\sqrt{v} \ll |\, a_0 \,|$ we get a more stringent condition. We have checked with numerical simulations that in the case $\sqrt{v} \gtrsim |\, a_0 \,|$ the QDT describes correctly anomalous fluctuations when (5) holds.

3. Anomalous fluctuations

It is well known that in the decay of an unstable state fluctuations, $\sigma = \langle I^2 \rangle - \langle I \rangle^2$, exhibit an anomalously large maximum [7]. The same kind of behaviour is found when sweeping across the instability point if $v > D$. However, when $v \approx D$ the maximum disappears (see Fig.1). In this situation the intensity follows adiabatically the steady–state value associated with the instantaneous control parameter [2].

We characterize anomalous fluctuations by the time at which the maximum occurs, t_m, and by the value of this maximum, σ_m. We have found by performing numerical simulations of (1) for $B = 1/2$, $D = 10^{-3}$ and $a_0 = -0.5, -0.05$ that t_m coincides with the time of maximum rate of growth of I, T, whenever $v \gg D$. This can be easily understood in the framework of the QDT by considering that the largest amplification of

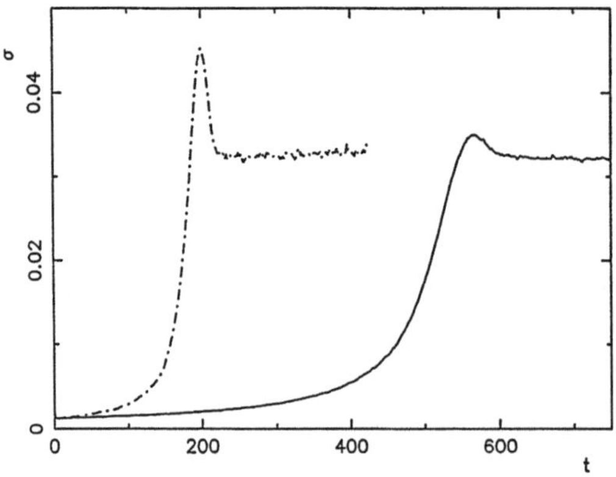

FIG. 1. Fluctuations vs. time for $B = 1/2, D = 10^{-3}, a_0 = -0.5$ and v: $-\cdot-\cdot-$, 3.10^{-3}; —, 10^{-3}.

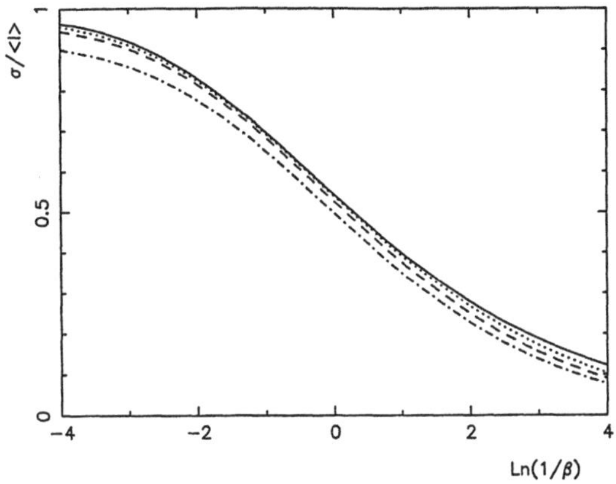

FIG. 2. Relative fluctuations vs. ln β^{-1} for the scaling function: —, and $B = 1/2, D = 10^{-3}, a_0 = -0.5$ and v: $- \cdot - \cdot -$, 5; $- - -$, 50; $\cdots$, 500.

the initial fluctuations occurs at T. We have found that for $a_0 = -0.5$ and $1 < v < 10^3$, T is well defined (relative fluctuations smaller than 0.06) and it scales with $v^{-1/2}$. This scaling is in agreement with experimental data [5] and with deterministic analysis [6] for large v. An explanation of these results can be given in the following way. When $v > 1$ the QDT is valid (see (5)). From (4) we get the following condition for T [6]

$$\left[(1+\beta)\exp[-x^2]\int_{-\sqrt{v}\,\bar{t}}^{x} du\, \exp(u^2)\right]^{-1} = x\left[3 - \left(1 - 4/x^2\right)^{1/2}\right]/2, \quad (6)$$

where $x = \sqrt{v}(T - \bar{t})$ and

$$\beta = \left[\sqrt{v}\exp(v\bar{t}^2)\right]/\left[2B \mid h_\infty \mid^2 \int_{-\sqrt{v}\,\bar{t}}^{x} du\, \exp(u^2)\right]. \quad (7)$$

When $\mid a_0 \mid < \sqrt{v}$ large variations of v or h_∞ in β can be compensated with a small variation of x. Therefore, $\sqrt{v}\, T \approx x$ is nearly constant.

As concerns σ_m, we find $\sigma_m \sim \sqrt{v}$. This result can be explained in the same way as the scaling of T. The relative fluctuations are given from (4) by the following universal function

$$\frac{\sigma}{\langle I \rangle} = \frac{\beta\left[\int_0^\infty dz\, \exp(-z)/(z+\beta)^2 - \left(\int_0^\infty dz\, exp(-z)/(z+\beta)\right)^2\right]^{1/2}}{\left[\int_0^\infty dz\, z\exp(-z)/(z+\beta)\right]}.$$

In Fig.2 we show that on increasing v this scaling function is approached.

Financial support from Dirección General de Investigación Científica y Técnica (Spain), Project PB-87-0014 is acknowledged.

References

[1]P. Mandel, in *Instabilities and Chaos in Quantum Optics II*, edited by N. B. Abraham, F. T. Arecchi and L. A. Lugiato (Plenum, New York, 1988), p. 321.

[2]G. Broggi, A. Colombo, L. Lugiato, and P. Mandel, Phys. Rev. A **33**, 3635 (1986).

[3]M. C. Torrent and M. San Miguel, Phys. Rev. A **38**, 245 (1988).

[4]H. Zeghlache, P. Mandel, and C. Van den Broeck, Phys. Rev. A **40**, 286 (1989).

[5]W. Scharpf, M. Squicciarini, D. Bromley, C. Green, J. R. Tredicci, and L. M. Narducci, Opt. Commun. **63**, 344 (1987).

[6]P. Mandel, Opt. Commun. **64**, 549 (1987).

[7]M. Suzuki, Adv. Chem. Phys. **46**, 195 (1981).

[8]F. de Pasquale, P. Tartaglia, and P. Tombesi, Phys. Rev. A **25**, 466 (1982).

[9]B. Caroli, C. Caroli, B. Roulet, and D. Saint James, Physica **108A**, 233 (1981).

Detection of Weak Optical Signals in a Laser

J.M. Sancho[1] *and J.I. Jiménez Aquino*[2]

[1]Departamento de Estructura y Constituyentes de la Materia, Universitat de Barcelona, Diagonal, 647, E-08028 Barcelona, Spain
[2]Departamento de Física, Universidad Autónoma Metropolitana-Iztapalapa, 09340 Mexico, D.F. Mexico

Abstract. Recently it has been shown that very weak optical signals can be detected effectively by a laser used as a supergenerative detector. The physical mechanism is the dynamical process of the relaxation of a laser from an unstable state to a steady stable state. We present an analysis of this process based on the study of Non-Linear Relaxation Time (NLRT).

1. Introduction

The transient dynamics of a laser relaxing from an unstable state to its steady stable state seems to be a very good physical mechanism for detecting very weak optical signals, as was pointed out recently [1]. The experimental process explained in Ref. [1] goes as follows . The laser is periodically switched on and off for time intervals of length T_c. The laser is prepared with an initial intensity $I_0 = 0$, which is an unstable state, and it evolves to its steady intensity which is different to zero. The dynamical evolution of the laser towards its steady state is very sensitive to the presence of an external field.

In Fig. 1 we show two time evolutions of the mean output intensity of the laser during that interval of time. The quantity A_0 is the area under the curve of the mean intensity in the absence of external signals, and the area A_e corresponds to the case when they are present. The ratio between the two areas, A_e/A_0, is called the receiver output and this is the quantity we are interested in, because it depends on the presence of an external field.

Our purpose here is to calculate the explicit dependence of this quantity on the parameters of the system and particularly on the internal fluctuations and the external signals.

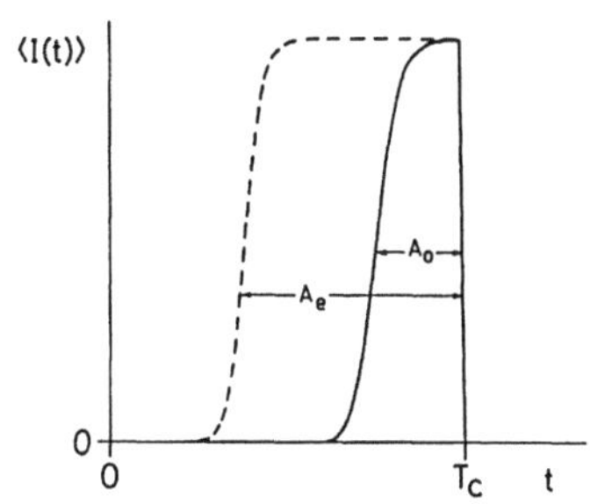

Figure 1. Time evolution of the mean intensity during the interval of time T_c.

Springer Proceedings in Physics, Vol. 55
Nonlinear Dynamics and Quantum Phenomena in Optical Systems
Editors: R. Vilaseca, R. Corbalán

Up to now two methods have been proposed to study the detection of very weak optical signals: one refers to studying the time evolution of the mean output intensity [1-a] (Fig. 1) and the second one looks at the Mean First Passage Time (MFPT) of this intensity [1,2]. We will study the first method. The second one has been analyzed in Ref. [2].

2. A Laser Model and the NLRT Approach

We use the same model of the laser as proposed in Ref. [1]. This model is characterized by the following Langevin equation for the complex, scaled, dimensionless laser field, E, in the presence of an injected external signal, E_e,

$$\frac{dE}{dt} = -kE + \frac{FE}{1 + \frac{AI}{F}} + q(t) + k_e E_e \, , \tag{1}$$

where

$$\langle \, q^*(t) q(t') \, \rangle = 2\epsilon\delta(t-t') \, . \tag{2}$$

$I = |E|^2$ is the laser intensity, q(t) is the white noise describing the internal fluctuations of intensity ϵ and k, k_e, A, F are the other parameters of the model.

The relation between the receiver output and the NLRT can be obtained from the definition of the NLRT associate with the mean intensity of the laser $<I(t)>$, which is given by

$$T_e = \int_0^\infty \frac{<I(t)> - <I>_{st}}{<I(0)> - <I>_{st}} dt \approx \int_0^{T_c} \frac{<I(t)> - <I>_{st}}{<I(0)> - <I>_{st}} dt \, . \tag{3}$$

The approximation is valid if $T_c \geq 1.5 T_0$, where T_0 is the NLRT in the absence of the field. From this expression it is possible to write the areas A_0 and A_e as functions of T_0 and T_e respectively. Since $<I(0)> = 0$ the receiver output is then

$$\frac{A_e}{A_0} = \frac{T_e - T_c}{T_0 - T_c} = 1 + \frac{T_0 - T_e}{T_c - T_0} \, . \tag{4}$$

The NLRT theory has been proposed in references [3-5]. However, we will use the explicit results deduced in Ref. [4] which tell us that the NLRT has the form

$$T_e = < t >_{QDT} - \frac{1}{2(F-k)} + \Delta C_{NL} \, , \tag{5}$$

where the term $\Delta C_{NL} = 7/12(F-k)$ comes from nonlinear deterministic contributions ($\epsilon \to 0$) and $<t>_{QDT}$ is the mean passage time that should be calculated with the Quasi-Deterministic Theory (QDT) using the linear approximation of the model (1). The formal solution can be expressed as

$$E_i(t) = h_i(t) e^{at} \, , \tag{6}$$

where $a = F\text{-}k.\ h_l(t)$ plays the role of an effective initial condition because for long times it transforms into a gaussian random variable $h_l(\infty)=h_l$. Then the marginal probability of the modulus h is

$$P(h) = \frac{h}{\sigma} I_0\left(\frac{ch}{a\sigma}\right) e^{-\frac{1}{2\sigma}(h^2+(c/a)^2)} , \tag{7}$$

where $c=k_e \mid E_e \mid$, $\sigma=\epsilon/a$ and $I_0(z)$ is the modified Bessel function of zero order [6]. Using this probability function, the MFPT for the intensity to get the prescribed value $I_r=F(F\text{-}k)/KA$ is then

$$< t >_{QDT} = \frac{1}{2a} < Ln\left(\frac{I_r}{h^2}\right) > \simeq < t >_0 - \frac{1}{2a}[E_1(b\alpha) + \gamma + Ln(b\alpha)] , \tag{8}$$

where $E_1(z)$ is the integral exponential function [6], γ is the Euler constant, $\alpha=(2\sigma)^{-1}$ and $b=c^2/a^2$. Here $<t>_0$ is the MFPT in the absence of the external signal which is calculated with the same method

$$< t >_0 \simeq \frac{1}{2a} Ln\left(\frac{aI_r}{2\sigma}\right) + \frac{\gamma}{2a} . \tag{9}$$

3. Results and Comments

Finally we get for the receiver output

$$\frac{A_e}{A_0} = 1 + \frac{E_1(b\alpha) + \gamma + Ln(b\alpha)}{2a\ (\ T_c - T_0\ )} . \tag{10}$$

Here the functional dependence on the external field is through $b\alpha$, which is explicitly given by

$$b\alpha = \frac{k_e^2\ |E_e|^2}{2a\epsilon} . \tag{11}$$

This variable is the ratio between the intensity of the external signal versus the intensity of the internal noise, which is the relevant quantity to focus on when one is interested in this kind of problem.

In Fig. 2 we plot the receiver output versus this variable for the values of the parameters of Ref. [1]. For large values of $b\alpha$ there is a logarithmic dependence which means that the deterministic motion is dominant [2]. This is not the regime we are interested in because signals are large and probably they could be detected by other standard techniques. For small values of $b\alpha$ there is no effect and the signal cannot be detected. For $b\alpha \simeq 1$ the laser works as a detector of very small signals. One can see that when $b\alpha = 1$ the effect is larger than 10% and it could be clearly seen with a not very large number of realizations.

We would like also to note that it is possible to study the effect of the fluctuations of the external signal here assumed constant. Its contribution to the

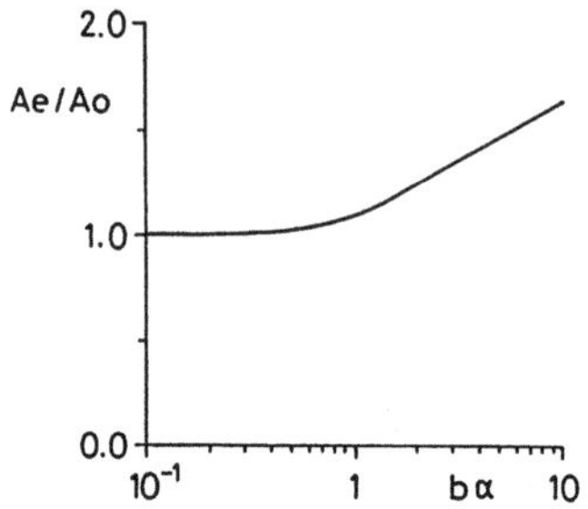

Figure 2. The receiver output versus the parameter bα obtained from Eq. (10).

receiver output is through a renormalization of the intensity of the internal fluctuations in a quantity ($\epsilon + \epsilon'$). This effect will be manifest as a positive vertical shift in Fig. 2.

Our result (10) has been compared satisfactorily [7] with the numerical simulations of Ref. [1-a].

We acknowledge the European Economic Community (Project No. SC1.0043.C(H)), the Dirección General de Investigación Científica y Técnica (Spain) (Project Nº PB87-0014) for financial support. J. I. J-A also acknowledges Consejo Nacional de Ciencia y Tecnología (México) and Consejo Superior de Investigaciones Científicas (Spain) for a grant.

References

[1] a) G. Vemuri and R. Roy, Phys. Rev. A 39, 2539(1989): b) I. Littler, S. Balle, K. Bergmann, G. Vemuri and R. Roy, Phys. Rev. A41, 4131(1990)
[2] S. Balle, F. de Pasquale and M. San Miguel, Phys.Rev.A41, 5012(1990)
[3] J.I. Jiménez-Aquino, J. Casademunt and J.M. Sancho, Phys.Lett. A133, 364 (1988)
[4] J. Casademunt, J.I. Jiménez-Aquino and J.M. Sancho, Physica A156,628 (1989)
[5] J. Casademunt, J.I. Jiménez-Aquino and J.M. Sancho, Phys. Rev. A40, 5905 (1989)
[6] M. Abramowitz and I.A. Stegun, Handbook of Mathematical Functions (Dover, New York 1972)
[7] J. I. Jiménez-Aquino and J. M. Sancho, Phys. Rev. A43,589 (1991)

Index of Contributors

Springer Proceedings in Physics

Managing Editor: H. K. V. Lotsch

MIX
Papier aus verantwortungsvollen Quellen
Paper from responsible sources
FSC® C105338
FSC
www.fsc.org

If you have any concerns about our products,
you can contact us on
ProductSafety@springernature.com

In case Publisher is established outside the EU,
the EU authorized representative is:
Springer Nature Customer Service Center GmbH
Europaplatz 3, 69115 Heidelberg, Germany

Printed by Libri Plureos GmbH
in Hamburg, Germany